Innovations in Materials Processing

Edited by
Gordon Bruggeman
Army Materials and Mechanics Research Center
Watertown, Massachusetts

and

Volker Weiss
Syracuse University
Syracuse, New York

PLENUM PRESS · NEW YORK AND LONDON

Library of Congress Cataloging in Publication Data

Sagamore Army Materials Research Conference (30th: 1983: Lake Luzerne, N.Y.)
    Innovations in materials processing.

    (Sagamore Army Materials Research Conference proceedings; 30th)

    "Proceedings of the 30th Sagamore Army Materials Research Conference, on in-
novations in materials processing, held August 1-5, 1983, in Lake Luzerne, New
York"—T.p. verso.
    Bibliography: p.
    Includes index.
    1. Materials—Congresses. 2. Metal-work—Congresses. I. Bruggeman, Gordon.
II. Weiss, Volker, 1930-   . III. Title. IV. Series: Sagamore Army Materials
Research Conference. Sagamore Army Materials Research Conference proceedings;
30th.
TA401.3.S23   1983   620.1′1   84-22817
ISBN 0-306-41839-8

Proceedings of the 30th Sagamore Conference, on Innovations in Materials
Processing, held August 1-5, 1983, in Lake Luzerne, New York

©1985 Plenum Press, New York
A Division of Plenum Publishing Corporation
233 Spring Street, New York, N.Y. 10013

Innovations in Materials Processing

SAGAMORE ARMY MATERIALS
RESEARCH CONFERENCE PROCEEDINGS

Recent volumes in the series:

This book is to be returned on or before
the last date stamped below.

PREFACE

The Army Materials and Mechanics Research Center in cooperation with the Office of Sponsored Programs of Syracuse University has been conducting the Annual Sagamore Army Materials Research Conferences since 1954. The specific purpose of these conferences has been to bring together scientists and engineers from academic institutions, industry and government to explore in depth a subject of importance to the Department of Defense, the Army, and the scientific community.

This 30th Sagamore Conference, entitled Innovations in Materials Processing, has attempted to focus on the inter-disciplinary nature of materials processing, looking at recent advancements in the development of unit processes from a range of standpoints from the understanding and control of the under-lying mechanisms through their application as part of a manufactur-ing sequence. In between, the classic link between processing and materials properties is firmly established. A broad range of materials are treated in this manner: metals, ceramics, plastics, and composites. The interdisciplinary nature of materials processing exists through its involvement with the basic sciences, with process and product design, with process control, and ultimately with manufacturing engineering. Materials processing is interdisciplinary in another sense, through its application within all materials disciplines.

The industrial community (and the Army as its customer) is becoming increasingly concerned with producibility/reliability/ affordability issues in advanced product development. These concerns will be adequately addressed only by employing the full range of disciplines encompassed within the field of materials processing. This point was emphasized at the outset of the conference by noting that the development of processing methods to yield products with the desired performance characteristics cannot be divorced from the design of products which take advantage of available processes and process economics. The materials processing innovations discussed in these proceedings are vital components in the overall process of producing reliable, affordable products.

    We wish to acknowledge the assistance of Ms. Karen Kaloostian
of the Army Materials and Mechanics Research Center and Mr. Robert
Sell of Syracuse University in all the many arrangements needed
before and during the conference. We wish to especially acknowledge
the efforts of Ms. Mary Ann Holmquist of Syracuse University, who
not only worked diligently on conference arrangements but also
did yeoman service in assembling these proceedings for publication.

                              The Editors

SAGAMORE ARMY MATERIALS

RESEARCH CONFERENCE PROCEEDINGS

Recent volumes in the series:

20th: Characterization of Materials in Research Ceramics and
Polymers
Edited by John J. Burke and Volker Weiss

21st: Advances in Deformation Processing
Edited by John J. Burke and Volker Weiss

22nd: Application of Fracture Mechanics to Design
Edited by John J. Burke and Volker Weiss

23rd: Non-Destructive Evaluation of Materials
Edited by John J. Burke and Volker Weiss

24th: Risk and Failure Analysis for Improved Performance and
Reliability
Edited by John J. Burke and Volker Weiss

25th: Recent Advances in Metals Processing
Edited by John J. Burke, Robert Mehrabian, and Volker Weiss

26th: Surface Treatments for Improved Performance and Properties
Edited by John J. Burke and Volker Weiss

27th: Fatigue - Environment and Temperature Effects
Edited by John J. Burke and Volker Weiss

28th: Residual Stress and Stress Relaxation
Edited by Eric Kula and Volker Weiss

29th: Material Behavior Under High Stress and Ultrahigh Loading
Rates
Edited by John Mescall and Volker Weiss

30th: Innovations in Materials Processing
Edited by Gordon Bruggeman and Volker Weiss

CONTENTS

SECTION IV

PROCESSING FROM THE LIQUID STATE

SECTION V

PROCESSING OF PARTICULATES

SECTION VI

ADVANCES IN MACHINING TECHNOLOGY

SECTION VII

SURFACE TREATMENTS

AFFORDABILITY THROUGH TECHNOLOGY:

BRIDGING THE GAP

Frederick J. Michel

U. S. Army, Headquarters, Development and Readiness
Command
5001 Eisenhower Avenue
Alexandria, VA 22333

INTRODUCTION

Affordability has become a major concern to the Defense Department and to the Army in particular. A whole new generation of
weapons systems was started on the drawing boards in the early
1970's, and we are just now in the process of transitioning them to
Procurement and Production. As we are transitioning these new
systems to production we are also experiencing substantial cost
growths.

The Army's procurement budget has more than tripled over the
last five years, going from $5 billion in FY78 to $17 billion in
FY83, and we still can't buy all the hardware we want and need.

In each quarter of the year, the Army is required to send
Selected Acquisition Reports to Congress on 14 major systems. On
those 14 systems, a comparison of the costs originally estimated
in then year dollars with the current cost projections in then
year dollars indicates nearly a threefold increase. Specifically,
the original early '70's cost of $30 billion has risen to today's
price tag of $82 billion. Of course it must be realized that the
capabilities requirements may have changed; and, of course, inflation has its effect. However, cost growth is certainly a major
element.

The U.S. Army Development and Readiness Command (DARCOM) is
procuring more than 1000 line items of equipment, of which one
third are categorized as high priority. If the 14 systems are used
as an indicator for all Army systems, it is obvious why affordability
has become a major concern as we transition our new weapons to
production.

Figure 1 presents the major causes of problems experienced in transitioning a newly developed system to production. The causes of these transitioning problems can be grouped into those derived from inadequacies or shortfalls related to the design phase (for example, going into production with incomplete designs or designs which are not producible) and those caused by factors related to the factory operation (for example, inadequate planning or use of inefficient manufacturing processes and equipment). Each of these can have a major impact on cost.

The problems of transitioning are due to a lack of adequate planning for production <u>early</u> in the development cycle. What is needed is a concerted production planning and engineering effort.

Toward this end, General Donald R. Keith, Commanding General of DARCOM, has placed high priority on improving the transitioning process of weapon systems from R&D to production. General Keith has expressed it thus: "There are no activities in weapons systems acquisition that demand greater attention than those directed toward assuring effective transition of developed hardware into efficient production."

What I hope to do here is to point out where the opportunities are which can influence this effort.

- PRODUCT COMPLEXITY/DIFFICULT TO MANUFACTURE

- UNEXPECTED PROBLEMS IN MATERIAL AVAILABILITY   DESIGN

- TECHNICAL DATA PACKAGES (TDPs) NOT VALIDATED
  OR STABILIZED

- INADEQUATE PLANNING FOR PRODUCTION

- ENTERING PRODUCTION WITH TOO MANY UNKNOWNS
  IN MANUFACTURING PROCESSES AND TECHNOLOGY   FACTORY

- INEFFICIENT PRODUCTION PROCESSES

- AGE OF FACILITIES/EQUIPMENT

1. Causes of Transitioning Problems

PRODUCT COSTS

As previously noted, the ultimate cost of an item is influenced by activities which are related to both the design of the product and the manufacturing operations which are needed to make it.

During the design phase, the engineer is called upon to achieve certain performance objectives and, at the same time, to make a number of tradeoffs in order to come up with a product which is reliable, maintainable and producible. In order to do that job, the technology must either be available or be developed during the design phase so that the designer can come up with a product with the optimum set of characteristics. Further, he must complete the design job in accordance with a schedule so that the manufacturing side of the house has time to get ready to produce it.

As it relates to the manufacturing operation, the production cost is influenced by the adequacy and completeness of the production planning (how the product will be produced), and selection and operation of the production resources (manpower, processes and facilities used to produce the product). The stability of the production has a major influence on the production cost.

Indeed, the production planning and engineering effort is not a small one. The production engineer must consider the manufacturing technology available, the capital equipment available in his company, what new equipment is needed, the layout of the manufacturing operation, flow of materials, designing tools and equipment, working up the detailed instructions on how to produce the product, and so on.

In order for this job to be completed and proven out in time to start production, these activities must be performed in conjunction with the design effort. Consequently, just as the technology must be made available to the designer to design the product, the technology to produce the product must be made available to the manufacturing operation up front.

As an item is being designed and developed, both time and cost are being sunk into the product. Consequently, when a change is made, it consumes both the time and money originally expended plus the time and money it takes to make the change. This results in an exponential curve of cost vs. time as items get nearer and nearer to production, as shown in Figure 2. This indicates that the introduction of producibility considerations must take place when the pencil first touches the drawing board or as early as possible thereafter.

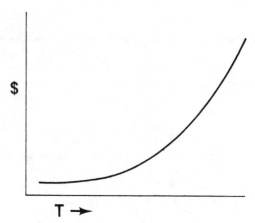

Figure 2.   Cost of Design Change as a Function of
Time During Development Cycle

COST INFLUENCES OF ENGINEERING vs PRODUCTION

Approximately 60% of all opportunities to reduce cost of the
average commercial product is typically determined by the activities of the design engineer, with 40% of the opportunity in the
production side of the house. The ratio is even higher in favor of
the engineer's influence when considering the higher technology of
defense products.

Consequently, the greatest influence on cost can be exerted by
making producibility technology available to the designer, and, as
noted previously, making it available early in the design process.

On the other hand, a major portion of the cost can still be
influenced by attention to the production side of the house. Here
approximately 70% of the manufacturing cost of a product is direct
labor and overhead. Material represents about 30% of the total
manufacturing cost.

Traditionally, the practice has been to look at ways of reducing direct labor in order to come up with cost savings. However,
with the advent of CAD/CAM and automation, our factories are
becoming more and more capital intensive, and direct labor, at
9-15% of the total cost, is not the primary target for cost reduction. Overhead is becoming the cost driver more and more, and
therefore is also the area where significant cost reduction
opportunities lie.

Factory floor cost is a direct measure of the productivity of
a manufacturing operation. As shown in Figure 3, technology

accounts for around 50% of productivity gains possible in the
operation. Saying this another way, one half of the 40% mentioned
earlier (as being that opportunity which the manufacturing operation
has to reduce cost) can be affected by technology improvement.

Consequently, from the preceding discussion, it is seen that
technology can have a major impact on cost, and therefore affordability, of defense products. The effect on this influence can be
enhanced considerably by placing new technology in the hands of both
the design engineer and the manufacturing engineer early in the
development process.

CONSTRAINTS TO NEW TECHNOLOGY DEVELOPMENT

The development of a new technology is constrained by a number
of "real-world" factors not only in Government but in the private
sector as well.

Financial Resources: Financial resources are always limited. There
never seems to be enough money to go around, no matter the endeavor,
and when the work to be done is perhaps speculative in nature, that
makes the problem worse.

Scientific Manpower: A definite lack of scientific talent, particularly as it relates to manufacturing, constrains technology
development even further.

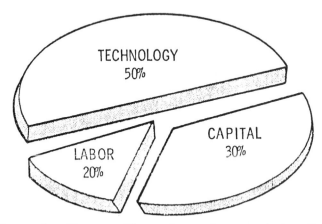

Figure 3. Factory Floor Costs-Opportunities for
Productivity Gains

Investment Payback Philosophy: Those in the position of deciding
among various competing potential developments have taken the
position that they want to see a tangible return for their dollar
and, in many cases, that must be a <u>fast</u> return as well. Long range
technology requirements are often neglected for want of a short
term return.

Gap Between Scientific Community and the Factory: Finally, a major
gap has existed in the past and still exists today between the
developers of new technology and the application engineers. It is
indeed a challenge to bring together the researcher and the factory
floor, as embodied in the manufacturing engineer.

BRIDGING THE GAP--TECHNOLOGY APPLICATION

In Japan, we have seen a concerted government/industry/academic
effort to bring new and improved technology to the factory floor.
The research we see is specifically aimed toward capturing a focused
segment of the marketplace. Application is the key word in the
Japanese philosophy.

In the United States we have always had the ability to come up
with new technological developments. From the preceding discussion
we have seen the impact which technology can have on the producibility and affordability of our products. What is needed in the
U. S. in order to bridge the gap between technology development and
its application is a systems approach toward identifying the opportunities, putting the players from the scientific and manufacturing
communities together and developing and applying those developments.

We must ensure that the R&D community works on <u>real live
problems in a cooperative effort</u> with the manufacturing engineer
who has the responsibility to introduce it on the factory floor.
The manufacturing engineer in turn has to get shop management involved early on to make it a party to the project, give it adequate
time to plan for the installation and start up the necessary
familiarization and training of the workers.

This philosophy can indeed be implemented successfully. A
case in point is the Robotics Institute at Carnegie Mellon. The
Westinghouse Defense Center at Baltimore wanted to automate several
operations in the production of electronic equipment; CMU agreed
to take on the assignment.

So far, the benefits are that CMU has assignments for graduate
students and instructors that require <u>real</u> solutions; Westinghouse
has the benefit of talent that is operating on the cutting edge of
technology, who in the laboratory develop solutions in cooperation
with its own manufacturing engineers; and the graduate students
get the opportunity to gain experience in problem solving which

will lead to tangible results. As a result, everyone's a winner.

Similar arrangements can be found at Rensselaer Polytechnic Institute's Center for Manufacturing Productivity and at Purdue, where about 30 professors are teaming up with Cincinnati Milacron, Control Data, Cummings, Ransburg and TRW in its Computer Integrated Design, Manufacturing and Automation Center to integrate the various components which will go into the factory of the future.

The National Academy of Engineering has also recognized the need to bring the scientific and the manufacturing communities closer together. The academy has established the Manufacturing Studies Board in an effort to do this.

SUMMARY

In summary, we realize that affordability can only be achieved by attacking our cost problems at all levels. Technology is a major influential factor at our disposal.

Recognizing the application is key to technology development. In order to effectively make technology work for us in achieving affordability it is required that we concentrate our efforts in those areas that are major cost drivers and potentially have the greatest pay-off. That is the challenge we face.

MANUFACTURING AND PRODUCTIVITY

Nam P. Suh

Massachusetts Institute of Technology

Cambridge, MA 02139

SUMMARY

One of the most significant problems currently facing the
United States is the competitiveness of its manufacturing industry,
which must be enhanced within the next several years. In order to
increase the productivity, the value added (i.e., selling price)
must be maximized and the manufacturing cost must be minimized.
Both of these depend on the quality of products and processes,
which in turn depend on the design decisions made during the product
and process development. The only means of assuring good design
is to use scientific principles for rational decision making such
as the synthesis Axioms. These manufacturing and productivity
issues are discussed by analyzing the margin of profitability of
the discrete mechanical parts manufacturing industry and by present-
ing the concepts and examples of the design Axioms. In addition
to these short-term issues, there are long-term issues related to
manufacturing and productivity, such as availability of natural
resources, which may only be resolved through technological
breakthroughs.

INTRODUCTION

During the last decade the United States has been keenly aware
of the need to increase its productivity in its manufacturing
sector of economy. Numerous articles have appeared stressing
various causes, remedies, and potential problems we may face if no

---

*Some of the materials presented in this paper are taken out of a
 book to be published, authored by N. P. Suh, who holds the
 copyright.

9

major actions are undertaken. It has also been quite popular to
visit factories in Japan to understand the methods and means
employed by Japanese firms to increase their manufacturing productivity. The debate and discussions that have ensued have been a
positive factor for taking remedial actions in many segments of the
manufacturing community in the U.S. Indeed, the innate strength
of the U.S. is its ability to face its problems and devise solutions.

The purpose of this keynote paper is to highlight the technological factors that have had the most significant effect on
productivity by distilling the critical problems and by addressing
only the central issues at the risk of overgeneralizing complex
problems in simple terms. An effort will be made to focus on the
issues that may shape the productivity problems of the 1980's and
1990's. A few remarks are also made on other related issues that
can significantly affect our future standard of living, such as the
availability of strategic elements, development of manufacturing as
a discipline, and the future industrial structure.

In order to understand the boundaries within which the U.S.
economy must respond, we will first consider the basic driving
forces that will dictate changes in the field of manufacturing. We
will then show that rational decision making ability is central to
increasing the future productivity and to dealing with fundamental
problems imposed on us by limited resources. Rational decision
making methods will be illustrated using the axioms for synthesis
developed at MIT [1, 2]. Emphasis will be given to the application
of these axioms to innovation of new products, processes, machines,
and systems. In the final analysis, major technological innovations
that can add greater values to our products and minimize the manufacturing cost are the only means by which we can maintain our
high standard of living.

DRIVING FORCE FOR TECHNOLOGICAL CHANGE AND ITS EFFECT ON
INDUSTRIAL DECISION MAKING

Technological changes that occur can be driven by many different factors, e.g., to satisfy defense needs, social aspirations,
environmental factors, or purely political ends. However, the most
influential factor on a long-term basis is the margin or profitability, since profits finance the future growth of economy [3].
The margin of profitability is a measure of the return on investment,
which may be defined as

$$\text{Margin of profitability} = \frac{\partial P}{\partial \$_i} \tag{1}$$

where $\underline{P}$ is the total profit over the product (or the process) life
and $\$_i$ is the total investment made.

The profit $\underline{P}$ may be expressed as:

$$\underline{P} = \$_p \cdot N - MfgC - OC \tag{2}$$

where $\$_p$ is the selling price of a product, N the number of products sold, MfgC the total manufacturing cost, and OC the other costs such as sales, R&D, and administrative cost. No business or firm can survive for very long if the margin of profitability, $\partial \underline{P}/\partial \$_i$, is negative. Therefore, on an individual firm basis, most investment decisions will be made with the view of maximizing the margin of profitability. Equations (1) and (2) state that the margin of profitability can be maximized by maximizing the total sales income and minimizing the manufacturing cost and other costs. Since the productivity is proportional to the ratio, $\$_p \cdot N/(MfgC + OC)$, one of the best ways to increase productivity is to produce goods which can command high price without affecting sales volume.

The selling price of a product is controlled by the perceived values that are formed in the minds of the customer, by the functions a given product performs, the quality of the product, the ingenuity of ideas behind the product, and by the reliability of the product. Many of the factors that affect the perceived value are directly governed by the design decisions made in configuring the product and the manufacturing processes. Therefore, unless the product can command a high selling price (that is, unless we can make products with high value added), one cannot improve productivity by lowering MfgC and OC alone. In mature industries the selling price of goods is essentially fixed. Therefore, the relative competitive position is established by the ability to produce the good at the lowest possible cost.

PRODUCTIVITY IMPROVEMENT STRATEGY

The overall criterion for judging productivity improvement strategies is provided by Equation (1). However, when the technological content is so low, (i.e., everyone can make the same product) that the selling price is more or less fixed, the criterion may be reduced to the following:

$$\frac{\partial}{\partial \$_i}(MfgC) < 0 \tag{3}$$

That is, we have to strive to lower the manufacturing cost. Since the manufacturing cost consists of the labor cost, LC, materials and energy cost, MC, information cost, IC, and capital cost, CC, Equation (3) may be expressed as

$$\frac{\partial}{\partial \$_i}(MfgC) = \frac{\partial}{\partial \$_i}[LC + MC + IC + CC] < 0 \qquad (4)$$

A strategy for minimizing the manufacturing cost can be developed by examining the relative magnitude of each one of the terms of Equation (4) [3].

It turns out that in large segments of the mechanical parts manufacturing industry, the direct labor cost (excluding the overhead) is only 5% to 25% of the manufacturing cost. In contrast the materials cost (defined as the cost of the components and raw materials coming into a plant) is anywhere from 50% to 75% of the manufacturing cost. Also, the information cost including most of management functions at all levels is high, being anywhere from one to three times the direct labor cost. Capital cost is difficult to quantify because it is affected by the amortization period, inventory management, capital equipment cost, and the interest. In many industries in the United States, the capital cost is lower than the direct labor cost, justifying the introduction of automation to reduce the labor content.

Since the direct labor cost is a surprisingly small fraction of the discrete parts manufacturing cost of industry, any effort to reduce only the labor content cannot have a major impact on the manufacturing cost. That is, one-to-one substitution of direct labor with automated equipment, such as a robot, cannot be justified unless the introduction of the automation provides secondary or subsidiary benefits such as the reduction of the information cost, more reliable part production, and improvement of the quality of the product. Clearly the recent fervor created around the robotics technology can hardly be justified in terms of the long-term implications of the productivity improvement strategy.

This is not to say that automation is not desirable, but rather to say that automation must be introduced with the view of improving the overall manufacturing operations. For example, flexible manufacturing systems (FMS) may reduce the in-process inventory and reduce the capital investment cost if the product lives change frequently. In manufacturing certain electronic components, the only way that reliable and dependable products can be manufactured has been through complete automation. However, it does not make sense to introduce a few robots here and there, leaving everything else alone and hope for productivity increases through the reduction of direct labor cost!

The fact that the materials cost is such a large fraction of the manufacturing cost in discrete parts manufacturing industry is counter intuitive. Obviously if a given plant could buy natural ores and make everything itself, the fraction of the materials cost would be smaller. That is, if a plant can be vertically integrated completely in its manufacturing operations, the materials cost would be a smaller fraction of the manufacturing cost. However, vertical integration does not guarantee lower manufacturing cost, because certain components that go into the final product cannot be made at a competititve price unless the lot size is sufficiently large. When the lot size is small, it is necessary to purchase these components from outside firms that specialize in manufacturing those specific items. Consequently, the degree of vertical integration, and hence the fraction of the materials costs vary from industry to industry. In certain electronics industries, vertical integration can be more readily achieved than, for example, in the machine tool industry. For this reason, the materials cost is a large fraction of the total manufacturing cost in many discrete parts manufacturing industries.

If materials consumption could be reduced by 20%, the corresponding productivity increase would be equivalent to eliminating the entire direct labor cost. Yet few systematic or rational efforts are being undertaken to deal with the materials issue. There are some major gains to be made through the use of rational design of processes and products. Rational designs imply effective use of materials through the reduction of reject parts, less dependence on materials removal processes, and through the optimum use of materials in meeting the functional requirements.

Currently, information cost is high since people are needed to manage information related to all facets of industrial operations. As a product becomes complex and as the unit of operations become large, the information data base increases and often is incomplete. Therefore, decisions must be made intuitively and in an ad hoc manner.

In many industrial operations there are two kinds of information: structured and unstructured. Unstructured information is the information that operators of machine tools and managers intuitively know to make the right decisions based on their years of experience on the job, whereas the structured information is that which can be clearly stated in explicit statements and can be quantitatively accounted for.

RATIONAL DESIGN AS THE BASIS FOR INCREASING PRODUCTIVITY

In the previous section, it was pointed out that the best way to increase productivity is to increase the value added by proper design and innovation of important new technology. It was also stated that the possibility of reducing the materials cost is often overlooked as an opportunity for major productivity improvement. In considering all these aspects of the productivity issues, rational design stands out as a single most important factor in determining productivity.

Another equally important implication of the foregoing analysis is that research and development must be an integral part of the U.S. industrial posture, since new innovations increase the value added to products. Indeed, basic research must be done in all areas of science and technology in order to create social and industrial culture that nurtures innovation, since the source of major new ideas cannot be identified a priori. The notion that R&D is very important has been pointed out by Dennison [4]. According to his statistical analysis, R&D has been the most important contributing factor to the U.S. productivity increase during the past five decades.

There seems to be evidence that economy is driven by technology rather than vice versa. Figure 1(a) shows how the technological level changes after a major new innovation (e.g., steam engine) is introduced. It appears that a large number of innovations have reached maturity in about (50 ± 20) years as shown in the figure. The net economic impact in terms of new jobs created and new wealth added to the economy is the differential of the curve shown in Figure 1(a), as shown in Figure 1(b). At the early stage of the technological development it is difficult to discern which one of the new technologies would constitute major innovations. Futhermore, economic impact is small because only limited economic activities take place during the early stages of major new technological development. Once the technology reaches maturity, the industry based on the mature technology will do well just to maintain its existing level of business. The maximum impact on the economy is made when the growth rate is most rapid and when the margin of profitability is large. The curve shown in Figure 1 (c) which plots productivity versus time has nearly the same shape as Figure 1(b). The value added becomes smaller and smaller as the product matures. During the early stages of the development, the value added may be large but the production volume and yield are low and the manufacturing cost high.

According to this view, the long term economic activity is given by the superposition of the curves shown in Figure 1(b) in the time domain. There must be sufficient overlaps between the curves to have a nearly constant plateau with an upward slope. Therefore, the prerequisite for a healthy economic climate on the continuing basis appears to be periodic introduction of new technological innovations.

The foregoing argument reinforces the notion that we, as a nation, must continue to devote significant resources to R&D if we are to have a bright economic future.

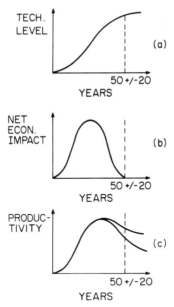

Fig. 1. Relationship among technological level,
net economic impact, and productivity
due to a major innovation.

SCIENTIFIC APPROACH TO MANUFACTURING

One of the major problems of the manufacturing field has been
that we rely too much on empiricism in all aspects of operation, be
it the design of the product, the design of the process or the
operation of the manufacturing system. The common problem with an
empiricism-dominated field is that the existing knowledge cannot be
stated in a generalizable form, or transmitted to others in an
effective way without requiring the accumulation of a similar experience base. However, if we are to make the field of manufacturing
into a discipline, we need to establish the science base. Since
the most important, all encompassing, element that transcends the
complete spectrum of the manufacturing field is synthesis, we have
to make the subject treatable in a scientific way. Then, the issue
becomes "How can we transform the decision making process involved
in synthesis into a science?"

In spite of the importance of innovation in increasing manufacturing productivity, the process of innovation has been treated
purely as a creative process that cannot be enhanced through
scientific principles. However, this type of thinking misses the
key aspect of the innovation process that idea generation is only
one of its steps. Before good ideas can be generated, one has to
define the essence of the problem in terms of functional requirements. Once the ideas are generated, we must be able to eliminate
bad ideas from good ideas. The final step in the innovation
process is the implementation of the best idea. In the absence of
these four steps of the innovation process, no innovation can be
forthcoming. Except for the idea generation itself, all other
aspects of the innovation process should be able to be treated on
a scientific basis.

The author has claimed for some time now that we can treat the
design and synthesis process scientifically using the design axioms
[1-3, 5-7]. Axioms are absolute principles that are required in
making rational design decisions for products, processes, systems,
and software. Axioms have played the crucial role in the development of modern science. Euclid's geometry, Newton's laws, thermodynamic axioms, and even Einstein's theory of relativity are all
based on axioms. Indeed, the history of science is the history of
axioms. Axioms enable us to generalize the observed behavior in
terms of explicit statements. Axioms cannot be proved but are
assumed to be valid as long as there are no exceptions or counter
examples. For example, Newton could not prove that F = ma precisely, nor could he prove that the action is equal to reaction.
Indeed, the concept of force can only be defined in terms of his
axioms, i.e., we cannot measure the force directly. Thermodynamic
axioms are the same. No one has proved that the energy is conserved.
Although we are now accustomed to think that energy and entropy are
physical quantities, they are simply conceptual quantities defined
by the first and the second law of thermodynamics. We cannot
measure _energy_ directly, but must infer that energy is based on
other primary measurements such as the fall of a dead weight from
a known height. The synthesis axioms, if they are correct, should
contribute to the development of manufacturing science as these
axioms have done in the past.

We can say that there are axioms that govern the synthesis
process, since there are good designs and bad designs, and all good
designs have common characteristics. Indeed, the design axioms
were generated by the author in 1977 by reviewing all the steps that
the author had taken that had made a significant improvement in the
quality of products, processes, and systems. Initially, the author
came up with 12 "hypothetical axioms" which were reduced to 6
"hypothetical axioms", and 6 corollaries through the discussion with
the author's colleagues at MIT (Bell and Gossard [1]), and later
only to 2 axioms after undertaking serious research on this topic.

The axiomatic approach to design and manufacturing differs philosophically from the current trend in manufacturing, which is to rely on large computers with all the required data base to arrive at the solution. While computer-integrated manufacturing has its place in modern manufacturing, the entire process of using computers can be simplified if key decisions are made a priori without depending on an exhaustive search. Furthermore, computers cannot be intelligent without relying on basic decision making rules. A very fundamental limitation of computer based technology is given by Bremmerman's limit [8], which in essence states that even if we make the entire earth into a single, most efficient computer, it cannot deal with more than 270 variables in a factoral design. Since manufacturing systems often involve a very large number of variables, a brute force approach to computer-aided manufacturing may not be the most fruitful exercise.

AXIOMATIC APPROACH TO DESIGN

Definition of Design

The first step in design is to define a set of functional requirements in the functional domain that satisfy the perceived needs for a product (or a process or a machine or a manufacturing system). The second step is the creation of a physical being in the physical domain that satisfies the stated functional requirements (FRs). Therefore, the design process involves transformation from the functional domain to the physical domain. In this sense, design can be defined as the mapping process to transform the functional description of a product into a physical entity (see Figure 2). The designer specifies how the transformation occurs by specifying the design parameters (DPs), i.e., the geometric shape, the physical component, the processes, and the spatial and temporal relationship among them. The designer thus creates a set of information in the form of drawings, circuits, software, and/or equations that describe the transformation process.

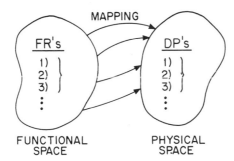

Fig. 2.   Design is defined as the mapping process from the functional space to the physical space to satisfy the designer specified FR's.

It should be noted that in the context of these axioms the word design is used to include all synthesis related activities: the design of products, processes, manufacturing systems, software, and even organization. This is consistent with the view that design transcends all subdisciplines of engineering and all of man's creative activities.

Design Axioms and Definitions

A design must satisfy both of the following two axioms:

Axiom 1   The Independence Axiom
           Maintain the independence of functional requirements.
Axiom 2   The Information Axiom
           Minimize the information content.

Axiom 1 states that the functional independence specified in the problem statement in the form of FR's must be maintained in the design of a solution, whereas Axiom 2 states that of the designs that satisfy Axiom 1, the one with the minimum information content is the best design among those evaluated.

In order to be able to use these axioms, one must adhere to the exact definitions of key words. In Axiom 1, the functional requirements (FR's) are defined as the minimum set of independent requirements for a product or process that completely characterizes the design objectives. That means that each one of the functional requirements in a set does not depend on other functional requirements, i.e., a functional requirement can be stated without consideration of any other functional requirements. The designer is free to choose any set of functional requirements, provided that the set of FR are self-consistent and minimal, in the sense that none of the FR's are redundant. The design that violates Axiom 1 is said to be a coupled design.

Functional coupling should not be confused with physical coupling which is often desirable as a consequence of Axiom 2. Integration of more than one function in a single part, may reduce complexity. Caution is required when attempting integration, however, so that the single part does not become more complicated than the several parts it replaces. These concepts will be clarified later using examples.

In addition to functional requirements, there are constraints which are defined as the bounds on acceptable solutions. By definition, a constraint is different from a functional requirement in that the constraint does not have to be independent of functional requirements of the design.

In Axiom 2 the word <u>information content</u> is defined as a
measure of complexity. It is related to the probability of a
certain event occuring when the information is supplied. Following
the information theory the information content is defined as

$$I = K \log_2(1/p) \tag{5}$$

where K is a constant and p is the probability. I is the information content given in bits.

Mathematical Representation of the Independence Axiom

In the preceding section design was defined as the mapping
process involving the transformation of functional requirements
(FR's) in the functional domain into design parameters (DP's) in
the physical domain. The functional requirements constitute
vectors FR's in the functional space, while the design parameters
constitute a vectors DP's in the design space. Then, the mapping
process may be represented as

$$\{FR's\} = [DM] \{DP's\} \tag{6}$$

where [DM] is called the design matrix. [DM] describes the transformation process.

Axiom 1 states that the design matrix must be a diagonal matrix
with all non-diagonal elements equal to zero. When this is the
case, each one of the FR's can be changed independently of the
other functional requirements by changing only one design parameter.
If a diagonal matrix is not possible, a triangular matrix also
satisfies Axiom 1, provided that the DP's are varied in a specific
sequence. In this case, one FR can be varied independently of the
other FR's. When [DM] is neither a diagonal nor a triangular matrix,
it has a unique solution which may or may not satisfy the specified
functional requirements.

A design that satisfies Axiom 1 by having a diagonal matrix is
called an <u>uncoupled</u> design, i.e.

$$\begin{Bmatrix} FR1 \\ FR2 \\ \vdots \\ FRn \end{Bmatrix} = \begin{vmatrix} X & 0 & \ldots 0 \\ 0 & X & \ldots 0 \\ \ldots \ldots \\ 0 & \ldots \ldots X \end{vmatrix} \begin{Bmatrix} DP1 \\ DP2 \\ \vdots \\ DPn \end{Bmatrix} \tag{7}$$

The design that yields a triangular matrix is called a "decoupled"
or "quasi-coupled" design, i.e.,

$$\begin{Bmatrix} FR1 \\ FR2 \\ FR3 \end{Bmatrix} = \begin{vmatrix} X & 0 & 0 \\ X & X & 0 \\ X & X & X \end{vmatrix} \begin{Bmatrix} DP1 \\ DP2 \\ DP3 \end{Bmatrix} \qquad (8)$$

All other designs are called coupled designs, i.e.,

$$\begin{Bmatrix} FR1 \\ FR2 \\ FR3 \end{Bmatrix} = \begin{vmatrix} X & X & X \\ X & X & X \\ X & X & X \end{vmatrix} \begin{Bmatrix} DP1 \\ DP2 \\ DP3 \end{Bmatrix} \qquad (9)$$

It should be noted that when the number of DP's is less than the number of FR's we always have coupled designs.

The significance of the Independence Axiom can be graphically illustrated for the two-dimensional case. When the physical space and the functional space are superimposed, the relationship between the FR axes of the functional space and the DP axes of the physical space falls into one of three cases as shown in Figure 3. In Figure 3(a), both of the FR axes and the DP axes are parallel to each other, whereas in Figure 3(b), only one of the FR axes is parallel to one of the DP axes. Figure 3(c) shows the case where neither one of the axes is parallel. For the parallel case shown in Figure 3(a), it is apparent that changing FR2 from A to B can be accomplished by changing DP2. This is the uncoupled case. The situation is quite different in the coupled case shown in Figure 3(c). In order to change FR2 from A to B, both DP1 and DP2 have to be changed. In the case of decoupled (or quasi-coupled) case shown in Figure 3(b), the functional state can be changed from A to B by first changing DP2 from the A level to the C level, and then adjusting DP1 to the correct value of FR1.

From the foregoing discussions of the two-dimensional case, it is clear that the merits of the design are determined by the following two factors:

1) Alignment of the DP axes with the FR axes.
2) Orthogonality of the DP axes with respect to each other.
   (Note that by definition the FR axes are always orthogonal to each other.)

Uncoupled designs are those with perfect alignment between the DP axes and the FR axes, and orthogonal DP axes. Even the design with orthogonal DP axes is a coupled design if these axes are not parallel to the FR axes.

Two metrics have been advanced to measure these two charac-teristics [10]. They are reangularity, R, and semangularity, S, which measure orthogonality and alignment, respectivley. They are defined in terms of the elements of the design matrix, DM, (some-times called the coupling matrix)

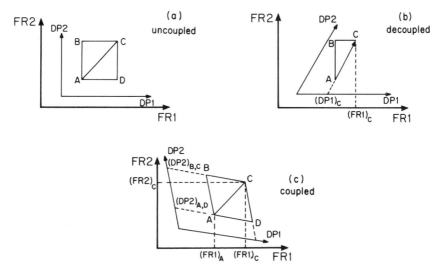

Fig. 3.   Graphical representation of the Design Mapping
          Procedure from the F- to P-Domain.   (a) uncoupled,
          (b) decoupled and (c) coupled system.

$$C_{ij} = \frac{Fr_i}{DP_j} \tag{10}$$

as

$$R = \prod_{\substack{1-1,n-1 \\ j=i-1,n}} \left\{ 1 - \frac{\left(\sum_{k-1,n} C_{ki} \, C_{kj} \right)^2}{\left(\sum_{k-1,n} C_{ki}^2 \right)\left(\sum_{k-1,n} C_{kj}^2 \right)} \right\}^{1/2} \tag{11}$$

$$S = \prod_{j=1,n} \left\{ \frac{\mid C_{ij} \mid}{\left(\sum_{k=1,n} C_{k_j}^2 \right)^{1/2}} \right\} \tag{12}$$

Figure 4 shows the plot of R and S in the two dimensional physical
space.   The uncoupled state corresponds to the case of R = 1 and
S = 1.

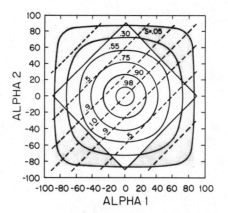

Fig. 4. Contours of semagularity (solid) and reangularity
         (dashed) as a function of alpha one and alpha two.
         Design parameters have not been selected properly
         if the design is represented in the shaded region
         of the figure.

In the underlined case the state can be changed from A to C by
first going from A to B and B to C, or going from A to D and then D
to C (see Figure 5(a)). The amount of information required is
exactly the same either way. Therefore, in this case, the information required to change the state is path independent, i.e.,

$$\Delta Iu_{AC} = \Delta I_{AB} + \Delta I_{BC} = \Delta I_{AD} + \Delta I_{DC} \tag{13}$$

where Iu_{AC} is the information associated with the uncoupled state
change. The decoupled case shown in Figure 5(b) indicates that the
information required is path dependent. Consider again the problem
of changing the state from A to C. If we first change DP1 to set
FR1 and then change DP2 to set FR2, FR1 will change from the original
set value. That is,

$$\Delta I^{dc}_{Ac \, \text{Path 1}} = \Delta I_{AB} + \Delta I_{BC} < \Delta I^{dc}_{Ac \, \text{Path 2}} = \Delta I_{AD} + \Delta I_{Dc} + \Delta I_{Ec} \tag{14}$$

where I^{dc}_{AC} is the information required to change the state from A to
C in a decoupled design. It follows that the coupled state requires
the most information and it is also path dependent.

The definition of information given in Equation (5) can be
generalized as the log of the range in which a given event can occur
and the acceptable tolerance of the event:

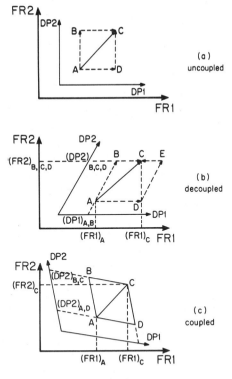

Fig. 5.   Graphical representation of the Design Mapping Procedure
          from the F- to P-Domain and illustration of path depen-
          dence of coupled and decoupled systems. Uncoupled system
          is path dependent when the state is changed from A to C.

$$I = K \log_2 \left(\frac{RANGE}{Tolerance} \right) \qquad (15)$$

Using the definition given by Equation (14), the information required to implement the designer specified tolerance by a manufacturing system with its own specific capability can be defined [9].

     Consider the problem satisfying the designer specified re-
quirement for the length of a bar, $(L \pm \Delta L_d)$, using a machine
system that can cut the bar within $(L - \Delta L_a)$ to $(L + \Delta L_b)$. The
probability distributions for these tolerances are assumed to be
uniform within the tolerance band as shown in Figure 6. When the
bar is cut so that its length is within the shaded region where

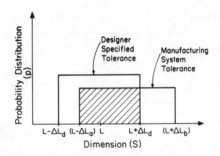

Fig. 6. The uniform probability distributions of the designer
         specified tolerance range and the manufacturing system
         tolerance range. The manufacturing system can, at best,
         make the bar fall in the shaded range, which is acceptable
         to the designer. However, there is a finite probability
         that the bar length will be outside the acceptable range,
         i.e., between $(L + \Delta L_d)$ and $(L + \Delta L_b)$. Note that the area
         under each rectangle is equal to 1, i.e., $\int p dS = 1$.

the designer's specification and the manufacturing system capability
overlap (i.e., between $(L - \Delta L_d)$ and $(L + L_d)$), it is an acceptable part. However, there is also a probability that it will be
outside the acceptable range, i.e., between $(L - \Delta L_d)$ and $(L + \Delta L_d)$.
When the shaded area lies completely within the designer specified
tolerance, the manufacturing system can cut the bar to the right
length everytime. Obviously in this case we do not need additional
information, i.e., $I = 0$. We have chosen the dimension of the bar
as an example, but the general concept is for other production
parameters such as hardness, cost, surface finish, etc.

Equation (15) may be re-written for the length problem as:

$$I = K \, Log_2 \left(\frac{\Delta L_a + \Delta L_b}{\Delta L_a + \Delta L_b} \right) \tag{16}$$

The coefficient K is taken to be equal to 1 so that the sum of the
information is also information [9].

Corollaries

From the two axioms of design, many corollaries can be
derived as direct consequences of the axioms. These corollaries
are sometimes more useful in making design decisions than the axioms
since they can be stated to be specifically applicable for a given
situation. Those experienced in design will find that some of the
corollaries have been used as design rules in the past, often
intuitively.

Some of the most important corollaries are:

Corollary 1)   Decouple or separate parts or aspects of a solution
                if functional requirements are coupled or become
                interdependent in the designs or processes proposed.

Corollary 2)   Conserve the resources (e.g., labor, materials, and
                energy) in fulfilling the functional requirements.

Corollary 3)   Minimize the number of functional requirements and
                constraints.

Corollary 4)   Integrate functional requirements in a single part or
                solution if they can be independently satisfied in
                the proposed solution.

Corollary 5)   Use standardized or interchangeable parts whenever
                possible.

Corollary 6)   A part should be a continuum if energy conduction is
                important.

Corollary 7)   Use symmetric shapes and/or arrangements if they are
                consistent with functional requirements.

Corollary 8)   Specify the largest allowable tolerance in stating
                functional requirements.

Corollary 9)   Seek an uncoupled design that requires less informa-
                tion if the proposed uncoupled design contains more
                information than coupled designs in satisfying the
                same set of functional requirements.

These corollaries are derived from one or both of the axioms.
For instance, Corollary 1 is a direct consequence of Axiom 1,
whereas Corollary 4 is derived from Axioms 1 and 2. Some of these
corollaries are self-evident, but some of these have much deeper
implications than they first appear to have. It should be noted
that even though it is not always specifically stated as part of the
corollary statement, these corollaries are valid only if they do not
violate the stated functional requirements.

Corollary 1 states that the functional independence must be
satisfied by decoupling if a proposed design couples functional
requirements. As shown by the examples given in the following
sections, decoupling does not necessarily imply that a part has to
be broken into two or more separate physical parts, or that a new
element has to be added to the existing design. Functional decoupling can be made without physical separation, although in some
cases it may be necessary. In fact, Corollary 4 states that as

long as the functional requirements are not coupled by physical
integration of parts, it should be done so as to minimize the geo-
etric information content.

The concept of decoupling functional requirements, so as to
maintain their independences, applies even when constraints are
imposed on the functional requirements by the laws of nature. Pressure and temperatures of a gas, for example, are inherently coupled,
and any adiabatic system which changes the pressure of a gas by
mechanical work must also change the temperature. However, if a
certain pressure and temperature were both functional requirements
of the system, an additional element (e.g. heat exchanger) must be
added to regain independence. The system is then decoupled and the
decoupled system is called a quasi-coupled system.

Corollary 2 is based on the fact that the more the resource
used, the larger is the information content, because more specifications are needed to describe how they are to be used. Furthermore,
the information content of a part with fixed tolerance increases as
the size of the part increases.

Corollary 3 states that as the number of functional requirements
and constraints increase, the system becomes complex and thus the
information content increases. This implies that the conventional
adage that "my design is better than yours because it does more than
it is intended" is wrong. Machines must be designed to fulfill the
precise needs defined by the functional requirements. Similarly,
a process which fulfills more functions than specified will be more
difficult to operate and maintain than that which only meets the
stated functional requirements.

Corollary 5 states a very well-known design rule. In order to
reduce inventory and facilitate the assembly operation, it is
necessary that odd parts (such as special screws, etc.) be replaced
with standard parts (such as standard screws and bolts). Furthermore, even the number of standard parts should be minimized so as
to decrease the inventory and simplify the inventory management.
That is, instead of using a variety of screw sizes, the use of a
few standard screws may reduce the manufacturing cost. Interchangeable parts also allow reduction of inventory and simplify
manufacturing and service operations.

Corollary 6 is somewhat ambiguous. It states that when heat
conduction through the part is important, it is easier to transfer
the heat by conduction if the part is made in one piece. This is
in contrast to making the part in two or more pieces and then incorporating other features to promote heat transfer. The latter is
likely to require more information.

Corollary 7 is self-evident. Symmetric parts require less

information to manufacture and to orient in assembly.

Corollary 8 states that tolerances should be as large as possible. As the tolerance is reduced it becomes increasingly difficult to manufacture a part so more information is required to produce parts with tight tolerances. Therefore, the tolerance should be made as large as possible, provided that the part is functionally compatible with other parts. The correct tolerance band is that which minimizes the overall information content. When tolerances are made too large, the overall information content will increase since the subsequent manufacturing processes will require greater operating information. This is the essence of Corollary 8.

Corollary 9 states that there is always an uncoupled design that involves less information than a coupled design. This corollary is a consequence of Axioms 1 and 2. If Corollary 9 were not true, either Axiom 1 or 2 is violated. The implication of this corollary is that if one proposes an uncoupled design that has more information content than a known coupled design, the designer should repeat the design process so as to uncover an uncoupled design that has less information content.

Some of these concepts are illustrated in the examples given in later sections. It should be noted that the existence of the design with the minimum information content cannot be proved, since we cannot define the best design using an absolute measure. We can only identify the best design among those proposed in terms of Axioms 1 and 2.

Simple Examples that Illustrate the Key Concepts

Two simple examples are given in this section to illustrate the concept of functional independence qualitatively.

After the concept of functional independence is emphasized here, additional examples will be given to illustrate the use of information metrics system defined by Equation (15). It should be emphasized that designs must satisfy both of the axioms. However, in normal design practice the functional independence (i.e., Axiom 1) is often satisfied first, followed by minimization of the information content (i.e., Axiom 2). In some cases, we have to develop several alternate designs that satisfy Axiom 1 and then choose the one that has the minimum information content. The supposition is that the design with minimum information content is easiest to manufacture, provided that all the right facilities and equipment are available. When the right equipment is not available, the information content involved in manufacturing processes must also be considered together with the information content involved in the product.

Example 1:

Design of a Refrigerator Door

Suppose that a refrigerator door must be designed and manufactured. The FR's of the refrigerator door are defined to be
the following:

> FR1 -- Minimize the heat transfer from the environment
> to the refrigerator so that the contents can be
> kept at $40 \pm 5^0 F$ with a minimum expenditure of
> energy.
> FR2 -- Provide access to the contents inside the refrig-
> erator.

One may add additional functional requirements to the above set of
FR's, but then the design problem is no longer the same.

In a conventional refrigerator, FR1 is satisfied by providing
a cooling system and an insulated enclosure. FR2 is satisfied by
a vertically hinged door. Is this a good design?

The design is not a good design in terms of the axioms. It
couples two functional requirements, thus violating Axiom 1. As
the hinged door on the refrigerator with the vertically hung door
is opened to remove the contents (i.e., FR2), the cold air inside
the refrigerator flows out and is displaced by the warmer room air,
thereby affecting the thermal requirement of the door (i.e., FR1).
The functions are coupled since the means for achieving the second
FR interferes with or compromises the means chosen to satisfy the
first FR. Is it possible to alter the design and develop an
uncoupled system?

An uncoupled design configuration is that of the top-opening
door which is often used in chest freezers. Since the door of the
chest freezer is the top surface of the enclosure, opening the door
does not allow the warm outside air to displace the cold air. The
thermal isolation of the refrigerator is compromised to a much
lesser extent than that of the conventional refrigerator when the
door is opened. If the tolerance specified in FR1 is satisfied by
the design, then the two FR's listed are uncoupled in the top-
opening horizontally hung door design. This means that the conventional refrigerator design is poor for the specified set of FR's.
However, if many other functional requirements are added to the
original set and if the tolerance of FR1 is increased, the conventional door may become an acceptable (i.e., uncoupled) design.

Example 2:

Bottle/Can Opener Design

Suppose now that we are interested in designing a manually
operated bottle/can opener. The functional requirements of the
device are
          FR1:   Open beverage bottles.
          FR2:   Open beverage cans.

A simple device that satisfies the above set of FR's is shown
in Figure 7. This is a very simple opener that can be made by
forming a sheet metal strip. In this design, the means for achieving
both FR's is contained in the same physical piece, and thus a
minimum information content is required to manufacture the device.
Are the functional requirements coupled? This design does not
couple FR's because the act of opening cans does not interfere with
or compromise opening bottles. The functional requirements are
coupled only if a functional requirement of the product is to open
bottles and cans simultaneously, which is not the specified FR's.
The functions have been physically integrated, but not functionally
coupled. Physical integration without functional coupling is
advantageous since the complexity of the product is reduced in
accordance with the second axiom.

Fig. 7.   Can and bottle
            opener.

USE OF THE AXIOMS IN DESIGN AND MANUFACTURING

In this section the use of the axioms will be illustrated using
real manufacturing problems as examples to clarify further the basic
concepts involved in the axioms and corollaries.

Design and Manufacture of Multi-lense Plate

Statement of the Problem:

XYZ Corporation must manufacture a multi-lense plate with 64
individual lenses as a component of a microfilm reader. The configuration of the lense plate is shown in Figure 8. The FR's and
the constraints (C's) of the multi-lense plate are as follows:

FR1:   The quality of each lense must be good within a few
       wavelengths of visible light.
FR2:   The distance between every pair and all pairs of
       lenses must be within ± 0.0005 inch (12×10^{-3} mm).
C1:   The manufacturing cost of the multi-lense plate
       must be less than $10.00 a piece.

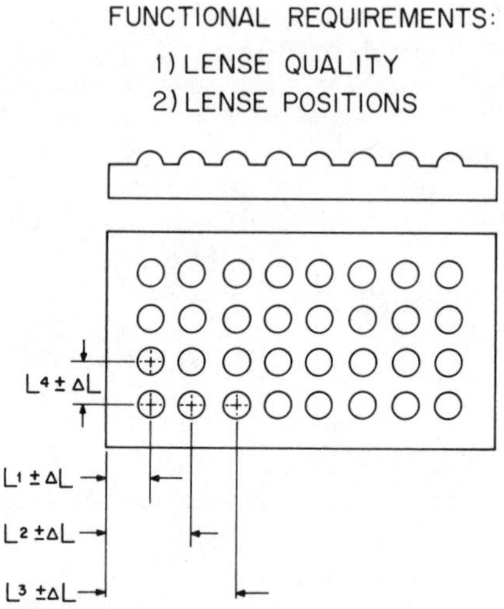

FUNCTIONAL REQUIREMENTS:

1) LENSE QUALITY

2) LENSE POSITIONS

Fig. 8.   Multi-lense plate.

In view of the cost constraint, XYZ Corporation decided that the
entire multi-lense plate should be made by injection molding a
thermoplastic called polymethylmethacrylate (PMMA) into a carefully
made metal mold.   The engineers of the corporation designed and
tested an injection molding system.   In their design, the molten
plastic was injected into a cold mold.   When the part solidified in
the mold, the part was removed from the mold.   After two years of
repeated trials and a great deal of expense, the engineers still
could not produce acceptable parts consistently and reliably.   They
found that if they re-machined the mold to correct a distance between lens cavities that was out of tolerance, then other dimensions
of the molded part, which had been within tolerance, changed to be
out of tolerance.

Solution:

The basic problem with their design is that it violates the
Independence Axiom (Axiom 1).   Every measurement within the part
depends on all the other measurements so that it is impossible to
change one part of the molded part without affecting the rest of the
part.

When the hot molten plastic is injected into the mold, the velocity distribution of the plastic behaves as shown in Figure 9. The axial velocity of the fluid component is highest near the center of the gap because of the wall friction. (The boundary layer is very thick). When the liquid particles reach the plastic front, the fluid motion changes its direction toward the surface. This is called the fountain effect. Furthermore, since the mold is kept at a temperature below the glass transition temperature, T_g, the injected plastic freezes. The plastic near the gate and near the wall freezes first and the thickness of the frozen layer increases proportionally with time. The temperature of the plastic front decreases as a function of distance from the gate. Furthermore, pressure acting on the plastic also changes as a function of the distance from the gate.

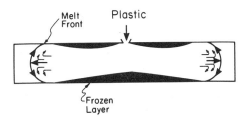

Fig. 9. Velocity distribution of molten plastic in a
         mold cavity during injection molding.

As the plastic freezes, its specific volume changes, that is, it shrinks. Shrinkgage of amorphous polymer can be represented by the state equation given by Spencer Gillmore [11] as:

$$(p + \pi)\ (v + \omega) = R'T \tag{17}$$

where   p = pressure
          v = specific volume
          T = temperature
          ω, R', π = material properties

So, when the initial pressure and temperature before freezing vary from location to location, the change in the specific volume of the plastic and thus the shrinkage will be position dependent. Furthermore, the volume changes are constrained by the presence of lenses. Residual stress and strain are established in the plastic due to these geometric constraints and non-uniform freezing rate in the mold. Therefore, if the molten plastic is simply injected through the end gate, the pressure-temperature history of the plastic differs from point to point. Furthermore, the dimensions between any pair of lenses are influenced by the residual stresses and the shrinkage that takes place elsewhere.

Therefore, the original design used by the XYZ Corporation is a coupled system. That is, the FR (functional requirement) vector and the PV (process variable) vector* are related by a design matrix which is neither diagonal nor triangular, i.e.,

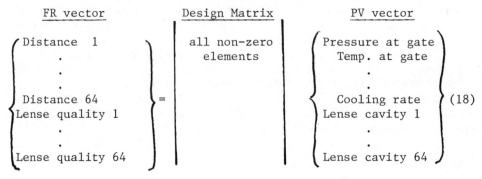

$$
\left\{ \begin{array}{c} \text{Distance \ 1} \\ \cdot \\ \cdot \\ \cdot \\ \text{Distance 64} \\ \text{Lense quality 1} \\ \cdot \\ \cdot \\ \text{Lense quality 64} \end{array} \right\} = \left| \begin{array}{c} \text{all non-zero} \\ \text{elements} \end{array} \right| \left\{ \begin{array}{c} \text{Pressure at gate} \\ \text{Temp. at gate} \\ \cdot \\ \cdot \\ \text{Cooling rate} \\ \text{Lense cavity 1} \\ \cdot \\ \text{Lense cavity 64} \end{array} \right\} \quad (18)
$$

FR vector   Design Matrix   PV vector

In the above equation one can recognize immediately that it represents a coupled system since the FR vector has more components than the PV vector. As discussed earlier, a coupled system does not satisfy Axiom 1.

Now we can decide based on Axiom 1 that the technique tried by XYZ Corporation is not a rational design and that we have to search for a new method. Furthermore, we know what the problem is: it is the coupling caused by non-uniform shrinkage of plastics throughout the mold. A coupled design may provide an acceptable part by accident, but normally it will produce an unacceptable part. In fact, we cannot even model the molding process accurately enough to specify each element of the design matrix.

The new solution we should seek is an uncoupled design. We have to eliminate the variable shrinkage of the plastic that occurs in the mold. A possible solution that is consistent with Axiom 1 is shown in Figure 10. It is a two-step process. The first step consists of making a preform by punching holes in a plastic or card-board sheet in regions where the lenses must be placed. The second step is to insert this preform into a mold and separately injection

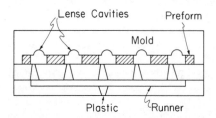

Fig. 10.   Example solution for multi-lense plate. Use preforms with holes.

* PV's are used in place of DP's when process design is considered.

mold individual lenses. In this process the location of lenses is
fixed by the position of the lense cavity in the mold, which is not
affected by the molding process. Also the quality of each lense is
independently controlled by the quality of the lense cavities.
Therefore, Axiom 1 is satisfied. The (FR) vector and the (PV)
vector are related as

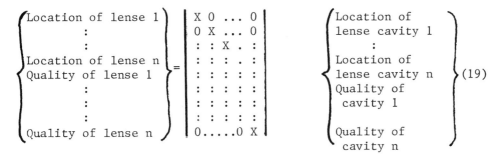

$$(19)$$

The design matrix of this design is a diagonal matrix indicating
that the function requirements can be met independently.

Going from the coupled design shown in Figure 8 to the uncoupled
design shown in Figure 10, the information content has decreased
significantly. The information required for the coupled design was
an accurate description of how much plastic shrinks, the exact mold
temperature, pressure and temperature of plastics, etc. This
information is much more than we can realistically provide for the
system. Even if we could have supplied all the necessary information, there was no guarantee that the system would have produced
good parts. In the uncoupled design, however, the information required is associated with the location and quality of each cavity,
assuming the temperature and pressure history for each molding
cycle could be controlled within specified tolerances.

It should be noted that the uncoupled design shown in Figure
10 is just one of many possible uncoupled solutions. When a large
number of uncoupled solutions are available, the optimal solution
among the known solutions is the one with the least information
content. If and only if we can conceive of all the possible solutions, which is clearly impossible, then we may talk about the best
solution.

Also note that we had to modify the design of the product in
order to make it meet the functional requirements of the product.
This fact may be stated as a theorem as follows:

Theorem 1:   Rational production of a product is not possible
             unless both the design of products and the design
             of processes satisfy Axiom 1.

The total information content of an uncoupled system for n-lenses may be written as:

$$I = \sum_{i}^{2N} (I_{location})_i + \sum_{i}^{N} (I_{\substack{lense \\ quality}})_i + I_{pressure} + I_{temp.} \qquad (20)$$

As a first approximation each of the information contents may be expressed as:

$$\sum_{i}^{2N} (I_{location})_i = \sum_{i}^{2N} \log \frac{\ell_i}{\Delta \ell} = \sum_{i}^{2N} \log \ell_i - 2N \cdot \log \Delta \ell$$

$$\sum_{i}^{N} (I_{\substack{lense \\ quality}}) = \sum_{i}^{N} \log \frac{R_i}{\Delta R} = \sum_{i}^{N} \log R_i - N \log \Delta R \qquad (21)$$

$$I_{pressure} = \log \frac{\Delta P_{machine}}{\Delta P_{overlap}}$$

$$I_{temp.} = \log \frac{\Delta P_{machine}}{\Delta P_{overlap}}$$

ℓ_i is the distance between a pair of lenses, $\Delta \ell$ is the tolerance between them, n is the number of lenses, R is the radius of curvature, and R is the tolerance associated with each lense cavity. $P_{overlap}$ denotes the overlap in pressure tolerance, i.e., between the designer specified tolerance and the machine (i.e., system) tolerance. $T_{machine}$ is the tolerance that can be achieved by the machine for the set temperature, and $T_{overlap}$ in the designer specified tolerances and the system tolerance.

Vented Compression Molding

Statement of the Problem:

The external tank of the Space Shuttle Program (see Figure 11) is made of aluminum which has to be insulated with a low thermal conductivity material to prevent the formation of ice and to limit excessive heat transfer to the liquid hydrogen and oxygen in the tank. The insulation material used in the Thermal Protection System (TPS) is a foamed composite consisting of silicone rubber and various fillers called SLA -- Super Light Ablative Material). For optimum thermal protection the insulation material must have a specific gravity of 1.6 to 1.7. In order to meet these needs, the functional requirements of the manufacturing process are specified as follows: The density must be 1.65 $\pm$ 0.05 and the thickness of the TPS layer must be 0.5 $\pm$ 0.05 inches.

FUNCTIONAL REQUIREMENTS:
1) THICKNESS
2) DENSITY

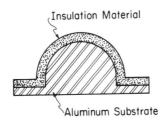

Fig. 11.   Thermal Protection System (TPS) for
the Space Shuttle's external tank.

The original process that was used to put the TPS material on
the aluminum substrate was rather cumbersome. The uncured material
had the consistancy of wet sand and, therefore, was very difficult
to distribute. Also, pressure had to be applied precisely to obtain
the specified density. The original process consisted of the
following steps:

    a) Distribute and pack by hand the SLA material.
    b) Cover the entire assembly with a polyethylene sheet
        and draw vacuum to apply uniform pressure
        (note: this process is sometimes known as vacuum
        bagging).
    c) Cure it in an oven.
    d) Machine off the excess SLA. (60 to 70 percent of the
        cured material was machined off using NC machines
        specially equipped with plastic cutters).

The management of the prime contractor asked MIT to review this
process and see if improvements can be made in the process so as to
lower the manufacturing cost.

The sequence of the manufacturing operation does not violate
Axiom 1, although it may involve many unnecessary steps. The
process is a quasi-coupled process (i.e., the design matrix is a
triangular matrix). However, the process seems to use excess information to satisfy the two functional requirements for density and
thickness of the TPS material.

Solution:

This problem was solved by McCree and Erwin (12) as part of the
MIT-Industry Polymer Processing Program. They noted that the FR's
of the process are :

FR1:   Control the thickness of the TPS to 0.5 ± 0.05 inches.

FR2:   Control the specific gravity of the molded material to 1.65 ± 0.05. (Since the density is a monotonic function of pressure, FR2 may be stated as control the pressure to 20 ± 0.05 psi.

Since there are two FR's there have to be a minimum of two process variables (PV's) to render an uncoupled design. In order to minimize information content as per Axiom 2, it will be desirable to develop a process that requires only two PV's.

A solution to the problem developed by McCree and Erwin is the Vented Compression Molding (VCM) technique shown in Figure 12. It consists of a mold which is made of wire net with a predetermined opening. When the wire net mold is pushed through the unpacked SLA material, a constant pressure is developed in the material in front of the moving net. The excess material extrudes through the opening of the net. The pressure is simply a function of the size of the net opening. When the vented mold reaches a pre-set distance away from the substrate, the excess material above the wire net is vacuumed away, leaving only the molded part. This simple process replaced all the complex manufacturing operations originally used.

The information content of VCM is associated only with the pressure and thickness control. If we denote the designer specified pressure and thickness as $p_o + p_d$ and $t_o + t_d$, respectively, and the system range as $p_o + p_s$ and $t_o + t_s$ for pressure and thickness, respectively, and if the probability is uniformly distributed (see Figure 13) the information content for the process may be written as

$$I = I_t + I_p = \log\left(\frac{p_s}{p_d} \right) + \log \frac{t_s}{t_d} \tag{22}$$

When $p_s < p_d$ (or $t_s < t_d$), the information required for the pressure (or thickness) is equal to zero since the overlap in the tolerances becomes equal to the system capability. It is clear that the system tolerance can be further related to the wire size and the net opening in order to compute the system tolerances if necessary.

For the VCM process the FR vector can be related to the PV vector by:

$$\begin{Bmatrix} \text{Thickness } t \\ \text{Density } p \end{Bmatrix} = \begin{vmatrix} X & 0 \\ 0 & X \end{vmatrix} \begin{Bmatrix} \text{mold gap control} \\ \text{wire net opening} \end{Bmatrix} \tag{23}$$

Without putting in detailed expressions for X's, it is obvious that this process is an uncoupled process, satisfying Axiom 1.

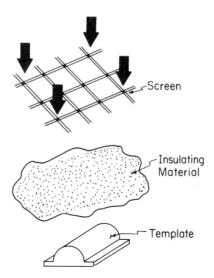

Fig. 12.   Vented compression molding.

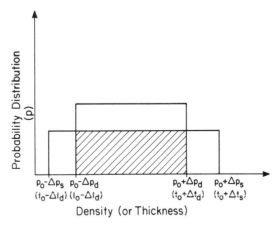

Fig. 13.   The probability distribution functions for the
           designer specified pressure range (and thickness),
           and for the manufacturing system tolerance range.
           Subscript d denotes the designer specified toler-
           ances, while subscript s denotes the system
           tolerance.

Robot Application in Assembly of Molded Parts

Statement of the Problem:

A major corporation in the U.S. has been manufacturing an instrument box by injection molding two halves of the box with impact grade polystrene and by bonding them together ultrasonically (see Figure 14). The original process used by the firm involved the use of conveyor lines on which the injection molded parts were dropped, orientation devices to properly line up the part in a gueuing line (a buffer), and the ultrasonic welding device.

FUNCTIONAL REQUIREMENT:

ASSEMBLE TWO INJECTION
MOLDED PARTS BY WELDING

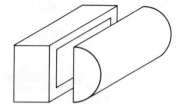

Fig. 14.   Robot and assembly.

One of the engineers reasoned that the process was not very efficient, since when the parts were in the mold their orientation and location were known. Once they were dropped on the conveyor belt, the spatial and orientation information was lost, thus requiring the addition of new information to orient the parts properly before welding. Therefore, the engineer suggested to his management that the company purchase a robot to pick up the parts from the mold, assemble and weld them together (see Figure 15). The engineer justified his recommendation as being more economical, since the use of the robot eliminated the need for the conveyor and the orientation device.

Solution:

The engineer's recommendation appears to be reasonable at the first glance. Indeed the company went ahead and implemented the concept. Contrary to their expectation, the productivity went down when the robot was used!

The problem stems from the fact that the new system based on the robot is a quasi-coupled system, as it will be shown later. The

entire production line goes down when any of the injection molding
machines break down or make imperfect parts. This problem was taken
care of in the original system by using the conveyor belt as a
buffer which gave sufficient tolerances to the injection molding
machines so that they could break down occasionally without affecting
the operation of the entire system.

The functional requirements (FR's) of the process are as
follows:

FR1:   Make the part A
FR2:   Make the part B
FR3:   Assemble A and B
FR4:   Weld the parts A and B.

The process variables (PV's) are:

PV1:   Injection mold part A
PV2:   Injection mold part B
PV3:   Use the robot to assemble
PV4:   Welding with the ultrasonic device.

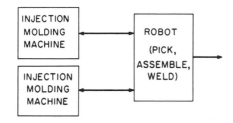

a) Proposed New Manufacturing Method

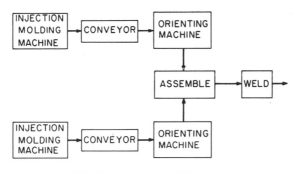

b) Original Manufacturing Method

Fig. 15. Assembly of injection molded part.

The relationship between FR's and PV's is

$$
\begin{Bmatrix} FR1 \\ FR2 \\ FR3 \\ FR4 \end{Bmatrix} = \begin{vmatrix} a_{11} & 0 & 0 & 0 \\ 0 & a_{22} & 0 & 0 \\ a_{31} & a_{32} & a_{33} & 0 \\ a_{41} & a_{42} & a_{43} & a_{44} \end{vmatrix} \begin{Bmatrix} PV1 \\ PV2 \\ PV3 \\ PV4 \end{Bmatrix} \tag{24}
$$

Since this is a quasi-coupled system, the PV's must be controlled in the sequence given in order to be able to vary FR's independently. from each other. If, for example, PV1 cannot be varied due to the failure of the injection molding machine, then FR3 and FR4 cannot be satisfied.

The information contents involved for FR3 and FR4 are large since conditional probabilities are involved. The information content may be expressed as

$$
\begin{aligned}
I = \log P_{11} &+ \log P_{22} + \log P_{33} + \log(P_{33}|P_{32}) \\
&+ \log(P_{33}|P_{31}) + \log P_{44} + \log(P_{44}|P_{41}) \\
&+ \log(P_{44}|P_{42}) + \log(P_{44}|P_{43})
\end{aligned} \tag{25}
$$

where P_{ii} is the probability associated with satisfying FRi by varying PVi, whereas (Pii/Pij) is the conditional probability of FRi occurring given the probability of satisfying FRi by PVj. Obviously when Pij is small because of the occasional malfunctioning of the injection molding machine, the information required becomes very large.

USE OF AXIOMS IN THE DEVELOPMENT OF INTELLIGENT MACHINES

It is clear that the factory of the future will make use of more intelligent machines which can make its own decisions. This will minimize the required data base and simplify the information flow between the central computer and individual machines. One of the basic requirements of the intelligent machine is that it must be an uncoupled system. Otherwise, the machine will be difficult to control and require continuous fine tuning. Such a machine cannot be reliable or dependable. The development of an intelligent brakeforming press will be used to illustrate the basic approach to intelligent machines.

The purpose of a brakeforming operation is to bend a sheet metal into a bent angle θ_f (see Figure 16). This is done by an experienced operator who chooses the right die and punch, and then sets the punch for the right displacement through a lengthy trial-and-error process. The difficulty in achieving the correct bend angle is due to the fact that the material properties vary and, therefore, the elastic spring back differs from sheet to sheet. The sheet must be overbent to an angle θ so that when the load is released, it springs back by an angle $(\theta - \theta_f)$ as shown in Figure 17.

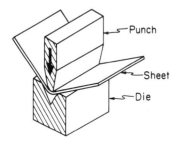

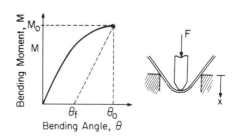

Fig. 16.   Typical brake-   Fig. 17.   Moment curvature relation-
           forming operation.   ship in sheet metal forming.

The functional requirement is that the bend angle be $\theta_f \pm \Delta\theta_f$, regardless of how the material properties vary. Based on the relationship between the bending moment, M, and the bending angle, θ, relationship shown in Figure 17, the functional requirement may be decomposed into the following three equivalent FR's as:

$$\{FR1\} = \theta_f \longrightarrow \{M_o, \theta_o, \tan^{-1}(EI)\}$$

Then, the problem becomes one of choosing the right design parameters (DP's) which yield a diagonal matrix. It is easy to see that M_o can be satisfied if the force exerted by the punch, F, can be measured for a fixed die shape, θ_o can be controlled by controlling the displacement of the punch X_a, and the elastic springback can be determined by monitoring the back-to-back displacement of the punch, ΔX_a, when the load is released from any M and θ. This can be represented mathematically as

$$\begin{Bmatrix} M_o \\ \theta_o \\ \tan^{-1} EI \end{Bmatrix} = \begin{vmatrix} X & 0 & 0 \\ 0 & X & 0 \\ 0 & 0 & x \end{vmatrix} \begin{Bmatrix} F \\ X_a \\ X_a \end{Bmatrix} \qquad (26)$$

Such an intelligent brakeforming machine was built at the MIT
Laboratory for Manufacturing and Productivity [11, 12]. The
machine's operating sequence was to bring down the punch and then
stop instantaneously and unload in the midst of the forming operation
to measure the springback. Once the springback is determined, the
relationship between Mo and θ_f can be used to calculate the final
punch displacement X_a. The machine produced the desired bend angle
within one degree regardless of the variations in the work material,
the first time and all the time, which is better than what the
experienced operator could produce.

The same concept can be used to develop other kinds of intelligent machines such as injection molding machines for plastics. The
concept, to reiterate the key points, is to define functional
requirements to be fulfilled by the machine and then to choose
design parameters that yield an uncoupled machine, i.e., a diagonal
matrix.

THE FIELD OF METALLURGY AND THE AXIOMS

The critical problem that can, in the future, significantly
affect the field of manufacturing and the society at large, is the
effective use of natural resources. As developing nations require
more natural resources to satisfy their increasing aspirations for
higher standards of living, the competititon for resources may be the
single cause for future conflict between countries. It is clear
that we cannot and should not consume materials at the current rate,
and that we cannot depend on the current availability of the strategic elements that are available only in other countries. There are
two important issues related to the use of natural resources:

1) Because the commercially available materials are much
   weaker and otherwise inferior than what is theoretically
   possible and because irrational design practices result in
   materials waste, we are using more materials than needed
   to accomplish a given task.

2) Because the field of metallurgy has been developed based on
   concepts founded on equilibrium thermodynamics (e.g.,
   phase diagram) and the kinetics of phase transformation,
   only certain elements can be used in making alloys, etc.
   For example, strategic elements such as vanadium and cobalt
   are used as alloying elements to control the kinetics of
   phase transformation rather than the chemical properties
   of the solid. In conventional metallurgy the microstructure
   and properties are controlled by following a precise recipe
   for the chemical composition, process variables, and
   process sequence.

These constraints on materials utilization have been created because the field of conventional metallurgy is a coupled system, where composition and microstructure cannot be independently varied.

In order to address the resources issues, we need to develop new concepts for decoupling the microstructural control of metals from the chemical properties. One such a process is called the Mixalloying process. In this section the Mixalloying process is described as a means of illustrating how the axioms can be used even in questioning a well accepted concept such as the metallurgy field.

The Mixalloying process is a new concept in metal casting and fabrication, which is designed to overcome the shortcomings of the conventional metallurgy and produce metals that have superior properties at low cost. The process was invented at the Massachu- setts Institute of Technology (MIT) by the author [14-17]. The Mixalloying process is unique and offers interesting possibilities in increasing strength, electrical conductivity, corrosion resistance, toughness, creep resistance and fatigue life without using expensive strategic elements. The improvements in these properties can theoretically be tailored to meet specific needs. We will first examine the ideal microstructure of metals and then describe how the Mixalloying process can generate the ideal structure.

Ideal Structure of Metals and the Mixalloying Process

Pure elements (e.g., iron, copper, nickel, zinc, and aluminum) are very soft. Therefore, alloys have been made to improve the strength and creep resistance of these pure elements by adding substitutional or interstitial alloying elements, and also by creating multi-phase structures through precipitation or dispersion hardening. However, these existing commercial processes cannot create the ideal structure where all the desired properties can be improved. For example, when the metal is hardened to increase its yield strength, the toughness is usually decreased. Furthermore, the precipitation hardening mechanism becomes ineffective at high temperatures since the precipitates dissolve in the matrix. In order to alleviate this problem, insoluble oxides have been in- corporated in the solid either through internal oxidation or by mechanical means in the form of dispersion hardened alloys, but these dispersion hardened alloys have very low toughness due to the large size of the particles and poor bonding between the matrix and the oxides.

The ideal two-phase structure is schematically illustrated in Figure 18. The microstructures shows a pure metallic matrix phase without any alloying elements, very small (100A) aluminum oxide particles that are bonded strongly to the matrix phase, and the spacing between the oxide particles of a few hundred to 1000 ang- stroms. In such a structure, the yield strength in shear, τ, is

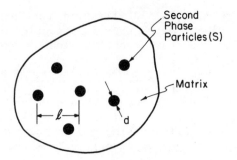

Fig. 18.   Ideal structure of two phase metals.

governed by spacing of the hard particles, ℓ, and the shear modulus, G, of the matrix phase, since the shear stress required to extrude dislocations by the Orowan mechanism is given by:

$$\tau = \frac{2Gb}{\ell} \qquad\qquad (27)$$

where b is the Burgers vector. In the case of iron matrix, G is 11×10^6 psi, and b is about 2A. If we let ℓ be 4×10^2 A, then the shear strength becomes 110,000 psi. Furthermore, since aluminum oxide does not dissolve in the matrix even at the melting point of the matrix, the material can have high strengths and excellent creep resistance even at these extremely high service temperatures. It can be shown theoretically and experimentally [18, 19] that when the particles are smaller than a critical size (of approximately 200 to 500 A), cracks cannot nucleate around these particles, because the energy criterion for crack nucleation cannot be satisfied. Therefore, this ideal material not only has high strength but also has excellent toughness. Furthermore, since the matrix phase consists of a single element, the electrical conductivity is nearly that of the pure element, whereas in commercial alloys conductivity is reduced normally by more than half due to the use of substitu-tional alloying elements. It should be noted that the hard particle must not be too small since dislocations will cut through them. That is, they should be large enough to block dislocations.

So far, no one could make these ideal materials because there has been no metallurgical process that could produce the ideal structure. The Mixalloying process is a technique that can most closely create the ideal structure.

When it is desired to make a high strength metal without the use of hard particles (or even with the use of hard particles), we can increase the strength by making very small grains so as to promote dislocation pile-up at grain boundaries. It can be shown that the yield strength, σ_y, of metals, is related to the grain size as

$$\sigma_y = \sigma_o + Kd^{-1/2} \qquad\qquad (28)$$

where σ_o and K are materials constants, and d is the grain size. Unfortunately, conventional metallurgical processes cannot easily produce small grain size of 1 μm or less, especially in the case of face centered cubic (f.c.c.) metals. If d can be reduced by a factor of 100, for example, from 100 μm to 1 μm, the yield strength can be increased by nearly a factor of 10.

The Mixalloying Process

The Mixalloying process for making the ideal Mixalloy structure consists of the following two basic steps:

a) Mixing and casting of liquid components to create a fine mixture of two solid phases and very small grains.

b) Chemical reaction between the two phases subsequent to the mixing and casting process.

There can be many variations and permutations of these basic steps.

Step 1:

Since the details of the mixing and casting process and the resulting microstructures of the Mixalloys are described in great deal, elsewhere [14-17] only the essence of this process will be given here. As shown in Figure 19, the Mixalloying process consists of two or more molten liquid reservoirs, the same number of conduits (pipes) through which the molten liquids are pumped into the mixing chamber, the same number of nozzles for injection of the liquid streams into the mixing chamber, and a chilled mold in which the

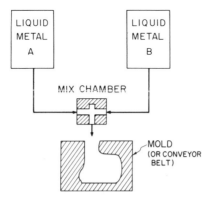

Fig. 19. Mixalloying process.

intimately mixed molten metals are cast. The liquid jets assume
turbulant motion due to their high Reynolds number as they emerge
from the nozzle. These turbulant jet streams impinge against each
other in the mixing chamber, creating a homogeneous mixture on
both macro- and micro-scale.

Through our research at MIT, we found that the size, ℓ_m, of
turbulant eddies (which may be thought of as rigid body motion of
liquid particles) decreases with the increase in the Reynolds
number, Re, based on the diameter of the nozzle, D, as [20].

$$\frac{\ell_m}{D} = A \cdot Re^{-3/4} = A \cdot \left(\frac{\rho VD}{\mu}\right)^{-3/4} \qquad (29)$$

Since, for liquid metals, the viscosity, μ, is very low and the
density, ρ, is very large, and since A is of the order of 1, small
eddies of the turbulant flow can be very easily created at low
head pressures. For example, in processing lead and tin, a 40 psi
head pressure can easily raise the Reynolds number to about 20,000,
making the eddy size, approximately 1 μm. The experimental results
agree quite well with the theoretical prediction given by Equation
29 [14, 15].

Now consider the case of two liquid streams A and B, impinging
against each other in the mixing chamber. Due to the large scale
mixing motion the eddies of A and B mix uniformly in the mixing
chamber. Although the turbulant motion decays rapidly, the eddies
maintain their size for a long time since the time constant for
coalescence is larger than the large scale flow of the liquids out
of the mixing chamber. When this liquid stream flows into the
chilled mold or is cast on a chilled plate, the eddy size is retained. Although the eddies near the center of the mold may
coalesce some during the solidification process, the growth of the
grains is not significant since, heat transfer time is more rapid
than the mass diffusion time.

When the liquid stream, A, has a significantly higher melting
point from that of B, the initial temperatures of A and B may be
chosen so that the eddies of the metal with the high melting point
(i.e., metal A) can be made to freeze almost instantaneously upon
their contact with the eddies of the opposing stream (i.e., metal
B) which is at a lower temperature. Then, the solidification of
the lower melting phase which surrounds the frozen particles of
higher melting phase can take place in the mold.

If A and B can form intermetallic compounds the interface of
A and B eddies will consist of intermetallic compounds.

The Mixalloys can be used after completing only this first step, since they have enhanced mechanical properties due to their small grain size. We can also make very different metals (and/or non-metals) with unique metallurgical microstructures and chemical properties, since any combination of elements can be used to make new alloys and compounds. However, more interesting results can be obtained by inducing chemical reactions between the constituents in phase A and phase B, which is the second step in the Mixalloying process.

Step 2:

Consider, as an example, the process of manufacturing very hard copper which is as conductive electrically as pure copper (which is very soft). Such a metal can replace expensive berylium copper. (Berylium copper is used extensively in the electronic industry due to its high hardness and reasonable electrical conductivity which is typically 40 to 50 percent of pure copper). This can be done by creating the ideal structure shown in Figure 18 through the use of slightly different alloys in stream A and stream B.

For this purpose let stream A be Cu/3% Cu_2O solution and stream B be Cu/2% Al solution (note that these solutions have a lower melting point than pure copper). Then, after impingement mixing, the two phases will be uniformly distributed in the solidified state as shown in Figure 19. If we heat the solid to a high temperature (say to half of the melting temperature of copper) to promote mass diffusion, aluminum from phase A will diffuse into B and oxygen from B will diffuse into A, aluminum diffusing faster. When aluminum meets oxygen, they will react to form aluminum oxide since the free energy of formation of aluminum oxide is lower than that of cuperous oxide. Once Al_2O_3 nucleate, these nucleated oxide particles will grow larger with the additional reaction and diffusion of Al_2O_3 to the nucleation sites. The final size of the oxide particles will depend on the number of particles nucleated, which depends on the size of turbulant eddies. As a result of the reaction, we obtain a solid with pure copper matrix strengthened by small aluminum oxide particles as per the following reaction:

$$Cu/Cu_2O + Cu/Al \rightarrow Cu + Al_2O_3$$

Properties of the Mixalloys

By using the Mixalloying process described in the preceding section, metals can be processed at low temperatures even though the final product may contain hard phases with high melting points. Different alloys can be made by choosing a proper set of alloys as illustrated below [23],

1) Welding Electrode (Cu/Al2 O_3)

2) Electrical Contact (Ag/CdO)

3) High Temperature Component (Ni/Y_2O_3)

4) High Specific Strength Components (Al/Mg_2Si)

5) Wear Resistant Surface (Co/WC)

Since the Mixalloys can be made soft until chemical reactions are induced to create hard particles, they can be easily machined or formed into shapes before hardening.

Since the mixalloys do not depend on the phase diagram (i.e., equilibrium thermodynamics) and the kinetics of phase transformation during the mixing stage, metals with extremely different properties can be manufactured that cannot be produced by conventional metallurgical methods.

Very large parts can be made to net shape, since the pressure involved is low, whereas powder metallurgical techniques are limited to small parts due to the extremely high compaction pressure. Continuous casting is also possible.

Metals with high strengths can be made even without using hard particles because of the ability of the Mixalloying process to produce very small grains.

Since some metals with small grains exhibit superplasticity, they can be formed into complicated shapes by a closed die forging process. The Mixalloying process may enable the creation of many superplastic alloys because of its ability to control the grain size.

MIXALLOYS, STRATEGIC ELEMENTS AND NATURAL RESOURCES

The foregoing example illustrates the possibility that the Mixalloying process may be used to create new alloys that have all the desired properties without having to use strategic elements, some of which are available only in politically volatile regions of the world. This may have a significant implication to the future need of these elements. Clearly there may be other processes which will enable us to produce alloys with desired properties using more abundant elements. In order to accomplish this goal, we need to overcome the constraints imposed on metallurgy by the phase diagram and the kinetics of phase transformation.

A process such as the Mixalloying process may have a profound effect on the demand for natural resources. It is a common aspiration of people all over the world to have a high standard of living,

which often means more material consumption. This aspiration for
improved living standards cannot be met by simply extrapolating
current technologies without the introduction of major technological
innovations. At the present time, the people in the U.S., which
constitutes 5 percent of the people in the world, consume 60 percent
of the world's resources. Clearly we must find means of satisfying
the material needs of the world by being able to use the natural
resources effectively and conservatively, but more importantly we
must devise means of utilizing more abundant elements that cannot
currently be used in making engineering and consumer products
because of the lack of suitable processes. On a long-term basis
this need to address the natural issue may be one of the challenging
tasks confronting the manufacturing community.

INFORMATION ISSUES IN MANUFACTURING

     It was stated earlier that the information cost constitutes
a large fraction of the manufacturing cost. The information cost
is high, because much of the existing information is wasted;
redundant information is continuously generated; the unstructured
information possessed by workers and managers is not well understood
and utilized; the system information cannot be fully managed and
optimized; and because intelligent machines and computers are not
generally available for manufacturing operations. This situation
creates the need for redundant layers of management. The problem
may be further compounded by the coupled organizational structure
where the overlapping functions make the ultimate decision making
process cumbersome.

     As defined earlier, information is related to probability of
a certain event occurring, such as meeting the production schedule,
developing a new device, and establishing a viable customer base.
In order to be within the functional requirements specified for
the event, we have to assign people to carry out the tasks involved
so as to improve the probability. For example, when the machine
tolerance is such that it is within the designer's specified
tolerance, the probability of meeting the functional requirement
is 1 and, therefore, no additional information is required. In
this case, we do not need an operator. However, when the tolerance
range of a machine is much larger than the designer's specified
range, we cannot simply rely on the machine capability to produce
the part within the specified functional requirements. In this
case, we need an operator to improve the probability of being within
the specified tolerance.

     In order to be certain that we have a best manufacturing
system, we have to satisfy the first axiom. That is, we must
create uncoupled systems. Then, we have to look for means of
minimizing the information content. If the system is a coupled
system, it will be an inferior system, requiring more information

than necessary. One of the theorems may be stated as follows: There are uncoupled designs that have less information content than coupled design.

Some of these information concepts have been used to develop a process planning system called PPINC [9].

Automation and Labor Issues

Automation of factories is coming, although the degree of and the rate of automation will differ from industry to industry. It will depend on competitive pressures. Automation will enable us to accomplish the following:

1) By enabling the more effective management of the manufacturing system, it will reduce information cost and the
   inventory cost.

2) It will force the manufacture of more precise and reliable
   parts by requiring more careful design procedures.

3) It will improve the quality of products and working life.

4) It will enable the manufacture of products that cannot be
   manufactured without automation.

Automation should not be attempted if the sole goal is to reduce the labor content since its impact on productivity will be small. Automation by itself is not going to provide the competitive edge to industrial firms for two basic reasons: the small fraction of the labor cost and the availability of automation technologies for mature technologies to anyone who is willing to pay for the techno-logy. In fact, some industries of the developing nations can only compete through automation because of the lack of skilled operators.

Conclusions

1) The maximum increase in manufacturing productivity will be
   possible when high value added products are made using
   efficient manufacturing technologies.

2) In this sense, rational design of products and processes
   will be extremely critical in establishing a competitive
   edge.

3) Rational design of products and processes cannot be
   achieved through ad hoc approaches. Rather, we need to
   establish the science base for the manufacturing field.

4) The most promising means of establishing the manufacturing

science base appears to be the axiomatic approach. The independence of functional requirement axiom (i.e., Axiom 1) and the information axiom (i.e., Axiom 2) are powerful scientific tools that aid the design decision making process. These axioms can be used to eliminate bad ideas from the early stages of synthesis.

5) From the axiomatic point of view, many conventional machines and processes are irrational, and therefore, alternate possibilities should be expored.

6) Even those well established areas such as metallurgy are suspect from the axiomatic point of view. Some of the shortcomings of the metallurgy field may be eliminated by using uncoupled processes such as the Mixalloying process.

7) Intelligent machines can be developed based on the axiomatic thinking.

8) Automation is necessary to produce quality products and to improve the quality of working life. However, the automation that is primarily designed to reduce the direct labor cost cannot be justified from both societal and technological points of view.

9) One of the major tasks ahead of us is to make Manufacturing a discipline. In order to achieve this goal, investment must be made in R&D and basic research.

References

1. N. P. Suh, A. C. Bell and D. Gossard, "On an Axiomatic Approach to Manufacturing Systems," Journal of Engineering for Industry, Trans. ASME, Vol. 100, 1978, pp. 27-130.
2. N. P. Suh, S. Kim, A. C. Bell, D. Wilson, N. H. Cook, and N. Lapidot, "Optimization of Manufacturing Systems Through Axiomatics," CIRP Annals, Vol. 31, 1978, pp. 383-388.
3. N. P. Suh, "The Future of the Factory," Robotics and Computer Integrated Manufacturing, Vol. 1, 1984.
4. E. Dennison, "Accounting for U. S. Economic Growth," 1929 - 1969, The Brookings Institution, Washington, D. C., 1974.
5. N. P. Suh and J. R. Rinderle, "Qualitative and Quantitative Use of Design and Manufacturing Axioms," CIRP Annals, Vol. 31, 1982, pp. 333-338.
6. J. R. Rinderle and N. P. Suh, "Measures of Functional Coupling in Design," Journal of Engineering and Industry, Trans. ASME, Vol. 104, 1982, pp. 383-388.

7. D. R. Wilson, A. C. Bell, N. P. Suh and F. Van Dyck, "Manufac-
    turing Axioms and Their Corollaries," Proceedings of North
    American Manufacturing Research Conference, Vol. 7, 1979.
8. H. J. Bremmerman, "Optimization Through Evolution and Recombina-
    tion," Self-Organizing Systems, Spartan Books, Chicago, 1982.
9. H. Nakazawa and N. P. Suh, "Process Planning Based on Information
    Concept," Robotics and Computer Integrated Manufacturing,
    Vol. 1, 1984.
10. J. R. Rinderle, Ph.D. Thesis, MIT, 1982.
11. R. S. Spencer and D. G. Gilmore, "Equation of State for Polysty-
    rene", Journal of Applied Physics, Vol. 20, 1949, pp.
    502-506.
12. J. McCree, S. M. Thesis, MIT, 1981.
13. K. A. Stellson, Ph.D. Thesis, MIT, 1981.
14. N. P. Suh, "Orthonormal Processing of Metals - Part I: Concept
    and Theory," Journal of Engineering and Industry, Trans.
    ASME, Vol. 104, 1982, pp. 327-331.
15. N. P. Suh, H. Tsuda, M. Moon and N. Saka, "Orthonormal Process-
    ing of Metals - Part II: Mixalloying Process," Journal of
    Engineering and Industry, Vol. 104, 1982, pp. 332-338.
16. N. P. Suh, U. S. Patent 4,278,622, July 14, 1981.
17. N. P. Suh, U. S. Patent, 4,279,843, July 21, 1981.
18. A. S. Argon, J. Im and R. Safoglu, "Cavity Formation from
    Inclusions in Ductile Fracture," Metall. Trans., Vol. 6A,
    1975, pp. 825-837.
19. K. B. Su and N. P. Suh, "Void Nucleation in Particulate Filled
    Polymeric Materials," Society of Plastics Engineers, 39th
    ANTEC, Boston, MA, May 1981.
20. C. L. Tucker and N. P. Suh, "Mixing for Reaction Injection
    Molding I. Impingement Mixing of Liquids," Polymer Engineering
    and Science, September 1980.
21. W. Giessen, Private Communications, 1983.

CLOSED-LOOP CONTROL OF SHEET METAL FORMING PROCESSES

David E. Hardt

Laboratory for Manufacturing and Productivity

Massachusetts Institute of Technology

ABSTRACT

The process of forming metal sheets involves large plastic deformation of the workpiece material. The outcome of these processes, therefore, are highly dependent upon the yield point and flow stress characterisitics of the material. To overcome the problems associated with uncertain and highly variable forming properties, a control strategy is presented that is based on closing the loop around the forming process and workpiece rather than just the processing machine. In this way the variations in material properties and the uncertainty of die-sheet interface forces can be overcome on a part by part basis. The details of such a process are discussed briefly in the context of conventional NC control and in detail for sheet metal forming. Examples of closed-loop control are presented for brakeforming, roll bending and general die forming. Specific experimental results for these cases illustrate that part control of sheet metal forming is quite feasible, and promises to improve the ultimate utility of such processes by reducing setup time, and by improving the part consistency.

INTRODUCTION

The need for improved control of sheet forming processes is motivated by the realization that such processes are often regarded as high volume-low tolerance operations. This results from the inherently uncertain forming characteristics of any metal, since yield point and flow stress characteristics can vary widely for the same alloy, and from the significant influence of die-sheet interaction forces, which are quite difficult to quantify. Traditionally two approaches have been taken to this problem, one

analytical and one purely empirical. In the former the details
of the mechanics of the forming operation and the properties of the
sheet are investigated and a forming model developed. However,
such models are usually limited in their utility for on-line process
control since they require calibration to give accurate prediction
and this calibration may be different for each part. Consequently
the most common form of control is purely empirical where the
forming process is performed, the result examined and modifications
made based on experience. The goal of closed-loop control is to
combine these two extremes into a process that consistenly produces
correct parts without regard to material property or processing
environment variations.

CLOSED-LOOP CONTROLLED PROCESSES

     A closed-loop controlled manufacturing process (or Part
Controlled Process) is one where the input commands to the processing machine are the desired part dimensions rather than commands
that relate only to machine functions or positions. To illustrate
this important distinction, and how it relates in particular to
sheet metal forming processes consider the control system block
diagram in Figure 1. This open-loop system shows that a manufacturing machine, which accepts machine related inputs (such as
position, speed or force) has an output that acts upon a workpiece
so as to produce the final output: the part. In a conventional
manual process the operator closes the control loop by measuring
part geometry (usually) and then manipulating the machine controls
until the actual geometry matches the desired. With numerical
control a hardware control loop is introduced and the machine output now becomes more predictable and is controllable from software
input. However, as shown in Figure 2, closing the NC loop still
leaves the workpiece transformation (from machine output to part)
outside any automatic control loop.

     For metal cutting processes such an ommission is not damning
because of close correspondence between machine output (e.g.
cutting tool position) and resulting part shape. As long as the
tool is sharp, the machine ways are stiff and all tool offset
dimensions known, machine control suffices for part control.

However, when this simple one-to-one correspondence does not apply, accurate part geometry control cannot be accomplished with simple machine or numerical control. Sheet metal forming is one such process, and the remainder of this paper will describe several specific methods for accomplishing part control based on geometry measurement as a primary feedback quantity. In this way the part control loop will be closed and the process inputs immediately become part-based rather than machine-based. This part-based input now facilitates communication with CAD system and obviates any open-loop calculations which are often used to force the marriage of the design and the manufacturing process.

Figure 1.   Open Loop Part Manufacture

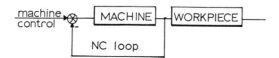

Figure 2.   Automatic Manufacturing Machine Control

SHEET METAL FORMING

Shapes are imparted to thin sheet metals in various ways. These
methods include brakeforming, for straight line bends, hydroforming,
for shrink and stretch flanging, roll bending, for producing
arbitrary two dimensional shapes, and die forming (in one of various
configurations) for general compound curvature parts. Regardless
of the process, the control input is usually displacement or strain
and the output is, of course, the resulting part shape. The
difficulty with sheet forming as opposed to machining operations
is that output shapes and input displacements are related by many
and often elusive variables, among these are intrinsic and extrin-
sic properties of the sheet material. Sensitivity of the output
shape to these properties varies with the process, but because
forming necessarily involves plastic deformations and the major
strains are usually introduced by bending, the yield point and
sheet thickness generally have the greatest influence on the process
outcome. If these variables were easy to hold constant or could
be easily premeasured, then the machine output-part shape transformation across the workpiece (see Figure 2) could be quantified
and used in a predictive manner. However, these quantities (as
well as the others listed below) are highly variable as demonstrated
by Nagpal et al. (1).

A first example is the process of brakeforming. In this
process, shown in an "air bending" configuration in Figure 3,
bending moments are induced by three point loading. The control
problem is one of determining die displacements (machine output)
that give the correct part angle when the sheet is unloaded. Not
only must the displacement-angle relationship be quantified, the
loaded-unloaded angle or springback must also be determined. The
bending characteristics of a sheet can be described using a moment-
curvature (M-K) constitutive relationship (see Figure 4). This
curve relates the bending moment to produce a particular radius of
curvature in the sheet (assuming a pure bending situation). As
with the stress-strain description of material mechanics, it shows
a linear elastic region, and a non-linear plastic region. Any
loading beyond the elastic moment limit will result in a permanent
curvature in the sheet. As can be seen from Figure 4, there is a
significant difference between the loaded curvature and unloaded
curvature for a given plastic moment (so called "springback"),
therefore the control problem in sheet metal bending is finding the
current loaded state of the sheet that will yield the desired
unloaded curvature. This is complicated by the fact that the
moment-curvature relationship of a sheet depends on the sheet
elasticity, yield point, strain hardening characteristic and the
sheet thickness.

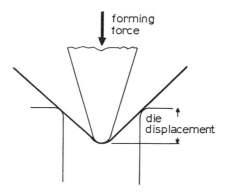

Figure 3.   Brakeforming with Air-Bending Dies

     To effect part control on such a process the machine displace-
ment-loaded angle and springback values must be predicted or
measured on-line. Several approaches have been taken to this
problem and they are distinguished by variable reliance on measurements versus a reliance on predictive models. Nagpal et al. (1)
took an open-loop approach by developing a finite element predictive
model that related unloaded angle to die displacement and calculated
springback. Exhaustive tests were then performed to predetermine
the material properties necessary for model calibration. They
demonstrated that such an approach can be quite successful but also
showed that this success is wholly dependent upon proper testing
of the infeed material.

Allison et al. (2) and Mergler et al. (3) took a closed–loop approach and directly measured and controlled loaded angle of the sheet. This reduces the problem to one of determining springback on–the–fly. Since it is known that springback always increases with angle, both methods converged upon the correct loaded angle by measuring the springback from an intentionally "underbent" state and then reloading the part using this springback as an estimate of the necessary overbend. The resulting systems accepted

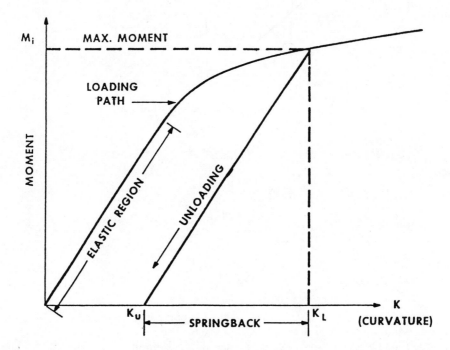

Figure 4. A Typical Moment Curvature Relationship

angle (part geometry) as the input and were as accurate as the angle transducer employed. However, these processes introduced iterations that result in a considerable increase in basic part production time.

A more fundamental approach combining both modelling and real-time measurements was taken by Stelson (4) who realized that the central variable in the bending process is the Moment Curvature (M-C) relationship for the sheet being formed. By measuring the center die force and displacement during deformation of the sheet (as shown in Figure 3) the moment-curvature relationship for the sheet was derived and used on-line to predict the springback and the correct die displacement to achieve the desired angle. This is an example of closed-loop identification of material properties and open-loop control of the process. Although less accurate than Allison's approach (+ 1.0 vs. + 0.2), the resulting process is as fast as the manual case and fulfills the CAD/CAM interface needs of an angle command input.

DIRECT SHAPE CONTROLLED PROCESSES: ROLL BENDING

In all of the above examples, the actual shape was controlled by estimating the springback and then predicting the correct machine commands to obtain the correct part shape. With the case of roll bending, we can demonstrate the use of "measurement" of the unloaded shape when still in the loaded configuration.

Consider the "three roll" bending process shown in Figure 5. The objective is to produce arbitrary two dimensional curves (which can be easily represented on a CAD system). Ideally the CAD/CAM design loop would contain the elements shown in Figure 5, where the central step is a shape or curvature controlled process. To effect such control, the unloaded curvature at each point on the sheet must be determined in real-time to insure fidelity to the part design. This had been accomplished in the following manner. Recall the moment-curvature diagram of Figure 4. This relationship illustrates that as a sheet is loaded with a bending moment, the curvature varies linearly until yielding, whereupon the curve changes slope and permanent deformation occurs. Unloading after this point occurs along a line that is parallel to the original linear portion, but shifted by the amount of permanent deformation. This illustrates that the unloaded curvature at a point can be determined from the loaded conditions by the relationship.

$$K_u(s) = K_1(s) - M(s)/EI$$

where

$$K_u(s) = \text{unloaded curvature}$$

$K_1(s)$ = corresponding loaded curvature

$M(s)$ = corresponding loaded curvature

$M(s)$ = elastic M-K slope (or beam bending stiffness)

s = position along the workpiece.

If $M(s)$ and $K_1(s)$ can be measured, and EI is determined either on or off-line, then the resulting unloaded curvature can be continuously calculated in real-time and used as a shape feedback signal. As shown in Figure 6, this now permits the use of an unloaded curvature servo loop. In this loop, the displacement of the rolls (the machine control) is governed by the unloaded curvature error. Notice that the input command to the control system is the desired quantity: the part curvature. Such a system was designed and tested by Hardt et al. (5). The results of experiments where circular rings were formed are shown in Figure 7. It can be seen that the desired curvature was maintained within measurement error (3%) regardless of part radius, sheet thickness or material type. Thus the goal of a simple, determinate process input (shape only) was achieved and such a system would need only a data link between the roll bending process control computer and a computer graphics system to provide the desired design loop automation.

DIE FORMING

When general three dimensional or compound curvature parts are required, a forming die is necessary to impart the appropriate strains on the workpiece. However, automatic control of such a process, while prone to the same problems cited above, is complicated by two additional factors: 1) The relationship between die shape and resulting part shape is extremely complex, because of the extensive and often non-unique in-plane material flows that occur; and it is not presently possible to develop a sufficiently accurate model for the machine output-(die shape) workpiece output (part shape) transformation even if finite element methods are used (see e.g. Wood (6)); 2) Forming dies are carefully machined sculptured surfaces and are therefore very costly to design and produce and modifications are very slow.

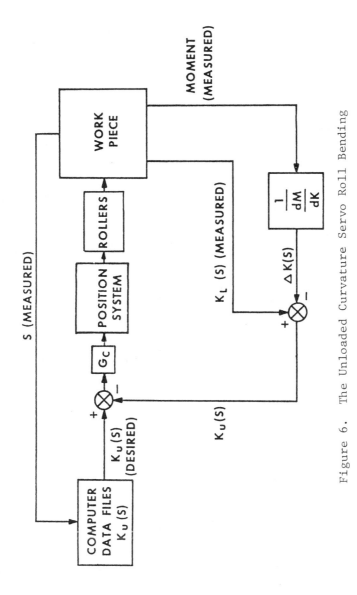

Figure 6. The Unloaded Curvature Servo Roll Bending

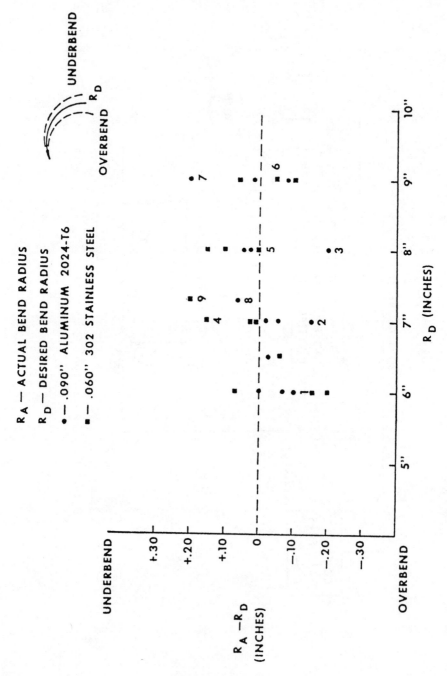

Figure 7. Radius Error for Roll Bending Experiments

To overcome these problems and still accomplish reliable shape control, Hardt and Webb (7) proposed the following solution. First, to provide a rapidly variable forming surface, they employed a "discrete die surface" that used 1/4" square elements whose relative positions could be varied under computer control (see Figure 8). That such a die can be used to successfully form parts has been confirmed (8.9) and a typical result is shown in Figure 9. With such a die, the control scheme shown in Figure 10 can be designed. The essence of this system is shape feedback, which is used to form a two dimensional shape of the discrete die until the correct part shape is achieved. Since no viable model of how to perform this adjustment can be developed, a simple shape iteration scheme was employed.

Figure 8. A Discrete Die Surface

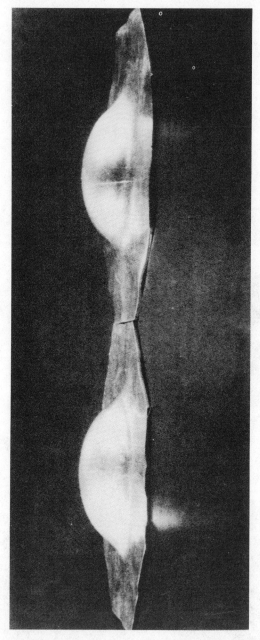

Figure 9.   Hydroformed Parts: on the left a continuous die was
used, on the right the die of Figure 7 was used.

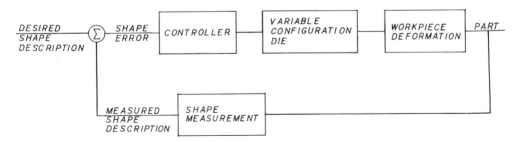

Figure 10. A Shape Control Loop for Die Forming

In this iterative control scheme no advance knowledge of sheet properties was required, rather it was simply assumed that deviations between a specific die shape and resulting part shape was caused.

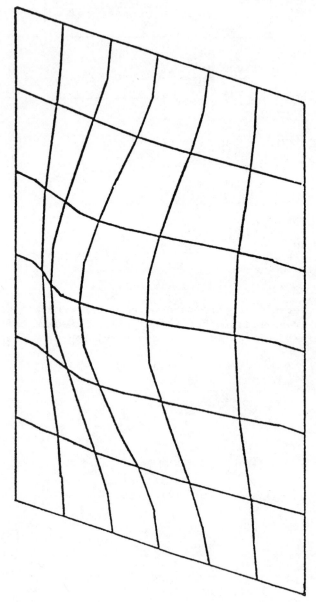

Figure 11.   The Desired Shape for the Forming Experiments

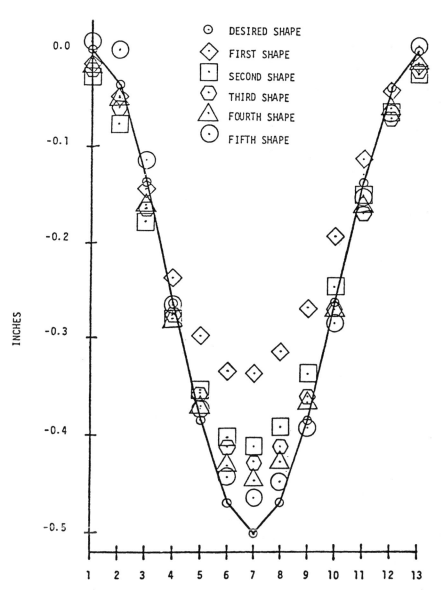

Figure 12.   Forming Results: cross section shape change for
             various iterations.

In a manner similar to the approach of Allison (2) and Mergler (3) this iterative scheme used the initial shape error as a direct estimate of the required die adjustment. Thus the forming cycle consisted of first setting the discrete die to the shape of the desired part. A part is then formed and the shape measured. This shape error, which can be expressed as a two dimensional array of elevation errors from a common datum, is in turn used to adjust the elements of the die a distance proportion to this elevation error. In this way regions of insufficient deformation are eventually "overbent" the correct amount and regions of excessive strain are reduced until acceptable shape error is achieved.

Experiments with this method, using a hydroforming configuration with a simple draw ring, showed that the shape illustrated in Figure 11 could be formed with reasonable accuracies in a limited number of forming iterations. A plot of the maximum sheet cross section for the initial set of iterations is shown in Figure 12. Subsequent forming cycles brought this cross section into close agreement with the desired shape.

CONCLUSIONS

The primary barrier to improved control of metal forming processes is uncertainty of material properties and process environment. This precludes exact predictive modelling and makes empirical methods slow and highly variable in outcome. Closed-loop control is ideally suited to such a problem, but application to metal forming operation is quite novel. Several examples have been presented where the essential forming properties of the workpiece are either implicitly or explicitly determined on-line and then used to control the process. It is expected that control methods such as these will lead to forming processes that are more predictable in performance and less costly to use for low part volumes.

REFERENCES

1. Nagpal, V., Subramanian, T. L. and Altan, T., "ICAM Mathematical Modelling of Sheet Metal Formability Indicies and Sheet Metal Processes". AFML-TR-79-4168, Nov. 1979, Air Force Material Laboratory Report, WPAFB, Dayton, Ohio.
2. Allison, B. T. and Gossard, D. C., "Adaptive Brakeforming", Proc. 8th North American Manufacturing Research Conference, 1980, pp. 252-256.
3. Mergler, H. W., and Wright, D. K., "Self-Adaptive Computer Control of a Ship Frame Bending Machine, Part II", NSF/RA-760394, March 1976.
4. Stelson, K. A., "The Adaptive Control of Brakeforming Using In-Process Measurement for the Identification of Workpiece Material Characteristics", Ph.D. Thesis, Department of Mechanical Engineering, Massachusetts Institute of

Technology, 1981.

5. Hardt, D. E., Roberts, M. A., and Stelson, K. A., "Material
    Adaptive Control of Sheet Metal Roll Bending", Proc. Joint
    Automatic Control Conference., 1981.

6. Wood, R. D., "The Finite Element Method and Sheet-Metal Forming",
    Sheet Metal Industries, August, 1981, pp. 561-575.

7. Hardt, D. E., and Webb, R. D., "Sheet Metal Die Forming Using
    Closed-Loop Shape Control", Proc. CIRP, 1982.

8. Hardt, D. E., Olsen, B. A., Allison, B. T., and Pasch, K.,
    "Sheet Metal Forming with Discrete Die Surfaces", Ninth
    North American Manufacturing Research Conference
    Proceedings, May, 1981, pp. 140-144.

9. Nakajima, "A Newly Developed Technique to Fabricate Complicated
    Dies and Electrodes with Wires", Journal of Japanese
    Society of Mechanical Engineering, March, 1968.

MODELING OF ORGANIC MATERIALS PROCESSING

Craig D. Douglas

Composites Development Division
Army Materials and Mechanics Research Center
Watertown, Massachusetts 02172

ABSTRACT

In an effort to reduce the empiricism involved in many polymer processing analyses and designs, this paper presents a finite element routine capable of predicting velocities, temperatures, stresses, and extents of chemical conversion in a variety of reactive polymer processes. The analysis accounts for diffusive, convective, and generation effects and also deals with transient situations.

After a brief introduction on the benefits of modeling polymer processes, the equations which govern nonisothermal reactive polymer flow are outlined. Using the equations, a finite element formulation is presented which results in a new element routine capable of analyzing complex flow domains. Program features include a penalty formulation to enforce fluid incompressibility, "upwind" weighting of highly convective terms, transient analysis capabilities, and iterative procedures to accomodate material nonlinearity.

The utility of the computer code is illustrated by sample analyses of representative industrial processes. These include pultrusion and reactive pressure flow which have parameters found in a number of other polymer processing operations. Material parameters as described in the literature are used and the effect of processing variables on the product internal temperatures and extents of chemical conversion are discussed.

Recommendations are made for future research and model development. These include alternative solution techniques to handle more complex geometries as well as the ability to model polymer behavior past the point of gelation.

71

INTRODUCTION

Quantitative analysis of polymer processing operations has been characterized by experimental measurement and observation. The development of more advanced materials and industrial practices has resulted in processes that cannot be fully understood by engineering intuition alone. A mathematical model such as the one described here is a valuable tool for the process engineer and allows him/her to identify trends or material/process interactions that might otherwise be overlooked.

Earlier modeling efforts consisted of solutions to exact equations with well-defined boundary conditions. A number of investigators have taken this approach (1-3). Research such as this established a firm foundation for analytical procedures based on continuum mechanics. The results of these investigations, although very useful, are limited to a family of problems where classical solution methods apply. As geometries, material behavior, and boundary conditions become more complex these classical approaches reach their limit of application.

Later attempts utilized numerical methods such as Runge-Kutta and finite difference approaches to gain a better understanding of more complex polymer interactions (4-10). This approach allowed a new class of problems to be analysed where material properties and boundary conditions are much more complex. Limitations of this method appear as the geometries increase in complexity. Despite this drawback, the finite difference method is a very powerful approach and still finds application in many problems analysed today.

The increased growth and acceptance of the finite element method, especially in structural analysis, has prompted use of this technique in fluid and thermal problems (11-17). The finite element method is similar to the finite difference method in that large and complex domains may be studied. Advantages of the finite element method include the capability to handle complex geometries, and since a global set of equations is solved simultaneously, local behavior of a variable is governed by the interaction of the total domain. Currently, a weighted residual formulation of the finite element equations is considered one of the best methods for modeling the complex boundary conditions and material properties so often encountered in polymer processing operations.

Numerical modeling techniques such as these should be used in conjunction with applied experimental methods. One important aspect of finite element modeling is it's ability to identify trends or interactions that may not be obvious when observed only by experimental methods.

The goal of this work is to develop a finite element routine capable of simulating various polymer processing operations.

Results of the model are then applied to specific processing operations. A uniform channel flow, typical of the entry region of a pultrusion process, is considered first. The effect of various die temperatures on the temperature and conversion profiles is shown. A second problem illustrates the effect of velocity changes on the temperature and conversion profiles in parabolic channel flow, such as occurs within the gate region of a resin transfer molding process. Finally, the value of these results are considered and suggestions for future model development are given.

REACTIVE PROCESSING

There are many different types of reactive processes, ranging from quiescent adhesive bonding to fast polyurethane cure used in automotive applications. Although most of these operations are characterized by low Reynolds numbers, the processes are complex and do not lend themself to simple analysis. They involve flow, heat transfer, and exothermic chemical reaction to form infinite molecular weight networks. Constitutive relations change dramatically during the process, with temperature affecting both reaction rate and physical properties of the material.

Reactive processes where flow is present can be found in a number of industrial applications. Three common examples are extrusion of thermosets, reactive injection molding, and pultrusion. Each of these have a characteristic velocity profile. In the case of extrusion, combined Couette (shear) and Poiseuille (pressure) flow is present. In reactive injection molding, Poiseuille flow is usually dominant throughout while in pultrusion, a uniform (plug) velocity profile exists. These velocity profiles are determined by the governing equations in conjunction with the applied boundary conditions and fluid properties, which are strong functions of temperature, molecular weight, shear rate, and other variables. The temperature in turn is affected by the viscous heat generation and the heat released or consumed by the reaction, and the reaction rate is also a strong function of the temperature. All of these variables interact and change in such a way as to make an intuitive grasp of the process almost impossible. There is obviously an advantage to being able to provide some sort of mathematical or numerical simulation of the process.

The equations which govern nonisothermal flow of a reacting polymer are derived in several texts on transport phenomena and polymer processing (1-2). Regarding velocity, temperature, and concentration of unreacted species as fundamental variables, the governing equations can be written as:

$$\rho \left[\partial\mu/ \partial t + U \cdot \Delta \right] = - \Delta P + \eta \Delta^2 U \qquad (1)$$

$$\rho C_p \left[\partial T/\partial t + U \cdot \Delta T \right] = QTK \ \Delta^2 T \qquad (2)$$

$$\left[\partial C/ \partial t + U \cdot \Delta c \right] = R + D \ \Delta^2 C \qquad (3)$$

The similarity of these equations is clear. In all cases, the time rate of change of a transported variable (velocity, temperature, concentration) is balanced by the convective or flow transport terms (e.g. $u \cdot \Delta C$), the diffusive transport (e.g. $D \ \Delta^2 C$), and the generation term (e.g. R).

The analyst seeks expressions for the space and time-dependent velocities, temperatures and concentrations which satisfy these equations and also boundary conditions which make real problems intractable: even if one were able to describe the boundaries mathematically, the resulting expressions would not likely be amenable to closed-form solution. In addition, many of the "constants" in the previous equations are often nonlinear functions of the problem variables. In reactive polymer processing, one might encounter such expressions as the following:

$$\eta = \eta^0 \gamma^{n-1} \ \exp \ [E_1/RgT] \ \exp \ (\beta\rho)(mw)^{3.4} \qquad (4)$$

$$Q = (\eta/2)(\ \gamma{:}\gamma) + R \ (\ \Delta H) \qquad (5)$$

$$R = -Km \ \exp \ (E_2/R_g t)c^m \qquad (6)$$

These expressions couple the governing equations and lead to a complicated set of equations. Equation 4 gives the relationship for viscosity as a function of shear rate, temperature, pressure, and molecular weight. The heat generation term given by equation 5 is composed of two separate parts: one being a contribution from viscous or shear heating and the second arising from a reaction heat which is usually exothermic and dominant, especially in the case of the polyurethane. The reaction rate as described by equation 6 also has an exponential dependence on temperature with the added complication that it is also a function of the local reactive group concentration C. Depending on the value of m, equation 6 not only couples the balance equations, but can lead to a nonlinear differential equation that must be solved.

The constants used in these three expressions can be found experimentally for the specific polymer system of interest. Macosko et al. have conducted experiments to obtain such data for the cure of polyurethane systems (18-19) while other researchers such as Hagnauer and Gillham have studied epoxy (20-23). The data supplied by these investigators are of extreme importance to model development and no simulation technique would be possible without them.

Reactive flow situations differ greatly depending on the material, geometry of the problem, and constitutive relationships used. It is often quite useful to define a set of dimensionless parameters to describe or characterize a particular problem of interest. Table 1 gives a list of dimensionless parameters with a short explanation of each and the material constants or variables that are used to evaluate each quantity.

Table 1. Dimensionless Groups sed in Reactive Flow

PARAMETER	DESCRIPTION	VARIABLES
Br(Brinkman)	viscous dissipation/heat conduction	η,u,K,T
N_d(Damkohler)	reaction heat/heat conduction	R,H,L,K,T
Pe_t(Peclet,temp.)	heat convection/heat conduction	u,L,ρ,C_p,K
Pe_m(Peclet,mass.)	mass convection/mass conduction	u,L,D
Re(Reynolds)	viscous forces/inertial forces	u,L,ρ,η
Pr(Prandtl)	viscosity/thermal diffusivity	ρ,η,C_p,K
Sc(Schmidt)	viscosity/mass diffusivity	η,α

Quantities such as those shown above are frequently used in problems concerning transport phenomena. Use of dimensional analysis greatly facilitates the solution process, especially when classical approaches are used. It also reduces the number of variables that must be followed thereby allowing one to easily get a "feel" for the problem. The analyst is tempted however to over-look specifics, especially when absolute values dominate the solu-tion as opposed to prescribed ratios.

Prior numerical modeling of reactive polymer flow problems has found its base in the finite difference method. A number of researchers have been active in the field (4,5,8,9,24,25,26). This work has provided the process engineer with further information to examine processes similar to those found in industry. Manzione and Osinski (27) recently reported some finite difference results. Here, enhanced color graphics routines were employed resulting in conver-sion maps that can be used as an aid in process optimization. Although very useful, this approach is still limited due to the nature of the solution method.

More recent efforts have been to use a finite element approach to examine the problem, and complex nonisothermal processes have

been studied (11-17, 28-31). Few of these, if any, include the
effects of a reacting polymer. One reference (32) uses the finite
element method to identify vitrified regions within a convection
cell, but does not emphasize the chemorheological phenomena occurring
prior to this point.

NUMERICAL MODEL

     The finite element method is now a well established tool used
in many areas of engineering analysis. The fundamental concept
assumes that any continuous quantity may be approximated by a set
of functions which have been discretized over the domain of interest.
The result of this approximation is a set of linear or nonlinear
algebraic equations that may be solved with the aid of a computer.

     A variety of methods exists by which the final set of equations
may be obtained, ranging from variational to weighted residual
approaches. Detailed descriptions are available in a number of
texts (33-38) and will not be treated here. Regardless of the
method used to formulate the finite element equations, the concept
of expressing problem variables in terms of an interpolation among
their values at various nodal points located on finite elements
within the problem domain is common.

     In this work a weighted residual approach is used and an
element routine has been developed to be compatible with a program
written by Taylor (39). Variations of the weighted residual approach
are used in the element routine to accommodate specific material
assumptions and problem characteristics. The formulation results
in a finite element routine capable of simulating two dimensional
flow of a reacting polymer. The program is capable of modeling nonlinear processes where mechanical, thermal, and chemical behavior
are coupled. It also has the added feature to analyse transient
problems.

     In its present form, the routine has been used primarily with
four-node bilinear elements. Excluding time, each node can be
associated with up to four degrees of freedom. These are:
velocities (in two directions), temperature, and concentration of
unreacted species.

     Fluid incompressibility is enforced via the penalty method
using selective reduced integration. Highly convective problems
are solved using an optimal upwind approach. The remaining element
contributions are a result of the conventional Galerkin weighted
residual approach.

A number of viscosity-state relationships can be employed.
These include: Newtonian, power-law, shear rate dependance,
Arrhenius temperature dependence, and effects from molecular weight
changes due to reaction. N^{th} order reaction kinetics are used,
with internal heat being generated by the combination of viscous
shearing and reaction energetics.

Execution of the program is extremely simple. Excellent pre-
and post-processing capabilities developed by Freese (40) allow
rapid generation and display of mesh data. These routines are
invaluable for interpretation of results where an intuitive grasp
of the problem is not always evident.

The program presently runs on a UNIVAC 1106 using double
precision arithmetic. An active column solution method is performed
in core, and all graphic representation of the data is displayed
on either a Tektronics 4014 or 4016 CRT.

The numerical model has a significant present ability to
simulate a wide range of problems in polymer processing. At the
same time, it is small enough to permit easy implementation in even
rather small processing facilities, and for quick familiarization
by process engineers.

PROCESS SIMULATIONS

The program's ability to simulate actual polymer processing
operations is demonstrated in this section. Simple processing
problems are considered, using a two-component thermosetting epoxy
system. Processing parameters typical of those found in industry
are employed. The examples are intended to show the response of
a reacting polymer system due to a variation in processing parameters such as temperature or pressure.

Thermal and kinetic parameters used in the model are shown in
Table 2. These values were determined experimentally by Price
(ref. 41) who studied a mixture of EPON 828/10%-mPDA and found it
to react relatively slowly. The data presented in this table were
used as input for the finite element model. The reader is cautioned
that these values are chosen only to demonstrate the program's
ability to model a process using experimentally determined values
having specific units. Thermosets used in commercial applications
are often more complex and their formulations are usually proprietary. If however, a manufacturer is able to supply the required
material properties for his formulation the process can easily be
modeled.

Table 2. Material Properties for DGEBA/10% mPDA from Price (41)

PROPERTY	VALUE	UNITS
Thermal Conductivity	0.00215	cal/sec-cm-K
Specific Heat	0.191	cal/gr-K
Density	1.1	gr/cm^3
Reaction Order	1.0	----
Heat of Reaction	116.7	cal/gr
Activation Energy	8720.0	cal/mole
Pre-Exponent	241.00	sec^{-1}
Gas Constant	1.980	cal/mole-K

Epoxy Pultrusion

The first reactive process considered is pultrusion of an epoxy composite. This is a continuous process used to manufacture products such as beams, truss members, or stiffeners. The operation consists of drawing a number of fiber tows from a creel through a resin bath and then into a heated die where the wetted fiber bundle cures producing a structurally sound part exiting from the die.

Thermal and chemical phenomena occurring within the die are of utmost importance to the process engineer. If the reaction proceeds too rapidly, especially in the case of epoxy, it is possible that the composite will bond to the die surface as has been observed by Tessier (42). This bonding problem results in a loss of production, damage to the part, and possibly the die.

Figures 1 and 2 present the results of a model for the entry region of a pultruder die. A plug (uniform) velocity profile equal to .25 cm/sec was enforced throughout the channel and the initial temperature of the resin mass was taken to be 273K. The material properties used for this model are found in Table 2. The entry region was 2.0 cm in length and 0.5 cm high. A modified Newton-Raphson technique was used to obtain the results shown in these figures. The solution converged to a set tolerance of $1. x 10^{-8}$ within 7 iterations.

In Figure 1, the thermal contours resulting from two different die temperatures are shown. When a wall temperature of 400K is applied, a boundary layer develops as cool material is pulled into the die. The material gradually increases in temperature and rises to a centerline value of 375K approximately 1.5 cm from the entrance.

When the die temperature is increased to 500K the thermal contours are significantly affected. In this instance, a centerline temperature of 375K occurs only 0.4 cm from the die entrance. At 1.5 cm, the centerline temperature has already risen to nearly 475K.

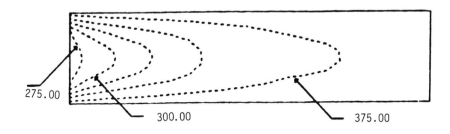

Tu=400

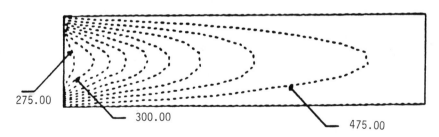

Figure 1.   Thermal Contours for Pultrusion of EPON-828/10% mPDA
            Resulting from Different Die Temperatures.

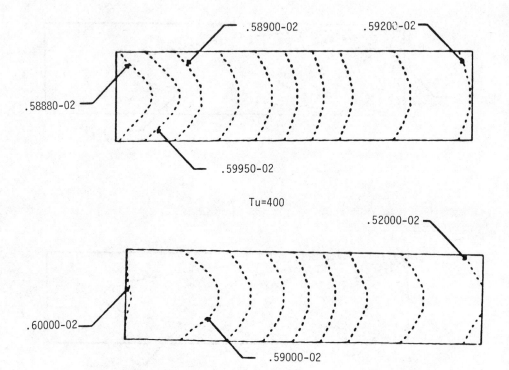

Tu=400

Figure 2.   Isoconversion Contours for Pultrusion of EPON–828/10
mPDA Resulting from Different Die Temperatures.

The predicted conversion contours for these two operating temperatures differed substantially. The inlet concentration was assumed to be .006 moles of reactive end-groups per gram, typical of a resin such as DER 332. The reaction proceeds quite slowly at a die temperature of 400K. Figure 2 shows that at 2.0 cm downstream, there is a centerline concentration equal to .00592 moles/gram, equivalent to a conversion value of 1.3%.

When the die temperature is increased to 500K, a much higher reaction rate is observed, with a concentration of .0052 moles/gram 2.0 cm from the entrance. This corresponds to a conversion value of approximately 13%. These two examples serve to illustrate the dramatic effect of die temperature on the thermal and conversion behavior within the composite during the pultrusion process.

Pressure-Driven Channel Flow

An extension of the pultrusion model just mentioned is pressure-driven channel flow, as might be encountered in a mold gate. Instead of material being drawn through the die at a constant velocity, a parabolic velocity profile develops due to an imposed pressure. This flow behavior is typical in many polymer processing operations.

One particular example is that of liquid injection or resin transfer molding (RTM). In this process a reacting mixture is introduced to a mold cavity by means of a runner and gate system. In the case of resin transfer molding, the cavity is pre-charged with a reinforcement. The reacting polymer then enters the cavity, impregnates the filler, cures, and the result is a reinforced component exhibiting superior mechanical properties.

The behavior of the polymer within the gate is critical to the outcome of the final part. For example, if there is excess heat present the polymer may react prematurely, causing blockage or an unwanted exotherm. Figures 3 and 4 show the results of a model for the gate region of a mold. These figures also show thermal and conversion profiles for two different cases.

In this example wall temperatures were maintained at a constant value and the maximum velocity was varied, as would result from an increase or decrease in injection pressure. The material properties used here are also taken from table 2., and the polymer was considered a Newtonian fluid with a constant viscosity of 500 Poise.

The thermal contours shown in Figure 3 illustrate the effect of velocity on the temperature field within the gate. At a centerline velocity equal to 1.0 cm/sec a pronounced boundary layer develops where the inner region of the gate is dominated by cool polymer over much of its length. When the velocity is reduced by

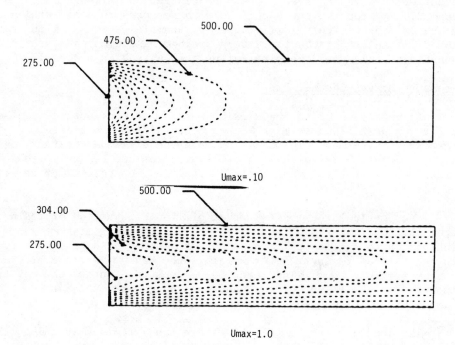

Figure 3. Thermal Contours in Pressure Flow of EPON-828/10% mPDA
           for Various Injection Velocities.

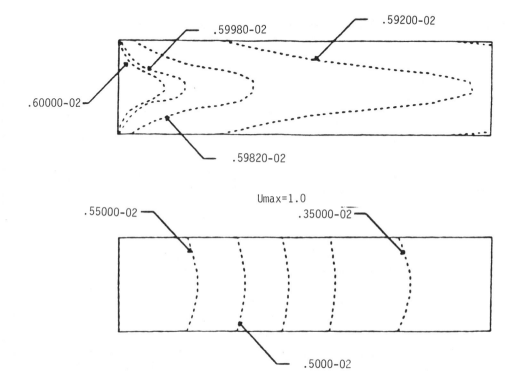

Figure 4.   Isoconversion Contours in Pressure Flow of EPON-828/10%
            mPDA for Various Injection Velocities.

an order of magnitude, internal temperatures approaching that of the wall occur less than 1.0 cm from the entrance. A case such as this could very likely cause problems such as premature gelation.

Figure 4 shows the conversion contours for these two flow situations. In the case where the centerline velocity is equal to 1.0 cm/sec centerline concentration values approach .00592 moles/gram approximately 2 cm from the entrance (again a conversion value of 1.3%). When the velocity is reduced to 0.10 cm/sec at the centerline, high values of conversion are observed and the boundary layer is lost. Here only .0035 moles/gram remain 1.5 cm downstream, a conversion of 41%! A process exhibiting this gate behavior could never yield a quality product because gelation is so far upstream. This example clearly demonstrates that pressure induced velocities play a dominant role in a process such as liquid injection or resin transfer molding.

These are only two of a number of polymer processing operations that could be used to demonstrate the program's ability to handle input parameters similar to those of an actual process. The next logical step would be to re-examine the RTM problem using non-Newtonian relationships where temperature and conversion values have an effect on the viscosity.

CONCLUSIONS

A numerical model has been developed which is capable of simulating reactive polymer processes similar to those found in industry. The most significant value of numerical results generated by the finite element program does not always lie in the ability to identify approximate spatial quantities within a domain, but in its ability to lead the analyst or engineer toward a better understanding of the total problem.

Once this experience is obtained, the engineer can draw upon it to make more intelligent decisions regarding operating input variables for a specific process in question. It is not difficult to imagine how information such as this might lead to such benefits as, improved productivity, better quality control, enhanced material properties, extended machinery life, etc..... in general a profit increase.

Finite element methods have already proven their value in structural mechanics, and it is only a matter of time before they are used as a tool in the field of reactive polymer processing.

Suggestions for Future Model Development

Reactive polymer processing provides the analyst with one of the most challenging problems to consider in manufacturing research today.

Judging from the available literature, the element mentioned in this paper appears to be the most comprehensive one available today for simulating reactive polymer flow processes.

Due to this relatively new application, the list of suggestions for future research appears almost endless. The following list initially suggests efforts germane to this work then tends toward a more general set of considerations.

- Addition of an out of core solution technique or installation on larger and more powerful computers, so that larger meshes may be treated

- Addition of anisotropic thermal conductivity (for pultrusion)

- Further emphasis on viscosity-state relations

- Upwind methods for higher order elements

- Addition of buoyancy effects

- Solid/fluid-fluid/solid transitions

- Viscoelastic effects

- Advancing front or free surface flow (liquid/gas)

- Gas/liquid interaction at the free surface (i.e. mold filling)

- Orientation parameters for chopped fibers

- Fluid/solid "stick-slip" effects

- Residual curing and temperature effects on internal stress distribution

- Shrinkage/curing phenomena

- Chemical/moisture degradation models (resulting from surface interactions)

- Solvent swelling models or solvent diffusion models (i.e. leeching of plasticizers)

- Stress induced diffusion or reaction

- Experimental verification for all of the above

REFERENCES

1. Bird, R. B., W. E. Stewart and E. N. Lightfoot, Transport
       Phenomena, John Wiley and Sons, Inc., New York, 1960.
2. Middleman, S., Fundamentals of Polymer Processing, McGraw-
       Hill Co., New York, 1977.
3. Crank, J. and G. S. Park, eds., Diffusion in Polymers, pp.
       85-100, Academic Press, London, 1968.
4. Rojas, A. J., Addabbo, H. E. and R. J. J. Williams, "The
       Flow of Thermosets Through the Nozzle of an Injection
       Molding Machine", Poly. Eng. and Sci., Vol. 22, No. 10,
       pp. 634-640, 1981.
5. Lindt, J. T., "Circular Couette Flow of a Polymerizing Fluid",
       Poly. Eng. and Sci., Vol. 21, No. 7, pp. 424-432, 1981.
6. Harry, D. H. and R. G. Parrott, "Numerical Simulation of
       Injection Mold Filling", Poly. Eng. and Sci., Vol. 10,
       No. 4, pp. 209-214, 1970.
7. Leonard, B. P., "A Survey of Finite Differences of Opinion
       on Numerical Muddling of the Incomprehensible Defective
       Confusion Equation", in T. J. R. Hues, ed., Finite
       Element Methods for Convection Dominated Flows, AMD, Vol.
       34, A.S.M.E., New York, 1979.
8. Domine, J. D. and C. G. Costas, "Simulation of Reactive
       Injection Molding", Poly. Eng. and Sci., Vol. 20, No. 13,
       pp. 847-858, 1980.
9. Manzione, L. T., "Simulation of Cavity Filling and Curing in
       Reaction Injection Molding", Poly. Eng. and Sci., Vol.
       21, No. 18, pp. 1234-1243, 1981.
10. Akay, G., "Stress-Induced Diffusion and Chemical Reaction in
       Nonhomogeneous Velocity Gradient Fields", Poly. Eng. and
       Sci., Vol. 22, No. 13, pp. 798-804, 1982.
11. Collins, B. R., S. M. Thesis, MIT, Cambridge, 1981, "Finite
       Element Analysis of Incompressible Viscous Flows by the
       Penalty Function Formulation".
12. Gresho, P. M. and R. L. Lee, "Don't Suppress the Wiggles-
       They're Telling You Something", Computers and Fluids,
       Vol. 9, pp. 223-253, 1981.
13. Roylance, D. and C. D. Douglas, "Finite Element Analysis of
       Nonisothermal Polymers Flows", Department of Materials
       Science and Engineering Research Report R82-1, MIT,
       Cambridge, 1982.
14. Reddy, J. N., "On Penalty Function Methods in the Finite
       Element Analysis of Flow Problems", International Journal
       for Num. Meth. in Fluids, Vol. 2, pp. 151-171, 1982.
15. Choo, K. P., Hami, N. L. and J. F. T. Pitman, "Deep Chemical
       Operating Characteristics of a Single Servo Extruder:
       Finite Element Predictions and Experimental Results for
       Iso-thermal Non-Newtonian Flow", Poly. Eng. and Sci.,
       Vol. 21, No. 2, pp. 100-103, 1981.

16. Reddy, J. N., "Finite Element Simulation of Natural Convection
        in Three-Dimensional Enclosures", A.S.M.E. TR 82-HT-71,
        1982.
17. Hieber, C. A. and S. F. Shen, "A Finite Element/Finite
        Difference Simulation of the Injection Molding Process",
        J. Non-Newtonian Fluid Mechanics, Vol. 7, pp. 1-32, 1980.
18. Brayer, E., Macosko, C. W., Critchfield, F. E., and Lawler,
        L. F., Poly. Eng. and Sci., Vol. 18, No. 5, pp. 382-
        387, 1978.
19. Lipshitz, S. D. and C. W. Macosko, "Kinetics and Energetics
        of Fast Polyurethane Cure", J. Appl. Polymer Sci., Vol.
        21, pp. 2029-2039, 1977.
20. Hagnauer, G. L. and I. Setton, J. Liq. Chromatogr., 1, 55, 1978.
21. Hagnauer, G. L. and D. A. Dunn, Materials 1980, SAMPE 12th
        National Technical Conference, Seattle, WA, pp. 648-655
        1980.
22. Gillham, J. K., "A Semimicro Thermomechanical Technique for
        Characterizing Polymeric Materials: Torsional Braid
        Analysis", AICHE Journal, Vol. 20 (6), pp. 1066, 1974.
23. Gillham, J. K., "A Generalized Time-Temperature-Transformation
        Phase Diagram for Thermosetting Systems", Office of Naval
        Research Report, NR 356-504, September 1979.
24. Lindt, J. T., "Flow of Polymerizing Fluid Between Rotating
        Concentric Cylinders", 8th International Congress on
        Rheology, Naples, 1980.
25. Lapidus, L., Digital Computations for Chemical Engineers,
        McGraw-Hill Book Company, New York, NY, 1962.
26. Pang, H. W. and B. H. Calude, "Modeling of a Modular-Polymer
        Process", Proceedings of Int. Sym. on Oilfield and
        Geothermal Chem., Stanford Univ., CA, pp. 307-324,
        May 1980.
27. Manzione, L. T. and J. S. Osinski, "Predicting Reactive Fluid
        Flow in RIM and RTM Systems", Modern Plastics, pp. 56-57,
        February 1983.
28. Mendelson, M. A., S. M. Thesis, "On the Numerical Simulation
        of Viscoelastic Flow", MIT, Cambridge, 1980.
29. Hami, M. L. and J. F. T. Pittman, "Finite Element Solutions
        for Flow in a Single-Screw Extruder, Including Curvature
        Effects", Poly, Eng. and Sci., Vol. 20, No. 5, pp. 339-
        348, 1980.
30. Phuo, H. B. and R. I. Tanner, "Some Coupled Flow and Heat
        Transfer Problems", Proceedings of the Int. Conf. on
        Finite Element Methods, 3rd, Univ. of NSW, Australia,
        pp. 731-746, 1979.
31. Hiber, C. D. and S. F. Shen, "A Finite Element/Finite
        Difference Simulation of the Injection Molding Filling
        Process", J. Non-Newtonian Fluid Mechanics, Vol. 7,
        pp. 1-32, 1980.

32. Gartling, D. K., "Finite Element Analysis of Convective Heat Transfer in a Pourous Medium", Sandia Natl. Labs TR# SAND-81-0463, 1981.
33. Zienkiewicz, O. C., The Finite Element Method, McGraw-Hill Co., London, 1977.
34. Huebner, K. H., The Finite Element Method for Engineering, John Wiley and Sons, Inc., New York, 1975.
35. Irons, B. and S. Ahmad, Techniques of Finite Elements, Ellis Honwood Series, John Wiley and Sons, Inc., New York, 1980.
36. Hinton, E. and D. R. J. Owen, Finite Element Programming, Academic Press, London, 1977.
37. Segerlind, L. S., Applied Finite Element Analysis, John Wiley, New York, 1976.
38. Clough, The Finite Element Method in Structural Mechanics, John Wiley, New York, 1965.
39. Taylor, R. L., Program FEAP, a finite element analysis program, found in reference 38, pp. 677-757.
40. Freese, C. E., "FEM.ABCDE" a finite element pre-post-processing, and analysis program, Army Materials and Mechanics Research Center, Watertown, MA, 1983.
41. Price, H. L., Ph.D. Thesis, "Curing and Flow of Thermosetting Resins for Composite Material Pultrusion", Old Dominion University, VA, 1979.
42. Tessier, N. J., Personal Communication, AMMRC, Watertown, MA, 1982.

WELD QUALITY MONITOR AND CONTROL SYSTEM

F. Kearney and Dawn Blackmon
U.S. Army Construction Engineering Research Lab.
P.O. Box 4005, Champayne, IL 61820

William Ricci
Process Research Div., Army Materials & Mechanics
Research Center, Watertown, MA 02172

ABSTRACT

A non-contact Weld Quality Monitor (WQM) system is being developed to detect, identify, and correct for deviations from established welding procedures and conditions which lead to weld defects in real time. The WQM continually measures with conventional transducers all primary process parameters such as current, voltage, and travel speed, and computes weld quality parameters such as heat input and weld bead geometry. In addition, the WQM monitors the spectral signature of the welding arc by means of a high resolution microprocessor controlled spectrograph. Here, the presence of weld pool and arc atmosphere contaminants, flux and shield gas effectiveness, arc energy input, and penetration/dilution into the base material can be determined. The spectral response from the welding arc and measurements of process parameters are then normalized, compared to preset operating limits, and processed in real time. Necessary adjustments to primary process parameters will be made by automated compensation devices to eliminate weld defects in real time. When necessary, the specific location of discontinuities will be provided to facilitate further inspection.

INTRODUCTION

During the welding process, changes in parameters, consumables, and the weld arc atmosphere can occur without the operator's knowledge. These changes may result in thermal damage to the base materials and defects (e.g., hydrogen induced cracking, porosity, embrittlement, lack of fusion and penetration) which seriously reduce the strength and service life of the welded joint. The

89

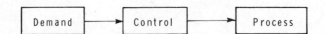

Figure 1.   Open loop control system schematic.

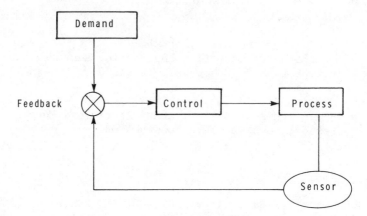

Figure 2.   Schematic of closed loop control system.

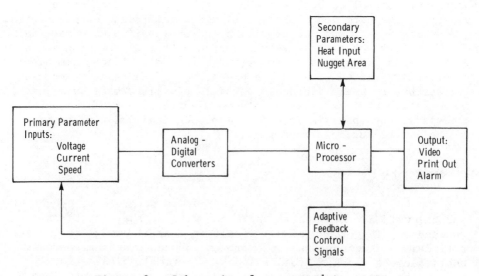

Figure 3.   Schematic of process data system.

cost of locating and repairing these defects constitutes a significant portion of the total weld fabrication cost. Prior attempts
have been made to develop techniques to quantitatively measure
welding conditions. These methods, however, often require direct
sensor-to-workpiece contact and are not considered suitable for
production environments because of sensor temperature limitations,
joint geometry limitations, and time lags which reduce the validity
and reproducibility of the information obtained.

Process control techniques have advanced rapidly in the past
decade, and as sensors, microprocessor technology, and artificial
intelligence methods advance also, it can only be assumed that the
rate of improvement in control technology will accelerate. The
welding process is capital intensive, prone to quality control
difficulties, and unpleasant for the human operator. Not surprisingly, a large effort is underway to automate welding and improve
weld process control.

In very simple welding controllers, an open loop exists, and
the process is controlled, in essense, by turning a knob. This
knob, which sets the "demand" signal, maintains the desired conditions (Figure 1). Most commercial welding power supplies, while
greatly improved in the past few years, fall into this category.
In a closed loop system (Figure 2), however, the controller utilizes
feedback from sensors that measure the value of certain quantities
of interest to improve process control. In the case of welding, it
is the primary process parameters, arc current, voltage, and travel
speed which are typically monitored. At CERL considerable attention
has been given to developing sensors capable of measuring various
welding parameters and resultant weld characteristics in real time.
This development has progressed on two fronts.

First, a process data system (PDS) (Figure 3)[1] has been
developed to measure arc current, voltage and travel speed. In the
PDS, a Hall effect device is used to obtain arc current values,
voltage is measured at the welding head, and one of a variety of
methods (tachometer, shaft encoder, etc.) is used to measure travel
speed. The analog output from these devices is digitized and the
data used by a microprocessor to determine whether the process is
being maintained within preset limits. Secondary information about
the weld quality, such as heat input and nugget area may also be
computed. The PDS generates this data in real time, and can be used
to stop welding, or alert an operator when an out-of-limits event
occurs; it can also be used to drive a feedback control system.

The second sensor system, which was developed in conjunction
with the Radio Research Laboratory of the University of Illinois,
is an optical data system (ODS) capable of detecting variations in
weld arc chemistry. During welding, loss of shielding gas, contaminated electrodes, contaminated shielding gas, or any of a number

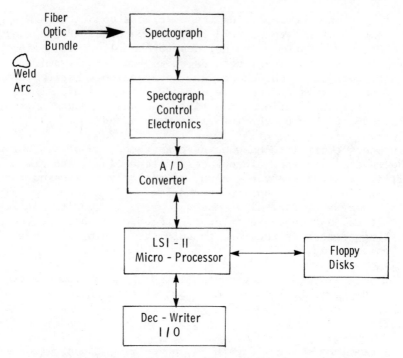

Figure 4.   Schematic of optical data system.

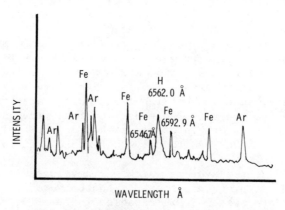

Figure 5.   Typical welding arc spectrum, centered at 6562 Å.

of other difficulties which may be associated with welding consumables can cause defects in weldments ranging from porosity to
cracking [2,3]. The detection of variations in weld arc chemistry
which are the cause of such problems is not a simple matter.
Because of the extremely high temperatures near the welding arc,
a remote sensing method is required.

Remote arc sensing has been accomplished by the development
of an opto-electronic method for observing the arc and noting
variations in its composition. The high excitation energies in the
arc plasma make spectroscopic analysis an attractive method of
obtaining qualitative and quantitative information about events
during welding. To this end, equipment capable of real-time
evaluation of the spectral features of the welding arc was developed
by the Construction Engineering Research Laboratory and the
University of Illinois [4-6].

EXPERIMENTAL TECHNIQUES

The opto-electronic system consists of a high-resolution,
microprocessor controlled spectrograph. A block diagram of the
system is illustrated in Figure 4. The optical radiation emitted
by the welding arc in the region from 300 to 1200 nanometers is
collected by a fiber optic bundle. The bundle, which is designed
to withstand the higher temperatures surrounding the welding arc,
is terminated at the entrance slit of an Instruments SA HR-320
Spectrograph/Monochromator. The HR-320, used as a spectrograph,
images a flat field spectrum onto a 1024 element linear diode array.
The photodiode array can be moved through the spectrum to obtain
samples with a band width of 60 nm. The resultant resolution of
this system is on the order of 0.6 nm. The photodiode array is
interfaced to a high-speed, analog-to-digital converter and a LSI
11/23 microprocessor. The spectral data, along with measurements
of the voltage, current and travel speed of the arc can be processed
or stored on floppy disks for later analysis.

As seen in Figure 5, the emission lines characteristic of the
various elements present in welding consumables (as well as
contaminants) may be observed using the equipment just described.
The intensity of an elemental emission line provides semiquantita-
tive information concerning the concentration of that element in
the welding arc. However, the welding arc is not a steady state
system and emission line intensities are not independent of insta-
bilities in the arc. But because so much iron and argon in the case
of argon shielded processes is present in the arc relative iron and
argon intensities and variations in the concentration of elements
of interest can be observed independent of changes in weld process
parameters. Accurate information about the concentration of these
elements may be obtained [7].

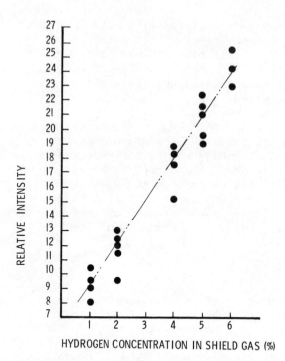

Figure 6. A linear rela-
tionship is found between
relative hydrogen line
intensity and hydrogen
concentration in
shielding gas.

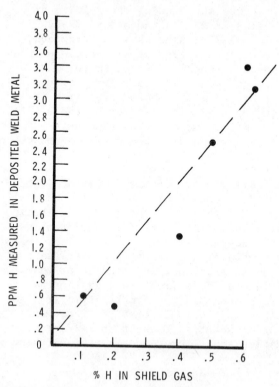

Figure 7. Hydrogen content
of shield gas vs. hydrogen
content measured in
weldment.

To accomplish this, an emission line characteristic of the element of interest must be identified. A suitable line must then be chosen for normalizing purposes. Finally, a series of welds must be made, under closely controlled, varying conditions so that a relationship between relative emission line intensity and elemental concentration may be obtained.

The possibility of detecting hydrogen in a weld is of particular interest because the presence of diffusible hydrogen in a steel weld can cause hydrogen induced cracking (often referred to as cold or delayed cracking) or embrittlement of the joint. Either of these phenomena can severely impair the mechanical properties of the weldment and necessitate costly repairs. Because of its wide economic significance, detection of hydrogen is of interest as an example of how the ODS can be used to detect quality problems with welding consumables [8,9].

Tests were performed to determine whether it would be possible to develop techniques for the detection in real time of the arc hydrogen concentration capable of causing hydrogen cracking in armor steel (MIL A 12560) used in the production of the Abrams tank. Two series of tests were performed to achieve this goal. Initial experiments were performed to relate hydrogen emission line intensity with the amount of hydrogen present. The second group of experiments correlated hydrogen emission line intensity with the amount of diffusible hydrogen present in the weld metal and the occurrence of hydrogen induced cracking in test welds.

Because MIL A 12560 steel has a carbon equivalent of 0.8, it is quite susceptible to delayed cracking problems [10]. The manufacturer's studies have indicated that dissolved H levels of less than 2 ppm may cause cracking. In the experiment, low concentrations of hydrogen gas were introduced into the argon shielding gas of a GMA weld and test welds were made and observed by the opto-electronic system previously described.

At the request of the Abrams tank's manufacturer, cruciform tests were used to detect cracking. Silicone oil immersion tests were used to measure the quantity of H present in the completed welds [11].

RESULTS AND DISCUSSION

As in previous experiments [12], a linear relationship was found between the concentration of H gas in the weld shielding gas and hydrogen line intensity (Figure 6). Since hydrogen emission line intensity is sensitive to hydrogen from any source, this has applicability to such problems as organic contaminants and moisture. In addition, it was found that hydrogen emission line intensity could be related to the occurrence of cracking, and loosely, to the

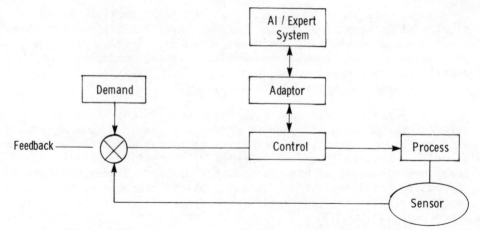

Figure 8.   Schematic of intelligent welding robot.

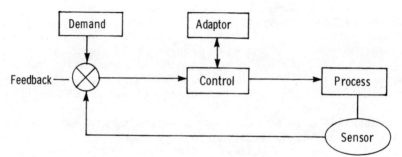

Figure 9.   Schematic of adaptive feedback control system.

Shield Gas Hydrogen Content	Type of Cracking
.1%	none
.2%	mild weld metal cracking
.4%	severe weld metal cracking
.5%	weld metal and underbead cracking
.6%	weld metal and underbead cracking

TABLE 1

type and severity of cracking which occurred (Table 1). As seen in Figure 7, a correlation between hydrogen emission line intensity and the hydrogen content of the resulting weldment was developed. While these data are preliminary results, it is clear that it is feasbile to use spectroscopic techniques to detect and measure hydrogen in the welding arc. Other consumables related welding difficulties which have proved to be measurable with this method include loss of flux in flux cored arc welding, and loss or reduction of shielding gas flow [12].

If improved sensors are to effectively enhance weld quality, better control is required. So, CERL's research efforts in this area have been directed towards adaptive feedback control, and a microprocessor based adaptive feedback controller has been developed for use with the PDS and ODS weld monitors. Adaptive feedback control differs from the control methods mentioned earlier in that it is capable of modifying the process to control function.

In control systems, the response to corrective control signals is described by transfer functions that are unique to a particular welding hardware configuration. Generally, these are complex functions of the form

$$K = A + iB$$

Each component in the control loop has a unique transfer function. For instance, in controlling wire feed speed, there would be two transfer functions: for the feed mechanism, a transfer function K_f, and for the servoamplifier driving the motor, the function K_a. The transfer function describing this section of the control loop would be

$$K_{wf} = K_a \cdot K_f$$

The operation of conventional control systems requires that the transfer functions be invariant [14].

Adaptive control systems are distinguished by their ability to compensate automatically for changes in system parameters and signal inputs. The wire feed system mentioned will serve as an example. If slippage occurs in the wire feed rollers, the transfer function K_f could assume a value beyond the controller's compensation range. In adaptive control, the system dynamics or control algorithms could be modified automatically to accommodate to this slippage (Figure 8). Because CERL's sensor systems are advanced, this adaptive control of welding has greatly increased flexibility.

The current work in real time weld monitoring and adaptive control forms the basis for an intelligent machine/robot welder which is under development. In the new system, an expert data base

using information acquired with the previously described sensor
systems will make intelligent decisions concerning process control
without human intervention (Figure 9). Work is underway on an
expert system capable of driving an intelligent robot welder. Two
major areas must be addressed. First, the ability to select the
correct solution to a given welding problem must be developed. That
is, the sensors must not only be able to detect an out of limits
event; the expert system must also be able to diagnose the cause
and correct the situation, or determine that a human operator must
be alerted. Second, a fitness-for-purpose criterion must be included. Thus, when a potentially defective length of weld is formed,
the significance of the defect may be assessed based on the ultimate
service conditions, so that an appropriate reaction (continue
welding, record defect location, stop welding, etc.) occurs.

The utilization of machine intelligence (expert system) is
feasible because of the opto-electronic sensors which form a computer
cognition system. The knowledge base for the system is being
developed through the specialized welding research in progress at
CERL. The inference machine used is a personal computer; presently
both IBM and Apple products are being used. Parallel processing
techniques will be used to couple the adaptive control and machine
intelligence functions. The hierarchies for the state space search
of the expert system are being developed to maximize response.
Current techniques indicate the system will operate adequately for
weld travel speeds of 20 inches per minute.

ARMY APPLICATIONS FOR THIS TECHNOLOGY

As previously mentioned, the WQM can be used to predict the
occurance of defects, such as hydrogen induced cracks, in armor
steel weldments made with solid and flux cored electrodes. Similarly, the embrittlement of reactive metal weldments such as those
based on Ti and Zr alloys can be predicted. Lack of penetration/
fusion in dissimilar metal weldments can also be determined and
adaptive control algorithms are presently being developed for the
real time in-process correction of these defects. One prime
application for this technology is in the joining of copper rotating
bands to ferritic based artillery shell bodies.

The application of the WQM is not limited to joining processes.
The WQM may also be used to control case depth and correct for surface melting in high energy beam surface modification processes such
as the laser or electron beam heat treating, cladding or alloying
of gear and bearing surfaces. WQM technologies can also be used
to monitor self propagating high temperature synthesis reactions
such as those used in the development of thermal batteries.

CONCLUSIONS

The results of ongoing welding research programs show that microprocessor based sensors of several kinds can be used to observe the welding process, and develop data concerning the weld as it is being produced. It is possible to correlate this data with the properties of the finished weld, making the use of microprocessor based control systems incorporating artificial intelligence a very attractive method of combined process and quality control; this is a result of the system's flexibility which offers a variety of responses (vary parameters, stop welding, note for later inspection) to the occurrence of an undesirable event. The effort to produce adaptive feedback control systems utilizing artificial intelligence should be pursued for welding since the process is very well suited to this type of control, both because of the cause and effect relationship between variation in process conditions and defects, and the human factors (operator discomfort, etc.) involved. The rapidly advancing state-of-the-art in all facets of the hardware and software required for this type of monitoring and control should make an intelligent welding robot a reality in the very near future.

REFERENCES

1. Kearney, F. et al., "Nondestructive Testing for Field Welds: Real Time Weld Quality Monitor - Field Tests," CERL Technical Report M-295, June 1981.
2. Carbon, K. W., Lawrence, F. V., and Radziminski, J. B., "The Introduction of Discontinuities in High Strength Steel Weldments," CERL Preliminary Report M-27, December 1972.
3. Lundin, C. P., "The Significance of Weld Discontinuities - A Review of the Current Literature," WRC Bulletin 222, December 1976.
4. Norris, M. and Gardner, C. S., "Microprocessor Controlled Weld Arc Spectrum Analyzer," RRL Publication No. 512, Technical Report, October 1981.
5. Norris, M. E., "Microprocessor Controlled Weld Arc Spectrum Analyzer for Quality Control and Analysis," CERL Technical Manuscript M-317, June 1982.
6. Gardner, C. S. and Kearney, F. K., "Electro-Optic System for Non-Destructive Testing of Field Welds," Proc. Conference on Developments in Welding Processes and Consumables, to be published.
7. Blackmon, D. R. and Hock, V. F., "An Opto-Electronic Technique for Identifying Weld Defects in Real Time," Proceedings SAMPE annual meeting, October 1983, to be published.
8. Graville, B. A., The Principles of Cold Cracking Control in Welds, Dominion Bridge, Co., LTD, 1975.
9. Evans, G. M. and Weyland, F., "Diffusible Hydrogen as a Criterion of the Quality of Welding Additives," Deutsch. Ver. fur Schweisstehnik, No. 50, p. 21-33, 1978.

10. Suzuki, H., "Cold Cracking and Its Prevention in Steel Weld-
     ing," Trans. TWS, Vol. 1, No. 2, 1978.
11. Ball, D. T., Gestal, W. J., and Nippes, E. F., "Determination
     of Diffusable Hydrogen in Weldments by the RPI Silicone
     Oil Method," Welding Journal, March 1981, 505-565.
12. Blackmon, D. R., Norris, M. E., et al., "Real Time Spectro-
     graphic Analysis of Hydrogen Concentrations in a Weld Arc,"
     Proc. 1st International Conference on Solutions to
     Hydrogen Problems in Steels, ASM, 1982.

HOT ROLLING SIMULATION OF ELECTROMAGNETICALLY CAST

AND DIRECT-CHILL CAST 5182 ALUMINUM ALLOY BY HOT TORSION TESTING

J.R. Pickens, W. Precht and J.J. Mills

Martin Marietta Laboratories
1450 South Rolling Road
Baltimore, Maryland 21227

INTRODUCTION

Much of the metal produced today is fabricated by rolling, extrusion, or forging at elevated temperatures. Consequently, it is desirable to optimize hot working parameters such as reduction schedule, rolling speed, allowable extrusion ratio, extrusion speed, and working temperature to maximize productivity. In addition, it is important to hot work a metal a) in a microstructural condition that lends itself to easy hot working, e.g., properly homogenized, and b) to a microstructure that gives desirable properties in the end product.

Commercial-scale ingots are often quite large, e.g., typically 40,000 lbs for aluminum rolling ingots, 100,000 lbs for steel. Consequently, it is impractical to work — and possibly scrap — many such ingots deformed under various commercial conditions to optimize a particular hot working operation. This has caused the need for hot working simulators, which can work relatively small amounts of material in a fashion that simulates commercial hot working. Furthermore, such simulators are extremely useful in establishing hot working parameters for newly developed alloys, which are costly and in limited supply.

In many respects the best hot working simulators are instrumented, scaled-down models of commercial units, e.g., an instrumented rolling mill or extrusion press. Unfortunately, such machines are expensive, and a typical metal working company would need one for each of the various commercial processes, e.g., a rolling mill, an extrusion press, and a forging press. This circumstance has created the need for relatively inexpensive, laboratory-scale mechanical test machines that can simulate hot-working. Several such units and techniques will be described herein.

REVIEW OF HOT WORKING SIMULATORS

Hot working simulation by hot tension testing is perhaps the simplest technique because a conventional tensile machine can readily be equipped with a furnace to deform specimens at elevated temperature.[1] This technique provides a measure of the material's resistance to deformation -- the flow stress (σ_o), which is of course in tension. Unfortunately, hot tension specimens often experience necking, which causes the local strain rate to vary and also complicates assessment of the material's working limits, or true strain-to-failure (ε_f).

Simulation by uniaxial compression[2-4] testing provides measurement of σ_o in compression using cylindrical specimens deformed between closing, lubricated platens. Friction between the flat faces of the specimens and the platens limits the uniform true strain to which the specimens can be worked. Friction constrains deformation on the flat faces causing the specimen to deviate from being cylindrical, and become "barrel" shaped. Concentric grooves are often machined into the flat faces to serve as reservoirs for lubricant, thereby extending the true strain to which the specimens can be deformed. Even with good lubrication, true uniform strains much above 1 are difficult, thereby limiting simulation capability for commercial processes involving large strains. The uniaxial compression simulator does not readily provide a means of measuring ε_f, and hot bend tests have been used to supplement the σ_o values obtained.

Simulation by plane strain compression[5-7] testing involves deforming a plate or sheet specimen by the advance of two opposing rectangular dies. The stress state in plane strain compression is the same as that during rolling so this technique is attractive for rolling simulation. However, to maintain uniformity of deformation, it is necessary to change die dimensions as the total deformation imparted increases, thereby interrupting the test. Moreover, the uniform strain that can be achieved is limited and strain heterogeneities often occur at the corners of the dies. As in the case of uniaxial compression, accurate measurement of ε_f is not readily achieved.

The hot torsion machine is, in our view, the most advantageous simulator.[8-11] Cylindrical dogbone specimens are deformed in torsion by rotating one end about the cylindrical axis while keeping the other end fixed. The technique achieves uniform deformation along the gauge section up to quite large strains.

Frictional problems are completely eliminated and true strains
large enough to represent commercial deformations can be
achieved. The surface stress on the specimens is calculated from
the measured torque and the equivalent tensile flow stress can be
calculated using the Von Mises criterion. The angular displacement of the rotating end can be converted to strain, and the true
strain-to-failure is readily obtained from the total angular displacement. Another advantage of hot torsion testing is that
interrupted deformation sequences can be performed up to large
strains, thereby simulating commercial hot rolling. Although
stress and strain vary with distance from the axis of rotation,
surface values can be used to assess σ_o and ε_f. Alternatively,
hollow tubular specimens can be used to reduce these stress and
strain gradients. The hot torsion machine can deform specimens to
strains high enough to simulate multipass rolling and assess
recovery and recrystallization times. Finally, the microstructures developed in hot torsion have been shown to correlate well
with those produced in hot rolling for several alloy systems.[12]

Martin Marietta Laboratories has the most sophisticated hot
torsion machine commercially available at this time. It was
developed by workability pioneer, C. Rossard,[13] and is
manufactured by Setaram, Lyons, France. We used this machine to
simulate the hot rolling of Al-4.5 wt%Mg alloy (5182) made by
electromagnetic casting (EMC) and direct-chill casting (DC). The
hot working data generated will be compared and discussed in
regards to observations of the alloys during commercial hot
rolling.

DESCRIPTION OF HOT TORSION SIMULATOR

The hot torsion system -- machine, interface, and computer --
has been described in detail elsewhere.[14] Briefly, the machine
consists of a lathe bed on which is located a 0-2000 rpm 5kW DC
motor, which drives a chuck, holding the specimen, through a gear
box and a fast-acting clutch-brake assembly. The stationary end
of the specimen is also mounted in a chuck which is rigidly
mounted to a stiff piezoelectric torque and axial force transducer. A furnace to heat the specimen is mounted on rails between
the chucks. The machine and its controls are interfaced to a
Digital Equipment LSI 11/23 computer with 256 kbytes of memory
through a CAMAC crate, an industry standard interface between
machines and computers. The computer, through software, controls
the motor speed and time of opening and closing of the clutch
brake. The program also collects time, torque, angular rotation,
axial force and temperature data every 1/60 s if required, and
tests for a predetermined strain or specimen break each sampling
interval. The system response is so fast that up to strain rates

of 10 s^{-1}, the computer can detect incipient failure quickly enough to stop rotation before the specimen has fully broken.

After completing a specimen run, the collected data are automatically converted to shear stress and strain using a modified Fields-Bakhoflen equation,[15] and using the Von Mises criterion, converted to equivalent tensile stress and strain. As mentioned above, the flow stress and strain-to-failure are also computed. The stress-strain data can be viewed on video graphics or plotted on a printer plotter.

MATERIALS

An ingot slice of commercial 5182 EMC was supplied by Kaiser Corporation, Pleasanton, CA. A similar slice of commercial 5182 DC was supplied by Martin Marietta Aluminum, Lewisport, KY. The materials were nominally of the same composition, particularly in Mg and Na contents, the latter known to promote edge cracking.[16-18] However, when the compositions, averaged in Table 1, were measured by both atomic absorption and the spark spectrographic technique from several locations on the ingot slice, the Mg contents were not identical. The 5182 EMC had about 0.5 wt% more Mg as well as a higher Na content.

Table I   Composition of 5182 DC and 5182 EMC Commercial Ingots (wt%)

Ingot	Si	Fe	Cu	Mn	Mg	Cr	Zn	Ti	Na
EMC	0.133	0.214	0.032	0.383	4.39	0.010	0.014	0.009	0.00020
DC	0.100	0.223	0.017	0.330	3.87	0.035	0.012	0.010	0.00015

EXPERIMENTAL PLAN

Torsion specimens of dimensions in Figure 1 were machined from each ingot slice, ∿6" below the top surface of the ingot, and ∿10" from the ingot sides. The long axis of each torsion specimen was machined parallel to the long axis of the ingot, which is the rolling direction.

Specimens were homogenized at a variety of temperatures and times corresponding to conditions reported by others on similar alloys. The effect of homogenization temperature at constant time, and homogenization time at constant temperature, on hot workability were specifically examined.

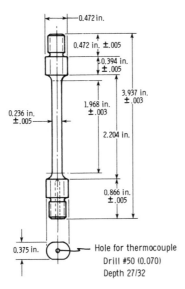

Figure 1.   Schematic of torsion specimen.

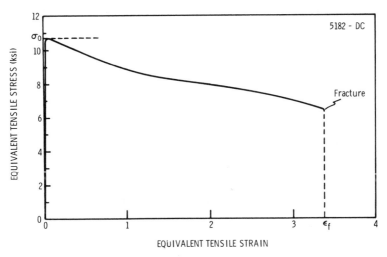

Figure 2.   Typical stress strain curve for 5182 showing how σ_o and ε_f are defined.

The mean-equivalent-tensile strain rate ($\bar{\dot{\varepsilon}}$) for a typical first pass on a commercial rolling mill was computed to be $1.2\ s^{-1}$ using the following equation:[19]

$$\bar{\dot{\varepsilon}} = \frac{2\pi}{60}\ N\sqrt{\frac{R'}{h_2}}\ \sqrt{\frac{1-r}{r}}\ \ln\left(\frac{1}{1-r}\right)\ \text{in}\ s^{-1}$$

where h_1 = entrance thickness
h_2 = exit thickness
N = angular speed of rolls in revolution/min
R' = deformed roll radius
r = reduction ratio = $\dfrac{h_1 - h_2}{h_1}$

The majority of the torsion testing was performed at the twist rate corresponding to a surface tensile strain rate of $1.2\ s^{-1}$. Specimens were deformed to failure at various working temperatures, $\sigma - \varepsilon$ data generated, and σ_o and ε_f computed from the data acquired by the computer. Unless specifically stated otherwise, the arbitrary deformation temperature of 920°F was used. In addition, the effect of working temperature was independently assessed for both as-cast and homogenized material.

We did not perform interrupted torsion testing to simulate pause times during commercial rolling, which has an effect on recovery. Our presumption is that ε_f during continuous deformation will be an approximate measure of resistance to edge cracking.

Microstructural observation and differential thermal analysis (DTA) of as-cast and homogenized material of each alloy were performed to aid in interpreting the hot working data.

RESULTS

A typical stress strain curve is shown in Figure 2. The elastic limit is exceeded quickly, and the peak stress (which we define as the flow stress, σ_o) is achieved. Softening occurs at higher strains until failure at ε_f.

Effect of Homogenization Time at 975°F (Working Temperature 920°F)

To compare the responses of the two materials to homogenization time, a typical homogenization temperature for an Al-4.5 wt% Mg alloy, 975°F, was selected from the literature.[20]

For as-cast material, σ_o was higher for 5182 EMC than for 5182 DC but ε_f was similar (see Figure 3). σ_o was fairly constant as a function of homogenization time at 975°F for each material

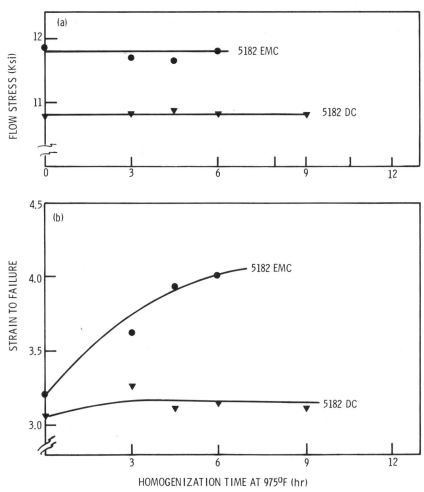

Figure 3.   Flow stress (a) and strain-to-failure (b) as a function
of homogenization time at 975°F. Working temperature is
920°F.

with the EMC being consistently higher (see Figure 3a). ε_f for
5182 DC did not improve significantly with homogenization time at
975°F but ε_f for 5182 EMC increased by 25% (see Figure 3b) over
the as-cast value after 6h. Thus, homogenization for up to 9h at
975°F does <u>not</u> improve the hot working behavior of 5182 DC, but
does improve that for 5182 EMC. It is clear that 5182 EMC
responds to homogenization at this temperature and 5182 DC does
not over the time span examined.

Effect of Homogenization Temperature at Constant Time of 6h (Working Temperature 920°F)

For a constant homogenization time of 6h, σ_o for each
material was largely insensitive to homogenization temperature
from 900-975°F, as shown in Figure 4, possibly decreasing slightly
for temperature greater than 1000°F. In addition, σ_o was consistently higher for 5182 EMC for any homogenization condition.

For a 6h homogenization, ε_f for 5182 EMC generally increased
with homogenization temperature from 900 to 1020°F (see Figure
5). In contrast, ε_f for 5182 DC did not improve up to 975°F, and
then increased by ∿15% at 1020°F. Moreover, throughout the entire
homogenization temperature range examined, the EMC material was
always more ductile at the rolling temperature used in this study,
(see Figure 5).

Effect of Working Temperature

For these tests, only 5182 DC was used. Material was tested
both as-cast and after homogenization at 1020°F for 6h. This was
the temperature found to maximize ε_f for 5182 DC.

For both as-cast and homogenized material, σ_o decreased with
working temperature from 800 to 1020°F (see Figure 6). Moreover,
the values for the as-cast and homogenized conditions fall in the
same population. For both conditions, ε_f increased with working
temperature from 880 to 920°F where it was maximum, then decreased
with working temperature up to 1020°F (see Figure 7). Throughout
this entire temperature range, the homogenized material exhibited
superior ε_f.

Microstructural Examination

The as-cast microstructures of 5182 EMC and 5182 DC are compared in Figure 8. The microstructure of the EMC material is
clearly more refined, exhibiting less coring, finer cell size, and

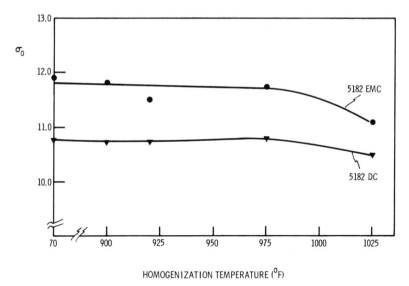

Figure 4.   Flow stress as a function of homogenization temperature
            for a constant time of 6 hours. Working temperature is
            920°F.

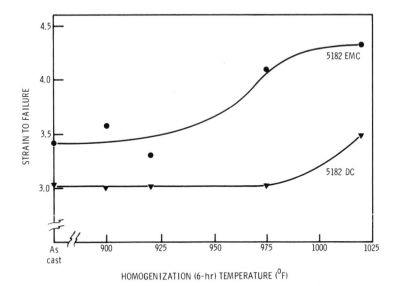

Figure 5.   Strain-to-failure as a function of homogenization
            temperature for a constant time of 6 hours. Working
            temperature is 920°F.

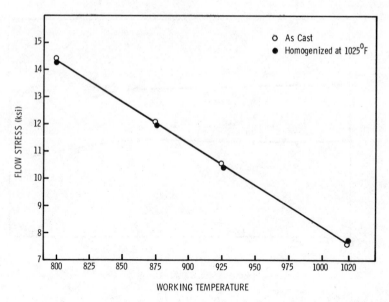

Figure 6.   Flow stress as a function of a working temperature.

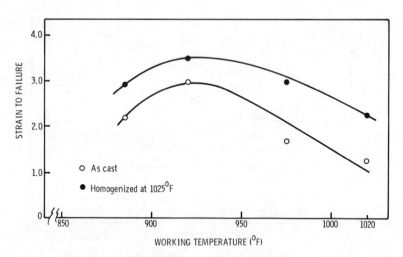

Figure 7.   Strain-to-failure as a function of working temperature
for as-cast and homogenized material.

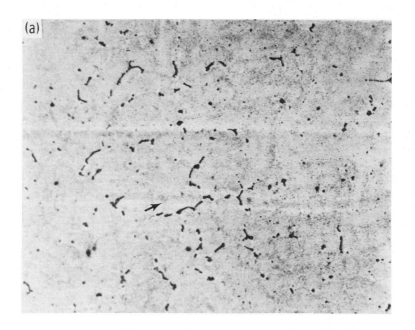

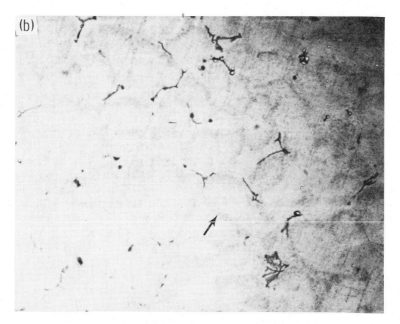

Figure 8. a) 5182 EMC as-cast b) 5182 DC as-cast (209X). Arrows
           indicate coring.

finer constituent particle size. Two types of constituents were observed (Figure 9): dark gray particles containing Mg and Si and light gray particles containing Fe, Mn, and Al, as determined by energy dispersive spectroscopy (EDS). The dark particles are most likely Mg_2Si, and the light particles (Fe, Mn) Al_6.[21]

Differential Thermal Analysis (DTA)

DTA was performed on 5182 DC only. Endotherms were observed between about 330-470°F, 650-870°F, and above 900°F. Each endotherm signifies a phase change. Metallographic analysis showed that this last endotherm was caused by eutectic melting between the constituent particles and the aluminum matrix. The onset of eutectic melting appears to occur just prior to the temperature where the peak in ε_f occurs (see Figure 10).

DISCUSSION

Each datum is the mean of at least two values. At times, the data exhibited scatter which we believe to result from compositional heterogeneities in the ingot from which the test specimens were machined. We later found that these heterogeneities, and the resulting scatter, could be greatly reduced by taking specimens from positions in the ingot corresponding to a solidification isotherm (generally the same distance from the ingot surface).

The flow stress for 5182 EMC was consistently 6-10% higher than that for 5182 DC which we attribute to the former's higher Mg content. Cotner and Tegart[22] measured σ_o as a function of Mg content in Al-Mg alloys at 930°F at a strain rate of 2.27 s^{-1}. From their work, we estimate that the 0.5 wt% difference in Mg composition between the materials should result in a 6-8% difference in σ_o, depending upon homogenization history. Thus, the difference in σ_o is approximately that expected for the Mg compositional difference between the two ingots, and not inherent in differences in the casting methods.

Homogenized 5182 EMC is clearly superior in ε_f, indicating that the EMC technology provides the alloy with an inherently greater hot ductility, and resultant greater resistance to edge cracking. The greater Mg content of 5182 EMC should decrease ε_f,[22] so the ductility advantage is perhaps understated. The improved response of the 5182 EMC to homogenization is probably a result of its finer cell and constituent size, resulting in easier diffusion paths.

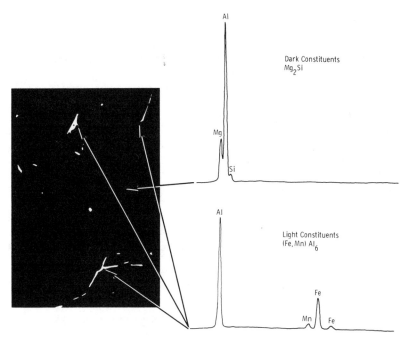

Figure 9.   Photomicrograph showing constituent phases in 5182 DC
            and EDS traces showing their elemental components.

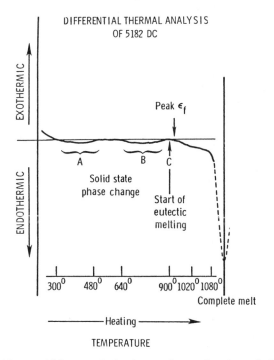

Figure 10.   Differential Thermal Analysis of 5182 DC.

The fact that ε_f passes through a maximum, followed by a <u>gradual</u> decrease as a function of working temperature, runs counter to the conventional wisdom of many plant metallurgists. They generally believe that hot ductility increases with working temperature and then rapidly decreases at the onset of melting. Indeed, Horiuchi et al.[23] found that ε_f measured in torsion increased monotonically for Al-2.5 wt% Mg alloy 5052 in the working temperature range 570-930°F. In addition, Coupry[24] found that ε_f for Al-6 wt% Mg increased with working temperature from 570 to 855°F. However, Thomson and Burman[20] found a peak in ductility measured by hot tension for an Al-4.4 wt% Mg alloy at 880°F. In addition, Cotner and Tegart[22] used hot torsion and found the highest ε_f at a working temperature of 1020°F for a high purity Al-5 wt% Mg alloy. Their result[22] is consistent with our findings because less eutectic melting of constituents would occur as the purity of the alloy is increased, resulting in a higher temperature for peak ductility. From the data collected in this study, we conclude that our hot torsion machine is extremely sensitive in measuring ε_f.

Observations of hot rolling 5182 EMC and DC have led to the conclusion that the EMC material exhibits less edge cracking.[25] This is true even when the liquated surface region (i.e., inverse segregation), which is present in DC but not EMC, is scalped prior to rolling. This observed resistance to edge cracking is entirely consistent with the ε_f data obtained, indicating that hot torsion testing correlates at least qualitatively with the commercial hot rolling of 5182.

CONCLUSIONS

1. 5182 EMC has better hot working characteristics then 5182 DC because of the former's greater hot ductility, indicating that 5182 EMC has better inherent resistance to edge cracking.

2. Flow stress decreases linearly with working temperature from 880-1020°F for 5182 DC.

3. The maximum hot ductility for 5182 DC occurs at a working temperature of 920°F, because eutectic melting between the aluminum matrix and the constituent particles begins at this temperature.

4. 5182 EMC is more responsive to homogenization that 5182 DC because of the former's finer cell and constituent size.

5. The greater hot ductility for 5182 EMC measured in hot torsion is consistent with the observations of its decreased edge cracking during commercial hot rolling when compared with 5182 DC.

ACKNOWLEDGEMENTS

The authors are grateful to Martin Marietta Aluminum for sponsoring and giving permission to publish this work. We appreciate A. Aromatorio's efforts in reviewing this manuscript, and the helpful comments of C. Brohawn and C. Mohr who supplied the test materials.

REFERENCES

1. P. Moore, Methods for studying hot workability: a critical assessment, in "Deformation Under Hot Working Conditions," ISI Publication 108 (1968), pp. 103-106.
2. J.L. Uvira and J.J. Jonas, Hot compression of Armco iron and silicon steel, Trans. Met. Soc. AIME 242:1619 (1968).
3. J.E. Hockett, Compression testing at constant true strain rates, Proc. ASTM 59:1309 (1959).
4. G. Fitzsimons, H.A. Kuhn, and R. Venkateshuar, Deformation and fracture testing for hot working processes, J. Inst. Met. 109:11 (1981).
5. J.A. Bailey and A.R.E. Singer, A plane strain cam plastometer for use in metal working studies, J. Inst. Met. 92:288 (1963-1964).
6. J.A. Bailey and A.R.E. Singer, Effect of strain rate and temperature on the resistance to deformation of aluminum, two aluminum alloys, and lead, J. Inst. Met. 92:404 (1963-1964).
7. G.R. Dunston and R.W. Evans, A technique for simulating the hot rolling of metals, Metallurgia 79:96 (1969).
8. F. Gatto, The hot torsion test for resolving extrusion problems, Light Metal Age 34:18 (1976).
9. M.M. Farag, C.M. Sellars, and W.J. McG. Tegart, Simulation of hot working of aluminum, ISI Special Report 108 (1968), pp. 60-67.
10. I. Weiss, J.J. Jonas, and G.E. Ruddle, Hot strength and structure in plain C and micro-alloyed steels during the simulation of plate rolling by torsion testing, "Proc. ASM Symposium on Process Modelling Tools," ASM publication, Cleveland, OH (1980).
11. I. Weiss, J.J. Jonas, P.J. Hunt, and G.E. Ruddle, Simulation of plate rolling on a computerized hot torsion machine and comparison with mill results, "Int. Conf. on Steel Rolling," Vol. 2, Iron and Steel Inst. of Japan, Tokyo, Japan (1980), pp. 1225-1236.
12. C.M. Young and O.D. Sherby, Simulation of hot forming operations by means of torsion testing, Tech. Report AFML-TR-69-294, Feb. 1970.
13. P. Blain and C. Rossard, Machines de torsion et de traction permettant la simulation au laboratoire des processus industriels de mise en forme par deformation plastique à chaud, "Lamiroirs," IRSID (1968), pp. 1-18.

14. J.J. Mills, K.C. Nielsen, and W. Merriam, Automating a hot
    torsion machine, in "Novel Techniques in Metal Deformation
    Testing," R.H. Wagoner, ed., TMS-AIME Conference Proceedings
    (1983), pp. 343-358.
15. D.S. Fields and W.A. Backofen, Proc. ASTM, 57:1259 (1957).
16. D.E.J. Talbot and C.E. Ransley, The addition of bismuth to
    Al-Mg alloys to prevent embrittlement by sodium, The Bulletin
    of the Bismuth Institute 23:1 (1979).
17. D.E.J. Talbot and C.E. Ransley, The addition of bismuth to
    aluminum-magnesium alloys to prevent embrittlement, Met.
    Trans. A 8A:1149 (1977).
18. C.E. Ransley and D.E.J. Talbot, The embrittlement of
    aluminum-magnesium alloys by sodium, J. Inst. Met., 88:150
    (1959-1960).
19. A. Gittins, J.R. Everett, and W.J. McG. Tegart, Strength of
    steels in hot strip mill rolling" Met. Technol. 4:378 (1977).
20. P.F. Thomson and N.M. Burman, Edge cracking in hot-rolled Al-
    Mg alloys, Mater. Sci. and Eng. 45:95 (1980).
21. Atlas of microstructures of industrial alloys, in "Metals
    Handbook," Vol. 7, Eighth Edition, ASM (1972), p. 244.
22. J.R. Cotner and W.J. McG. Tegart, High temperature deformation
    of aluminum-magnesium alloys at high strain rates, J. Inst.
    Met. 97:73 (1969).
23. R. Horiuchi, J. Kaneko, and A.F. Elsebai, The characteristics
    of the hot torsion test for assessing hot workability of
    aluminum alloys, ISAS Report No. 443, Vol. 35 (1), Tokyo,
    Japan, February 1970.
24. J. Coupry, "Utilization de la torsion à chaud dans l'étude
    des processus industriels de deformation, Rev. Alum.
    May:259 (1976).
25. Private communication concerning proprietary information with
    J.P. Faunce, Martin Marietta Laboratories, Baltimore, MD,
    April, 1981.

COMPUTER MODELING OF ENERGY INTENSIVE PROCESSES:

SOME FORD MOTOR COMPANY APPLICATIONS

C. Bagwell, R. Creese*, W. Evans, R. Fekete, C. Feltner,
J. Grant, R. Hurley, D. Piercecchi, H. Smartt**, and
S. Weiner

Ford Motor Company
Manufacturing Processes Laboratory
Research Staff
Detroit, MI   48239

INTRODUCTION

The oil shortage and subsequent price revolution of the 1970s focused attention on energy intensive industrial processes. After the initial housekeeping savings, the question became how efficiently can these processes be run? In order to answer this question, work was begun to develop computer models of such processes. The modeling applications discussed in this paper are: cupola melting of iron; steel slab reheating; and semi-permanent mold aluminum casting. All of the models developed have the potential to be used in developing automated process control strategies. However, the iron melting and steel reheat models were directed towards energy savings as the primary objective, while the aluminum casting model was aimed at improving casting quality. This paper provides an overview of some results achieved in applying these models rather than a discussion of model construction. Details of model construction can be found in the original publications[1-6]. The approach in this paper is to illustrate by example the use of computer models to provide further understanding and insights into three energy intensive processes. The models were used to define a manufacturing window, to study process modifications, and to provide a basis for automated process

---

*   Now at West Virginia University, Morgantown, WV
** Now at EG&G, Idaho Falls, ID

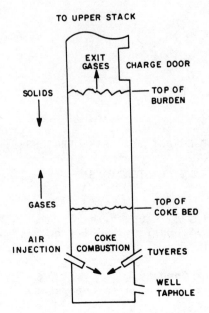

Fig. 1. Basic components of a cupola melting furnace.

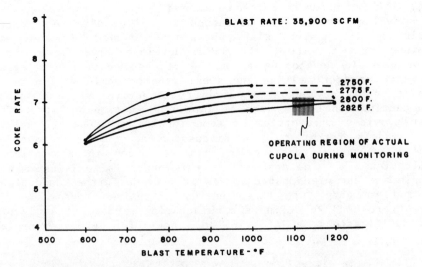

Fig. 2. Effect of blast temperature and metal temperature on coke rate for a 138-inch water-cooled cupola.

control. In one case, the aluminum casting model, an experimental
aluminum brake drum was made under computer control using a low
pressure casting machine.

Cupola Model

The cupola is a coke-fired shaft furnace used primarily for
the production of cast irons. The basic design is illustrated in
Figure 1. In the late 1970s, Ford Motor Company was melting 6300
tons per day (TPD) of cast iron using 13 cupolas in four found-
ries. The largest cupola had a capacity of 1300 TPD. The major
costs of cupola melting result from consumption of coke and
natural gas. The latter is used not only to preheat incoming air
but also to assist the combustion process for removal of carbon
monoxide from the stack gases.

The cupola model allows investigation of the effects of such
operating variables as blast rate, hot blast temperature, coke
quality, oxygen enrichment, furnace dimensions, and product chemistry and temperature[1]. Several applications of the model are
discussed including effects of blast temperature and oxygen
enrichment on coke rate, the ratio of iron melted to coke consumed. Portions of the model were also used to assess the feasibility of secondary air injection in place of afterburners to
reduce carbon monoxide emissions[2].

In order to validate the model, a monitoring program was in-
stituted. Metal temperature and composition, slag composition,
stack gas temperature and composition, and air blast rates and
temperatures were monitored directly. Melt rate and coke rate
were determined by monitoring the number and makeup of the
charges. The model was verified using data from four different
cupolas. These cupolas varied in size from 34" diameter to 138"
diameter. They also varied in number of tuyeres, blast rate and
temperature, and type of stack cooling.

Figure 2 presents model results illustrating the change in
coke rate with changes in hot blast temperature over the range of
316 to 649C (600 to 1200F). For each of the four curves indicated, all input conditions except hot blast temperature remain
constant. The effect of metal temperature is also indicated; each
of the curves represents a different iron temperature over the
range of 1510 to 1552C (2750 to 2825F).

The operating conditions encountered during monitoring of the
actual facility are approximated by the shaded area in Fig. 2. It
is apparent that these data represent a very small range of
potential operating conditions. Investigation of conditions
significantly different than the usual operating region involves

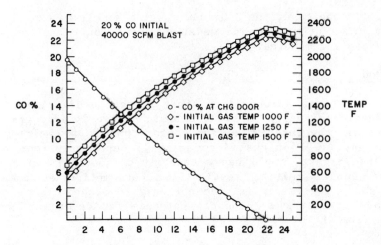

Fig. 3. Values of charge door stack gas temperatures and CO content for various initial injection point temperatures and injection amounts.

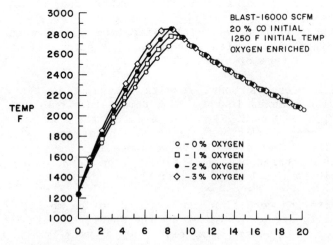

Fig. 4. Maximum temperatures obtained at injection point with various oxygen enrichment levels.

considerable risk if done with a production cupola, whereas the
model can provide estimates of these effects with no risk.

The results shown indicate that a lowering of blast temperature leads to a gradual deterioration in coke rate, especially at
lower blast temperatures. Also note that lowering the requirement
for final metal temperature leads to an improved coke rate.

The cupola model was also adapted to study the feasibility of
using secondary air injection to eliminate natural gas afterburners used to control CO emissions. Results on a laboratory scale
cupola indicated that the level of carbon monoxide in the off-gas
can be reduced to less than one percent using this method provided
a temperature of 1200°F is achieved[7].

Using an "Instant Combustion" approach, stack gas temperatures were calculated as a function of initial CO content and
amounts of secondary air for a cupola having a blast rate of
16,000 SCFM. The result of these calculations is shown in
Figure 3. As secondary air is injected, the charge door temperature rises and the residual CO content falls. At an injection
rate of about 9,000 SCFM, the peak temperature in Figure 3 is
attained. At this point, the residual CO content is zero.
Further increases in the amount of secondary air have a cooling
effect, since the hot gases are being diluted with ambient air.

If all the CO were burned, charge door temperatures would be
far in excess of the desired 1200 F. Figure 3 indicates that
1200 F can be reached by adding between 2000 and 3000 SCFM of air.
This would give a residual CO content of about 13 percent, which
would then ignite with the oxygen drawn in through the charge
door.

Additional calculations indicated that in order to minimize
the possibility of premature coke ignition it would be best to
keep the temperature at the charge door below 1300°F. Thus, the
problem became how to hold the temperature in the range of
1200-1300°F. To this end, the effects of oxygen enrichment and
secondary air preheating were also modelled. The effects of en-
riching the secondary air by up to three percent oxygen were
studied. The required total secondary air is reduced as the oxygen enrichment level is increased. However, as shown in Figure 4
the reduced injection amount has a negligible effect on maximum
temperature at the injection level. Thus oxygen enrichment of the
secondary air does not prevent the temperature from exceeding
1300°F. The results of preheating the secondary air were also
calculated to be negligible. Thus, the model calculations showed
that for this type and size of cupola the elimination of natural
gas afterburners by using secondlary air injection is feasible but

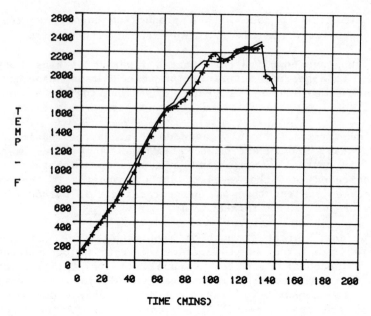

Fig. 5. Measured (+) vs. calculated temperatures 3/4" from top
slab surface; over a stationary support rail.

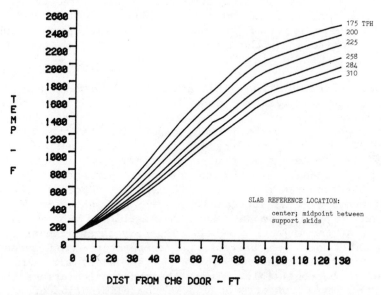

Fig. 6. Effect of changing production rate on the heating profile of a 7.75"-thick steel slab.

not practical unless a method is found to eliminate premature coke ignition.

In brief, the cupola model has been used to study a variety of actual or proposed operating conditions to improve energy efficiency without sacrificing product quality. As in all models, greater reliability should be placed on results obtained in the mid-range of model calibration/verification rather than on those obtained at extrema.

Steel Slab Reheat Model

Rouge Steel, a Ford Motor Company subsidiary is one of the ten largest steel companies in the United States. Its principal products are low carbon and high strength low alloy sheet steel. The current production facilities include coke ovens, blast furnaces, ingot molding facilities, soaking pits, rolling mills, and annealing furnaces. At the start of the hot rolling operation, there are three 125 foot long natural gas fired furnaces used to preheat the slabs prior to rolling. These furnaces have a capacity of 360 tons per hour (TPH) and can consume 1.1 billion Btu/hr. In the late 1970s, fuel efficiency was some 25-30%. This translated to a natural gas bill in excess of $30 million annually to support production of approximately 2 million tons of steel. Since the average weight of a steel slab is 18.5 tons, the rough cost of heating a slab to rolling temperature is slightly under $300.

As part of the overall effort to improve energy efficiency, a mathematical model of the steel slab reheating process has been developed[3]. The general objectives of the model are to provide a means of evaluating current operating practice and reheat furnace efficiency and to generate a basis for automatic furnace control. The specific short-term objective of the model is to determine the heat flux distribution, including useful heat flux to the steel, as affected by operating and design parameters, including heat input combustion air temperature, throughput, slab properties, and furnace geometry. In the longer term, the model was also designed to help define the on-line furnace control strategy which would be required for automated process control.

The mathematical model views the eight zone steel slab reheat furnace as longitudinally symmetric about its mid-plane. Each of the zones is treated as a separate system in evaluating radiative exchange. However, the eight zones are linked by the flow of flue gases. The model considers temperature variations across the furnace width and thermal properties of all insulation, and also includes convection and enthalpy flux in the heat flux balance. Heat transfer within the steel slab is treated as a transient problem using the finite difference method.

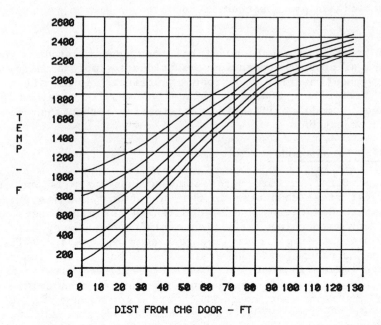

Fig. 7. Effect of slab charging temperature on the heating pro-
         file of a 7.75"-thick slab; 225 TPH at standard heat
         input; slab center between support skids.

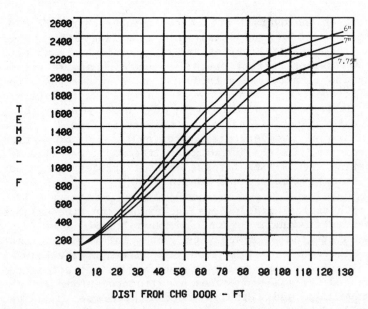

Fig. 8. Effect of slab thickness on the heating profile of the
         slab center; furnace time constant at 2.5 hours.

In order to verify the model, data were taken using the two thermal couples installed in each furnace zone as part of the fuel control system along with an additional 11 wall and roof thermocouples installed on one side of the furnace. Slab temperatures were measured by thermocouples imbedded in a test slab connected to an insulated recording device which rode with the slab through the furnace. A typical comparison of measured and calculated values of slab temperatures as a function of time is shown in Figure 5. Good agreement is obtained both in shape of the curve and in the absolute value of the temperatures.

The furnace throughput was varied from 175 to 310 TPH. The calculated slab heating curves vs. distance from the charge end for points between the support rails is shown in Figure 6 for the center temperatures. Thus the appropriate production rate corresponding to the desired discharge temperatures can be closely estimated for a given heat input pattern.

Slab charging temperatures of 75, 250, 500, 750 and 1000F were used at constant furnace heat input and a production rate of 225 TPH. Figure 7 illustrates the calculated results for the center of the slab for a 19.7 cm slab.

Slab thicknesses of 6", 7" and 7.5" cm were used at the same heat input pattern as in the other studies. In Figure 8, the calculated heating curves are given for a constant furnace time of 2.5 hours, giving production rates of 180, 210, and 235 TPH (165, 192, and 213 metric TPH) respectively.

In summary, the steel slab reheat process model considers furnace operating and design parameters including slab properties, heat input, combustion air temperature, throughput, and furnace geometry. Experimental and calculated results have been compared and the results of parameter studies of throughput, heat input, charging temperature and slab thickness summarized. Results of the model were used to effect a 5-10% reduction in natural gas usage.

Semi-Permanent Mold Aluminum Casting Process Model

Aluminum casting processes have traditionally been operator controlled. Using experience as a guide, the operator adjusts the fill parameters, coolant flows, die open time, and mold coating or lubrication to achieve casting quality while meeting the production schedule. In this complicated process, the operator uses feedback obtained from visual inspection of the casting. This section of the paper describes some recent results on computer control of low pressure casting. The point of departure from earlier studies is that the selection of the thermal set points

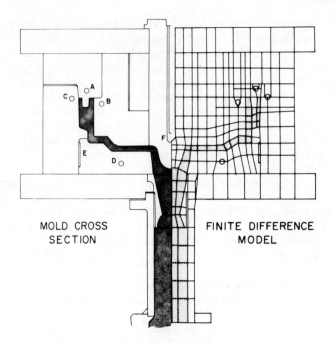

BRAKE DRUM LOW PRESSURE CASTING MOLD

Fig. 9.   Brake drum mold/casting nodal analysis.

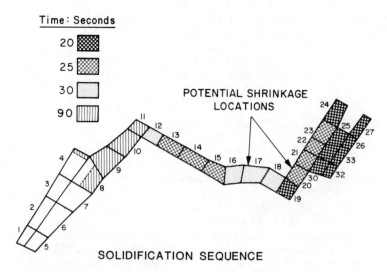

SOLIDIFICATION SEQUENCE

Fig. 10.   Solidification progression with time.

for die temperature control and die open is based upon a die thermal model.   The details of model construction and verification
have been presented elsewhere and will only be summarized in this
presentation[5,6].

The test casting for all of the work described is an experimental aluminum brake drum.   A cross section of the brake drum
mold is shown in Figure 9.   The cooling passages are designated A
through F.

The finite difference method of heat transfer analysis was
used to model temperature and heat flow in the mold/casting system
as a function of time.   In this method, the system to be modeled
is conceptually divided into a mesh of volume elements or "nodes."
Each node is then represented as a thermal capacitance which is
connected to neighboring nodes by finite thermal resistances.   The
division of the brake drum mold and casting into nodes is also
shown in Figure 9.   A total of 357 internal and surface nodes are
used in the model.

Two types of model verification studies were run.   In the
first series, the calculated results were used to generate the
predicted solidification times for various sections (nodes) of the
casting as shown in Figure 10.   The node numbers are shown for
reference purposes.   In the calculated solidification sequence,
nodes 16-18, and nodes 20-23 are predicted to solidify after the
nodes which feed them.   This implies that solidification shrinkage
might take place in these areas.   This prediction was verified on
X-ray examination of the castings.

While the prediction of the observed shrinkage is necessary
for model verification, it is not sufficient.   In addition, the
heat pick-up through the cooling passages (A-F) was also monitored.

The comparison of model and experimental data was made after
the mold had warmed up to operational (quasi-steady state) temperature.   For this paper, only the comparison for location (12)
is shown (see Fig. 11).   This particular location was chosen
because it is close to an observed shrinkage area.   On the whole,
the agreement between calculated and experimental data is fairly
good, both in terms of the temperature-time behavior and the absolute temperatures.   The greatest discrepancy noted is where the
model predicts a thermal arrest region not observed experimentally.   This can be attributed to two factors: undercooling in
the liquid and the small size of the thermocouple tip relative to
the node.   In the model, the node is held at constant temperature
until all the eutectic portion of the latent heat is dissipated,
giving average conditions over the volume, whereas the thermocouple responds to local conditions.   Lastly, the measured and

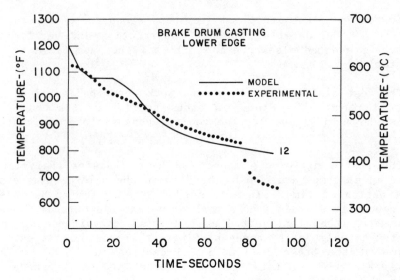

Fig. 11.   Casting temperature: thermocouple #12.

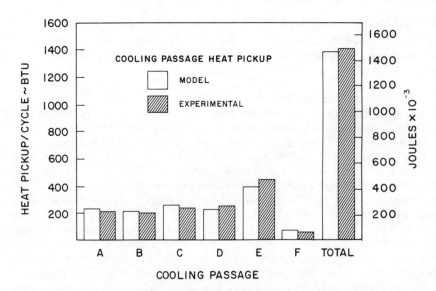

Fig. 12.   Mold cooling passage heat pickup.

calculated amounts of heat transported by the cooling passages during a casting cycle are compared in Figure 12.

Agreement is fairly good for all passages. The model predicts that 65% of the heat is rejected to the cooling lines, the remainder being transferred to the mold environment by radiation and natural convection. Note also the large amount of heat rejected by lines D and E. In the absence of other information, the high heat rejection at D would most likely be regarded as the cause of solidification shrinkage in the step area. Thus, an operator would probably attempt to reduce the shrinkage by slowing down the flow of air through these passages.

In parallel with model development and verification, work was initiated on computer control of die cooling and die opening for the low pressure casting machine/brake drum mold system. The existing thermocouples were used for feedback control. For mold opening the control thermocouple was located in the gate region.

System reliability and safety were enhanced by installing three separate thermocouples for die open control. In addition, the actuating relay was placed in series with the machine solidification timer and was under control of the pressure safety valve. Thus, the mold could not be opened until the pressure in the furnace was below 1 psig.

A strategy for controlling the air flow in the mold coolant passages was selected after consideration of the various factors which influence mold temperatures and mold and thermocouple response times to these factors. Cyclic average mold temperatures are influenced predominently by cooling level, overall cycle time, and mold open time. Superimposed on these average temperature levels are cyclic variations as the casting solidifies. The sensors are located in the mold such that they are fairly sensitive to this cyclic variation, rising and falling 100-200°F in a cycle. Typical data showing this response is presented in Figure 13 for thermocouple number 10. However, mold and thermocouple response to the other factors (i.e. cooling level) is relatively slow, as indicated in the Figure, taking place over a number of cycles. It was, therefore, decided to select high and low temperature limits within a cycle for each thermocouple to control coolant flow. Where possible, redundant thermocouples located at similar radial locations in the mold were used to increase system reliability.

As mentioned earlier ultimate selection of the temperature control limits for the mold was guided by the thermal model predictions. In the initial stages of control system implementation however, because of model unavailability, initial set points were selected based on experimental data obtained from earlier running of the mold. The "best" casting was selected from those produced

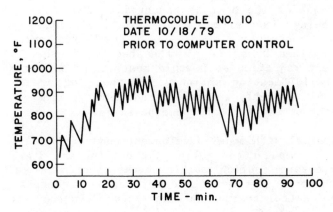

Fig. 13.   Typical temperature excursions.

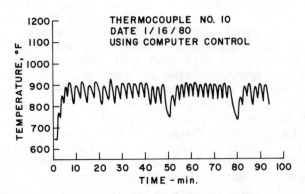

Fig. 14.   Typical temperature excursions under computer control.

during the previous run. Mold temperatures recorded during the cycle which produced this casting were then used to establish high and low temperature set points for the control system. The brake drum mold was then cycled under computer control and the actual temperatures reached were then recorded and analyzed. A typical plot of the data is shown in Figure 14. Comparing the results with (Fig. 14) and without computer control (Fig. 13) reveals a marked reduction in cycle to cycle temperature excursions under computer control. The overall reduction in standard deviation of the temperature during a typical steady state run prior to computer control was approximately 50%.

In addition to reducing the temperature range in any one cycle, computer control also significantly decreased the time required to reach steady state operation on start-up. Under computer control (Fig. 14), steady-state temperatures are achieved in the fourth cycle (~8 mins.) whereas 6 cycles (~17 mins.) are required without computer control. This beneficial effect is also seen when there is an interruption in steady state operation. When there was a delay at the 65 minute mark in the uncontrolled run, 6 cycles were required for restoration of the original operating conditions (Fig. 13). If, for any reason, the operator failed to shut off the cooling air during the delay the mold would be overcooled and additional scrap castings made to restore thermal balance. In the controlled run shown in Figure 14, two delays were deliberately introduced, one at 50 minutes and the other at 78 minutes into the run. As can be seen, thermal balance was obtained in both instances in the second casting after the delay.

The set points selected in the manner described above worked well in controlling temperature in some areas of the mold (as in Figure 14) and prevented mold cooldown in all areas during extended interruptions. However, in other areas of the mold maximum cycle temperatures exceeded the set points for extended time periods. This is due to the mold cooling system capacity being inadequate to maintain these areas at the desired low level over continual operation. This delineates the problem of selecting set points which, however desirable, may not be maintainable over extended operation. As described above, the initial set points were selected to correspond to measured mold conditions occurring during production of a particular casting. If, for example, this casting was the one from the fourth machine cycle shown on Figure 13, it can be seen that the measured mold temperatures will be lower than those physically maintainable under steady operation.

Although the use of computer control as shown did maintain thermal balance, or stabilized mold temperatures, and improve casting consistency, it did not eliminate shrinkage. This was particularly true of that found in the step area of the brake

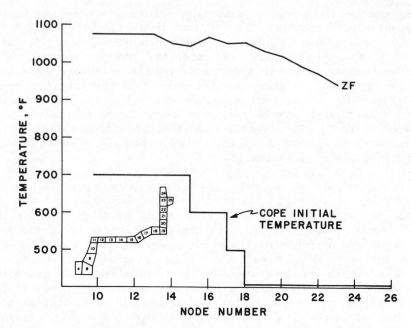

Fig. 15.   Temperature distribution with ideal initial cope temperature gradient.

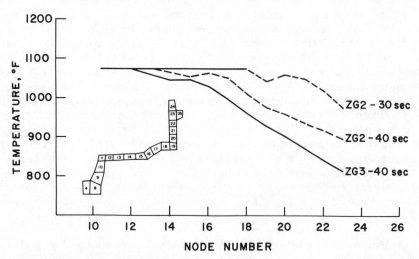

Fig. 16.   Temperature distributions with variable mold coating thickness.

drum. The existence of stable thermal conditions in the mold is desirable only if they correspond to those conditions which will produce quality castings.

The occurrence of shrinkage in the brake drum casting was consistent with the model calculations discussed earlier. In an attempt to alleviate the predicted shrinkage (and the actual shrinkage in the real mold) the model was exercised to determine the effects of various cooling alternatives on shrinkage in the step area. Most of the efforts were directed towards establishing a more favorable temperature gradient in the mold in order to achieve directional solidification. To this end, upper mold cooling (channels A, B, C) was increased while that in the lower mold (channels D, E) was decreased. The model was also used to assess the effects of longer cycle time and of higher metal temperature coupled with longer cycle time. Again no significant reduction in temperature inversion (shrinkage) was achieved.

In order to establish if further iterations on the cooling distribution would be worthwhile, a "best possible case" was evaluated. An assumed ideal temperature profile was imposed on the mold cope as shown in Fig. 15. This profile has a fairly steep gradient parallel to the casting face. Such a gradient would be difficult to achieve in practice, but serves to define the limiting effects of mold temperature distribution on the casting solidification sequence. As seen in Figure 15, a temperature gradient inversion is still present in the casting at nodes 14 and 15, and at node 18. Thus, it was established that variations in mold cooling alone could not overcome the effect of the greater section thickness of the step area relative to the bolt face area (nodes 13 and 14). This ability of the model to show that shrinkage in the step area could not be eliminated by changes in cooling practice with the existing die represents a significant improvement in ways of doing business.

After concluding that the shrinkage could not be eliminated by changes in cooling practice alone, attention was directed to retarding solidification in nodes 14 and 15 by use of different mold coating practices. The importance of mold coating practice in directional solidification has been verified both in high and low pressure casting. Accordingly, attention was directed to retarding solidification in this area (nodes 13 to 15) by use of thicker mold coatings. The results are shown in Figure 16. In one case, ZG-3, no inversion is found in the step area.

The coating thickness distribution suggested by case ZG-3 was evaluated on the actual mold. Thickness distribution was achieved by multiple coating layers in the various areas of the mold. Masking templates were employed to ensure accurate coating placement as well as reproducibility. Using this coating pattern

macroshrinkage voids in the step area of the casting were eliminated. Although some defects were still observed in castings from
the process, overall quality was significantly improved.

The semi-permanent mold aluminum casting model considers such
variables as mold cooling rates, aluminum temperatures, and mold
coating. The model was verified over the complete operating cycle
of a permanent mold system used to cast experimental aluminum
brake drums. After the model was validated, it was exercised to
define thermal set points and casting practice consistent with
shrink free castings. These calculated values were then input to
a minicomputer which was used to control die cooling and die open.
The control algorithm compares the readings from thermocouples
located in the die with the set points selected by the model and
makes the necessary adjustments.

Under computer control, cycle-to-cycle variability was
greatly reduced and the casting quality was more consistent.
However, by itself computer control of die cooling did not eliminate shrinkage porosity from the casting. This inability to
eliminate shrinkage porosity by mold cooling practice alone was
predicted by the thermal model. Following other predictions of
the model, castings of improved quality were made repetitively
under computer control after mold coating practice was modified.

SUMMARY

In summary, models of three energy intensive processes have
been developed, validated, and applied successfully to answer
operating questions. In one case a model has also been used to
improve process consistency and to provide direction for improved
product quality.

REFERENCES

1.   W. J. Evans, R. G. Hurley, and R. C. Creese, "A Process Model
        of Cupola Melting," AFS Transactions 80, 411 (1980).
2.   R. G. Hurley, W. J. Evans, and R. C. Creese, "The Potential
        for Combustion of Carbon Monoxide in Cupola Stack Gases
        Using Secondary Air Injection," Paper No. 80-34.7,
        presented at the 73rd Annual Meeting of the Air Pollu-
        tion Control Association, Montreal, Quebeck, Canada,
        June, 1980.
3.   W. J. Evans and C. D. Piercecchi, "Development of a Mathe-
        matical Model of the Steel Slab Reheating Process,"
        sumbitted for publication in AIME Transactions.
4.   J. W. Grant, "Thermal Modeling of Low Pressure Casting
        Molds," Paper No. 2-4 SDCE Low Pressure Casting
        Seminar, Troy, MI, October, 1979.

5.   J. W. Grant, "Thermal Modeling of a Permanent Mold Casting
         Cycle," Engineering Foundation, AIME, ASM Conference
         on Modeling of Casting and Welding Processes, Rindge,
         New Hampshire, August, 1980.
6.   R. J. Fekete, J. W. Grant, C. W. Bagwell and S. A. Weiner,
         "Verification and Use of Die Thermal Modeling for
         Computer Control of Low Pressure Casting," Proceedings
         of the Low-Pressure Tooling Seminar, International Die
         Casting Exposition and Congress, Cleveland, Ohio,
         June 1981.
7.   A. B. Draper, H. Choi, and J. W. Robinson, "Supplemental Air
         to Minimize Carbon Monoxide in Cupola Effluent," AFS
         Transacting 78, 487 (1978).

PROCESS MODELING OF P/M EXTRUSION

H. L. Gegel and J. C. Malas
AFWAL Materials Laboratory
Wright-Patterson Air Force Base, OH 45433

S. M. Doraivelu
Universal Energy Systems, Inc.
Dayton, OH 45432

ABSTRACT

A powerful and efficient thermoviscoplastic finite-element program called ALPID has been combined with advanced CAD/CAM methods for die design to permit mathematical and physical modeling of the extrusion of P/M materials. New constitutive equations and flow rules have been developed for porous materials as well as for the case of fully dense sintered and hot-pressed billets. These new equations were integrated into the analysis software through a material data base. The modeling results clearly show that the streamlined die geometry strongly influences the state of stress and uniformity of metal flow, which subsequently controls the quality of the product in terms of density, microstructure, and mechanical properties. Special attention is being given to the extrusion of RSR aluminum alloy powders as well as aluminum alloy powders containing SiC whiskers.

INTRODUCTION

Many non-conventional powder-metallurgy (P/M) alloys have been developed for aerospace applications, the intent being to achieve better combinations of strength, toughness, fatigue resistance and resistance to stress-corrosion cracking than in those alloys produced by ingot-metallurgy (I/M) techniques. Composite materials are being developed which incorporate SiC whiskers into the powder agglomerate for the purpose of increasing

137

the modulus of elasticity and the strength properties of aluminum alloys. Although these alloys have many attractive properties, they are difficult to fabricate by metal-working processes. A number of applications exist for these composite materials in aerospace structural components if extrusion or forging into complex geometries can be accomplished without fracturing the whiskers and changing the aspect ratio. The strength properties are strongly dependent upon the aspect ratio, while the effective modulus of elasticity is less strongly influenced by it beyond an aspect ratio of $\simeq 12$. Figure 1 shows the change in aspect ratio of SiC whiskers as a function of processing steps and Fig. 2, the dependence of the modulus of elasticity upon the aspect ratio.

The conventional method of extrusion using shear, conical, or parabolic dies is not satisfactory for this new class of materials. The internal shearing and turbulence of the material due to improper die geometry and control of metal flow cause fracture of the SiC whiskers, forcing them to be agglomerated. As a result components which are produced using conventional dies suffer from improper alignment of whiskers whose aspect ratio is reduced below the critical value of 12, as shown in Fig. 1. Thus, the modulus and strength properties of such products are markedly lower than expected values. Similarly other P/M alloys such as aluminum alloys (Al-8Fe-4Ce) and superalloys (Ni-base) suffer from defects related to discrete foreign particles and particle size distribution. These defects are also generated due to improper die design and control of metal flow during the deformation process. Therefore, innovative processing techniques and die-design procedures are needed for fabrication of these new alloys by metal-working processes such as extrusion and forging. Such innovation can be achieved by the development and appropriate use of the computer-aided engineering/computer-aided design/computer-aided manufacturing (CAE/CAD/CAM) concept which integrates material-behavior modeling, interface effects, geometric modeling, and analytical process modeling for generating data required for CAD/CAM of dies and process control. Using this concept, dies would be designed and optimized based on process simulation and manufactured using NC techniques. The process would be controlled by micro-processors using the optimized process parameters obtained from material-behavior modeling and process modeling.

Thus, process modeling of P/M metal-working operations becomes a key element in the development of a CAE system. A continuum-mechanics approach is being used to model different P/M metal-working processes utilizing the finite-element methods (FEM) for obtaining solutions. This requires a plasticity theory applicable to porous materials as well as constitutive equations with density as one of the parameters for studying densification during deformation.

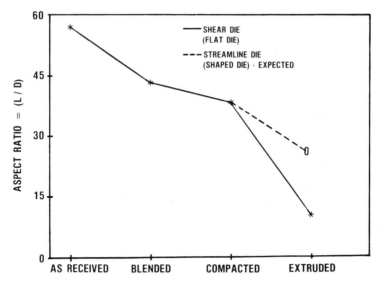

Figure 1.   Variation of Aspect Ratio of SiC Whiskers During
            Processing.

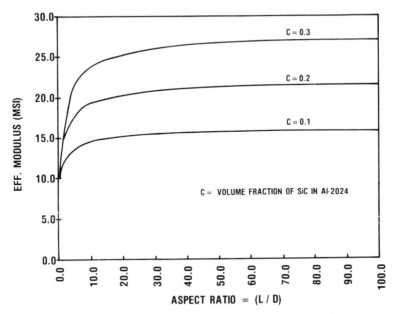

Figure 2.   Effective Modulus vs Aspect Ratio.

PLASTICITY THEORY -- POROUS MATERIALS

The design of billet-consolidation processes and of the dies used to forge or extrude porous P/M materials to full density requires a special yield function for development of the plasticity analysis. The plastic-flow behavior of porous P/M materials is more complicated than that of ingot materials because the hydrostatic component of stress influences the onset of plastic flow. The effect of hydrostatic stress is taken into account by considering a yield function of the form

$$AJ_2' + BJ_1^2 = Y_R^2 = \delta Y_o^2. \tag{1}$$

Here J_2' is the second invariant of the deviatoric stress, and J_1 is the first invariant or the hydrostatic component of stress. Y_o and Y_R are the yield stresses of fully dense and partially dense materials, respectively. A, B, and δ are functions of relative density.

Many researchers have determined these constants through heuristic arguments and the use of experimental results. Green [1] presented an analytical method which considered a uniform cubic array of spherical voids in a solid under states of stress corresponding to pure shear and hydrostatic compression. The results for these two stress states allowed determination of the two variables A and B. He assumed the stress distribution to be uniform in two directions through the minimum section of the array and the effect of the voids on the stress in the third direction to be negligible on the planes midway between the voids. Oyane, Shima, and Kono [2] determined A and B using more stringent assumptions. They found poor agreement between their theoretical and experimental results and, therefore, reported experimentally determined values in a later paper. Shima and Oyane [3] abandoned this analytical approach entirely and, instead, refined empirical relations obtained for their experimental values. Kuhn and Downey [4] also presented experimentally obtained values for these two variables. However, in the present study, the authors derived these variables taking into account the distortion energy due to the total stress tensor. The final equation obtained is

$$(2 + R^2)J_2' + \left(\frac{1 - R^2}{3}\right) J_1^2 = (2R^2 - 1) Y_o^2 \tag{2}$$

Theoretical curves along with experimental results for the uniaxial state of stress are shown in Fig. 3 for X7091 Al alloy. This figure not only shows good agreement between the theoretical and experimental results but also explains the influence of hydrostatic stress at various density levels. When full densification is achieved, the function automatically becomes the von Mises

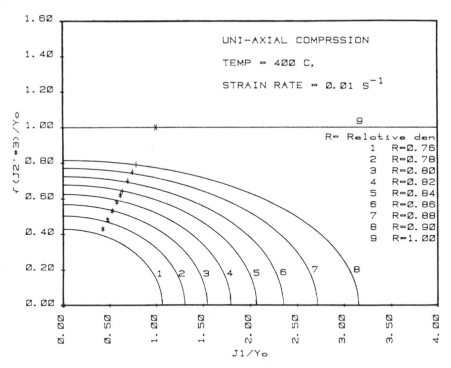

Figure 3.   Yield Surface in J_1 and J_2' Space.

yield function. Using this yield function, effective stress- and
effective strain-relative density relationships have been derived
and compared with the experimental results for the uniaxial state
of stress for the same alloy. Excellent agreement has been
observed between theoretical and experimental results, as shown in
Figs. 4 and 5. The theory is also applied to various P/M processes
such as frictionless closed-die forging, frictionless plane-strain
compression, and hydrostatic compression. These results are
presented in Figs. 6 and 7. The results show that beyond 95
percent of theoretical density, it is difficult to obtain full
densification in hydrostatic and closed-die forging by application
of pressure alone since the pressure required for full densification increases beyond that obtainable by commercial presses.
Therefore, this suggests that pressings should be made at a
temperature and strain rate which will accelerate diffusion and
permit full densification to be obtained.

In order to study the effect of strain rate upon densification during hot working, a constitutive model has been developed
by the authors for CT 91 (X7091) Al alloy. Uniaxial compression
tests have been conducted at different temperatures and strain
rates, as shown in Fig. 5; and using these results, the hot-
working temperature has been selected. The values of flow stress

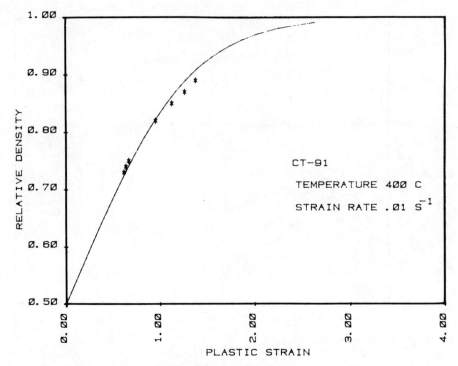

Figure 4.   Relative Density vs Plastic Strain.

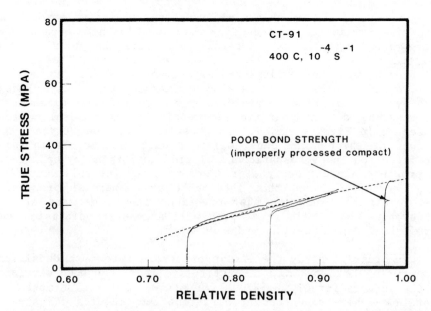

Figure 5.   True Stress vs Relative Density for Uniaxial State
of Stress.

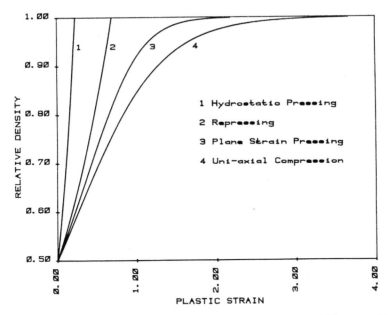

Figure 6.   Relative Density vs Plastic Strain for Different Processes.

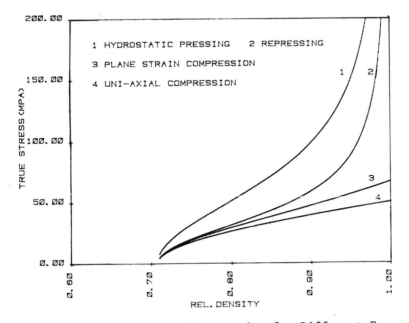

Figure 7.   True Stress vs Relative Density for Different Processes.

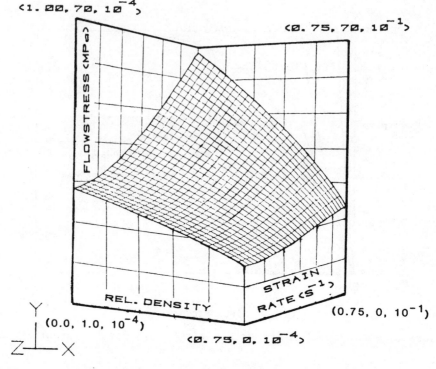

Figure 8.   Constitutive Behavior Model for X7091 Al Alloy.

are plotted in three-dimensional coordinates as a function of density and strain rate using the computer, as shown in Fig. 8. This data base has been used in the model for studying geometric hardening (densification) during various metal-forming operations for X7091 Al alloy.

ANALYTICAL PROCESS MODELING

Several methods are available for the analytical modeling of metal-working processes. However, the FEM has become popular for the solution of metal-working problems due to the rapid development of computers and numerical methods. Many studies have been conducted to solve various metal-working problems using this method. These were reviewed and discussed in detail by the present authors in another paper [5]. In all these studies the von Mises yield criterion has been used since the material under

consideration is always fully dense; however, for P/M applications
the influence of hydrostatic stress must be taken into account.
In the current study the method developed by Kobayashi and Oh [6]
has been used for predicting the metal flow through streamlined
extrusion since the starting billets are fully dense material.
However, a program called ALPID will be modified to incorporate
the yield function presented above for future studies which may
consider the use of porous billets. From this point of view, the
relevant equations required for modification are summarized below.

Equilibrium conditions, neglecting body forces:

$$\sigma_{ij,j} = 0 \tag{3}$$

where σ_{ij} is the stress tensor and ,j denotes partial differentiation with respect to j.

Compatibility conditions:

$$\dot{\varepsilon}_{11} = 1/2 \ (\frac{\partial V_1}{\partial X_1} + \frac{\partial V_1}{\partial X_1})$$

$$\dot{\varepsilon}_{12} = 1/2 \ (\frac{\partial V_1}{\partial X_2} + \frac{\partial V_2}{\partial X_2})$$

$$\dot{\varepsilon}_{ij} = 1/2 \ (V_{i,j} + V_{j,i}) \tag{4}$$

and $\dot{\varepsilon}_v$ is not equal to zero; mass is assumed to be constant.

Boundary conditions:

$$\sigma_{ij} \ n_i = F_j \ \text{on} \ S_F$$

$$v_i = V_i \ \text{on} \ S_V \tag{5}$$

where n_i is the unit vector normal to the surface; $|fs|$ = frictional
stress with the proper sign.

Equations (3) - (5) can be put into the variational principle
as

$$\Phi = \int Y_R \left\{ \frac{2}{3(2 + R^2)} \ [(\dot{\varepsilon}_1 - \dot{\varepsilon}_2)^2 + (\dot{\varepsilon}_2 - \dot{\varepsilon}_1)^2 + (\dot{\varepsilon}_3 - \dot{\varepsilon}_1)^2] \right.$$

$$+ \left. \frac{1}{3(1 - R^2)} \ \dot{\varepsilon}_v \right\}^{1/2} + \int (\int f_s dV_s) dS - \int F_i V_i dS. \tag{6}$$

APPLICATION - EXTRUSION OF P/M 2024 Al-20 VOL.% SiC WHISKER
MATERIALS

Extrusion is a primary metal-forming operation which is
characterized by its ability to produce long structural shapes and
induce very large strains in the product in a single operation.
The former is useful in the manufacture of various shapes including
rounds (produced generally using shear dies), and the latter is
useful in conditioning of the billet to improve its workability
for further manufacturing operations such as forging. Use of the
extrusion process--particularly with reference to P/M materials--
has specific advantages, provided due consideration is given to
die design and control. The present investigation is concerned
with producing complicated structural components from difficult-
to-extrude 2024 Al-20 vol.% SiC whisker material.

Metal-Flow Simulation Using FEM

Constitutive relations were developed for this alloy by
compression testing small cylindrical specimens as a function of
temperature and strain rate. A MoS_2 lubricant was used to assure
homogeneous deformation at a strain rate of ~ 0.50 true strain and
temperatures of 400 to 550°C. The results are shown in Fig. 9 for
a strain rate of 10^{-2}/sec. Similar results were obtained for
different strain-rate values. These relations were integrated
along with the die geometry which was modeled by means of a method
developed by Gunasekera, et al., [7] for the metal-flow analysis.

Effect of Geometry on Round-to-Round Extrusions

The FEM program was employed in the modeling of metal flow
for round-to-round extrusions using a variety of die geometries,
the objective being to determine the optimum die geometry for
avoiding velocity discontinuities at the entry and exit of the
die. The die shapes were shear, conical, parabolic, and
streamlined surface (generated on the basis of radius or area).

The shear die is the most common one used in industry,
particularly for conventional aluminum extrusions. The design and
manufacture of this die are relatively easy. However, correction
for thermal expansion/contraction and die deflection due to
loading must be taken into account for proper dimensional control
of the product. The length of the die land is also important.
Figure 10(a) shows the velocity field for a 9:1 reduction using a
shear die; the arrows represent the magnitude and direction of the
velocity vector of various points in the deformation zone. The
results are compared with those obtained for other die geometries.

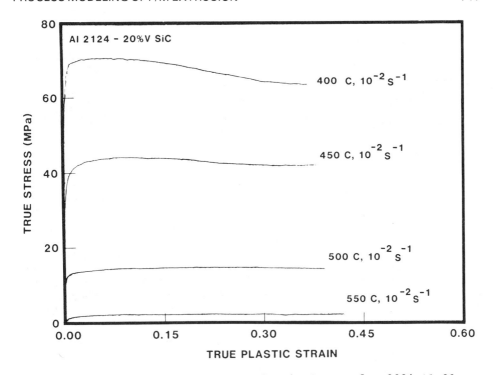

Figure 9.   True Stress vs True Strain Curves for 2024 Al-20
            Vol.% SiC.

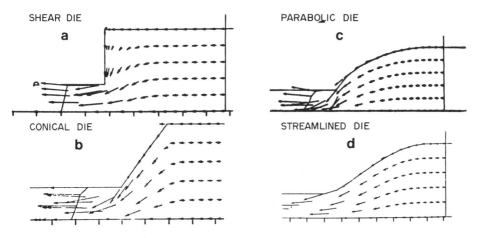

Figure 10.   Results of Analytical Modeling for Different Die
             Geometries.

The conical die, which is relatively easy to design and manufacture, is also fairly well known and is being used for conventional and hydrostatic extrusions. In this case material flow is more uniform than for the shear die. However, the conical die may induce some turbulence near the die exit due to the abrupt change of material flow at that point, as shown in Fig. 10(b).

The parabolic die has smooth entry; but the exit is sharp, creating turbulence and discontinuity in the velocity, as shown in Fig. 10(c).

Streamlined dies have smooth entry and exit, thus avoiding sharp velocity discontinuities at the exit or entry side, as shown in Fig. 10(d). Two types of streamlined dies are available, both based on cubic spline fits. The first is based on the radius (along the axis of extrusion) at the die entry and exit; the second is based on the cross-sectional area at the die entry and exit. The splines differ in shape for these two cases. Experience has shown that both types of streamlined dies have applications depending on the material used and other process parameters. However, the results of the analysis indicate that the die surface generated based on radius tends to produce a better combination of hydrostatic and shear stresses and homogeneous material flow which is required for the production of quality products.

On the basis of results on different die geometries and shapes (both analytical using FEM and physical using experiments), it was concluded that the optimum die geometry for round-to-round extrusions of the material tested is a streamlined die (a cubic spline) based on radius. Hence, for studying the effect of other variables such as die length, friction, and material, this die was chosen.

Effect of Die Length

In the past, optimization of die length (or semi-cone angle) for both wire drawing and extrusion has been undertaken using the minimum forming-force criterion. A typical force-vs.-die-length curve is shown in Fig. 11. The argument put forth is that a shorter die length reduces die/material frictional work but increases redundant work due to abrupt reversal of the velocity direction. A longer die length, on the other hand, produces a higher frictional work component but reduces redundant work. An optimum die length can be selected, as shown in Fig. 11. When extruding exotic materials or metal-matrix composites with fibers, this criterion may be inappropriate. In this case optimization

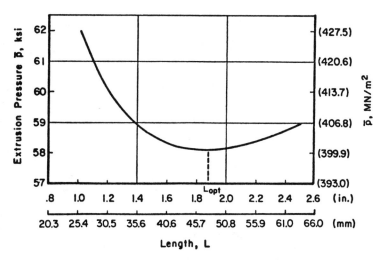

Figure 11.   Extrusion Pressure vs Die Length (Used for Selection
            of Optimum Die Lengths) from Ref. 8.

must be carried out on the basis of material flow rather than
minimum extrusion force, the objective being to maintain near-
homogeneous deformation.

Figure 12(a) shows the results of modeling using two different
die lengths. The 50-mm die creates turbulence near the die exit
and, hence, is unacceptable for producing a homogeneously deformed
product. The 75-mm die is adequate. Hence, for these given
conditions (i.e., 9:1 extrusion ratio of round to round, friction
factor m = 0.3), the aspect ratio of the die should be at least
one in order to achieve a near homogeneously deformed product.
The results of physical modeling also show that the metal flow is
more uniform when the 75-mm die is used, as can be seen in Fig.
12(b). For a higher extrusion ratio, a longer aspect ratio is
required. Experiments were also conducted using the material with
whiskers, and the effect of turbulence was observed in the 50- and
75-mm dies. In the former, the whiskers were broken into small
pieces and also were oriented in different directions as shown in
Fig. 12(c). This type of microstructure leads to poor mechanical
properties in the extruded bar. However, in the 75-mm die, the
turbulence effect was minimum and the whiskers were longer, as
shown in Fig. 12(c). This product exhibits better mechanical
properties and meets the design goals.

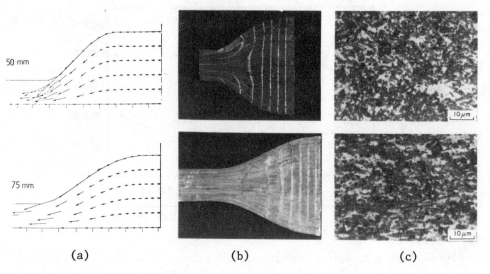

(a)   (b)   (c)

Figure 12.   (a)   Results of Analytical Modeling
              (b)   Results of Physical Modeling
              (c)   Microstructures of Extruded Product through
                     Streamlined Dies of 50 mm and 75 mm Length.

Effect of Friction

    In order to study the effect of friction on material flow,
the die length (for an extrusion ratio of 9:1) was fixed at 75-mm
(i.e., L/D = 1). Figure 13 shows the various stages of ram
position for two frictional conditions, m = 0.1 and m = 0.3, all
other variables being held constant. The analytical-modeling
results show very little differences in grid distortion and
velocity field. It is concluded that the effect of friction on
the material flow pattern is negligible for a small variation in
frictional conditions.

Effect of Extrusion Ratio

    Different extrusion ratios (9:1 and 9:4) were used, with all
other variables such as friction (m = 0.3) and die length (75 mm)
being held constant. As can be seen in Fig. 14, the material flow
was more uniform when an extrusion ratio of 9:4 was used for the
75-mm die.

    The effective strain distribution for this case (extrusion
ratio of 9:4) is shown in Fig. 15, which also shows that the
material undergoes uniform deformation when proper die length and
extrusion ratio are selected for the given frictional condition.
This result is important for the forging industry in determining

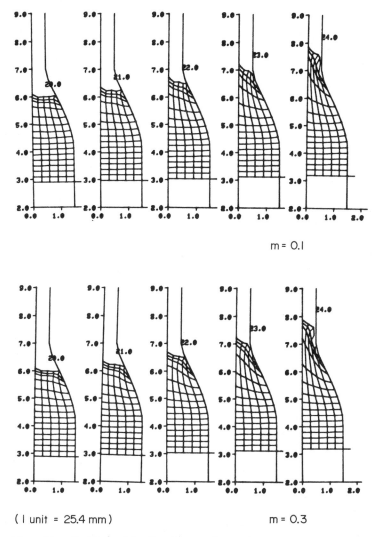

m = 0.1

(1 unit = 25.4 mm)   m = 0.3

Figure 13.   Metal-Flow Simulations Through Streamlined Dies for
             Different Frictional Conditions.

the amount of strain induced in the billet during conditioning of
the billet by extrusion.   A certain amount of uniform strain in
the product across the section is necessary in the billet to
obtain uniform recrystallization and, hence, to increase the
workability.   Thus, these results are highly useful for the metal-
working industry.

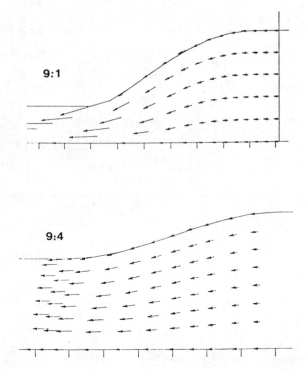

Figure 14.   Effect of Extrusion Ratio on Metal
             Flow (Die Length - 75 mm).

Computational Details Used for Solving the Above Problems

        In the current investigation, the input-data preparation was
done manually. However, in the future an automatic mesh generator
will be used. The incremental displacement of the upper die for
each step was 0.02 times the height of the original billet. The
velocity of the ram for the analysis was 25.4 mm/sec., and one of
the dies used was stationary. The measure of convergence repre-
sented by $||\Delta u||/||u||$ was chosen to be 10^{-4}. For all these
problems, four-node isoparametric quadratic elements were selected.
For displaying output data such as grid distortion and point
velocity field, a post-processor such as FEMGRA developed by Oh of
Battelle Columbus Laboratories was used.   MOVIE.BYU developed by
Brigham Young University was used for displaying the stress and
strain contours. The typical time required to solve these problems
for one run varied between one and five minutes CPU time in the
CDC Cyber 750 computer.

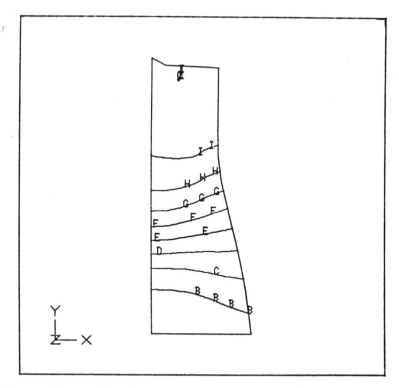

Figure 15.   Effective Strain Contour of the Product During
             Streamlined Extrusion (Extrusion Ratio 9:4, Die
             Length - 75 mm).

CAD/CAM of Die

Design and Data Transfer

   Once the die length for approximating streamlined flow has
been determined from the FEM analysis, co-ordinates of the die
surface are generated at various sections along the die length
using a fully interactive die-design package developed by
Gunasekera [5]. This package incorporates innovative concepts and
principles of die design discussed in detail in Ref. 8 and is run
on the PRIME 550 Computer. It requires input such as product
cross-section, die length, and type of cubical spline. Once these
data are fed into the program, the co-ordinates of the die surfaces for any complex shape are automatically generated.

Figure 16 shows a perspective view of the EDM electrode for the manufacture of a streamlined die for a T section. The die coordinates are output to a nine-track magnetic tape in ASC11 format for transfer to an Applicon Computer Graphics System which utilizes a PDP 11/34 computer. This transfer is necessary since the manufacturing group is not on the same computer system as the engineering-development group. This coordinate data transfer is rapid and requires no IGES compatibility. At this time the data are reformatted for graphics display to verify the correctness of data transferred.

Prior to the installation of the CAM system in the manufacturing shops, the surface coordinates were output to a printed medium that required interpretation and then transformed into templates which were used to check manual machining operations. This method proved to be time consuming, error prone, and incapable of maintaining the desired tolerances.

When designing the die, consideration must be given to the selection of the fillet radii since these radii determine the maximum diameter of the ball-end mill which can be used in the machining operation.

CNC Programming

The NC graphics software on the Applicon System is not capable of outputting a cutter-location (CL) file where complex surfaces exist. For this reason, a copy of the original data points must be transferred to the computer-operating-system (RSX) side of the graphics system. The data are formatted and compiled first using an F-Mill routine and then using Automatically Programmed Tools (APT) processing. F-Mill is a program which

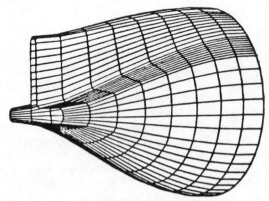

Figure 16. Perspective View of EDM Electrode
for the Manufacture of Streamlined
Die for T Section.

utilizes point data to develop surfaces not handled by conventional APT definitions. Once the surface normals are defined in this format, the F-Mill output file is used by APT as a data file for generating a CL file. This CL file is comprised of a series of tool paths offset tangentially from the surface by the value of the cutter radius.

Previous attempts to use B-Surf in the manufacture of the die in Fig. 17 proved unsuccessful due to inaccurate definition of the surface where abrupt changes to the geometry occur.

CNC Machining

Machining of the EDM electrode is accomplished on a Computer-Numerical-Controlled (CNC) milling machine capable of three-axis simultaneous motion. Paper-tape output from APT processing is loaded into the memory of the milling machine which then takes commands from the computer. Holding fixtures were designed as shown in Fig. 17 for transporting the electrode and fixture from the milling machine to the EDM machine for the proper orientation. This process reduces set-up time and ensures accurate alignment with the die block. This becomes especially important should the electrode become damaged during the EDM process and require removal for rework. A roughing electrode is used to remove the bulk stock rapidly, with little concern for die surface finish. The die is then heat treated, and a finish electrode is used to produce a smooth finish.

Extrusion

Using the die manufactured following the above procedure and a cylindrical billet of 2024 Al-20 vol.% SiC whisker material, the extrusion process was carried out under working conditions selected from process simulation to produce T sections. Two other extrusions were also carried out under the same working conditions for comparison purposes. For the latter experiments a conventional die and a die which was designed based on minimum-force calculations were used. The results are shown in Fig. 18 along with dies and respective products. Obviously, the extrusion process which made use of the die designed and manufactured following the CAE concept produced the best results.

CONCLUSIONS

Analytical process modeling has been used very effectively to evaluate various process parameters such as die length, die shape, and effect of friction. The limitation of the analysis is that in its present stage of development, the program can solve only two-dimensional axisymmetric plane-strain and plane-stress problems which involve fully dense P/M materials. The solution to problems

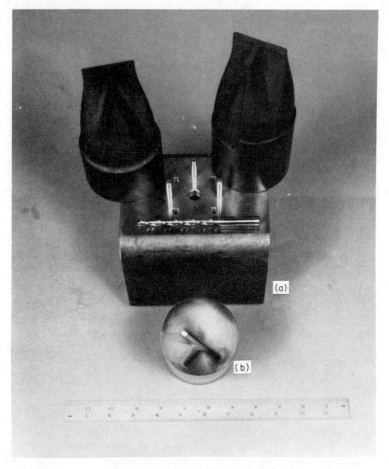

Figure 17.   (a)   Roughing Electrode and Finishing
                   Electrode with Fixture and End Mill.
             (b)   Finished Die, Ready for Extrusion.

which involve porous materials and three-dimensional non-
axisymmetric metal flow is in progress. However, for non-
axisymmetric extrusion, a "shape-complexity" factor can be
incorporated to define the degree of complexity of the product
geometry. This factor can also be combined with results of
two-dimensional or axisymmetric FEM analysis to provide
approximate predictions for non-axisymmetric extrusion.

It is extremely important to establish the criteria used for
optimizing a given process or die-design procedure. Different
criteria usually yield different optimum process parameters. In
the present investigation, the conventional minimum-force
criterion for the design of dies does not automatically guarantee

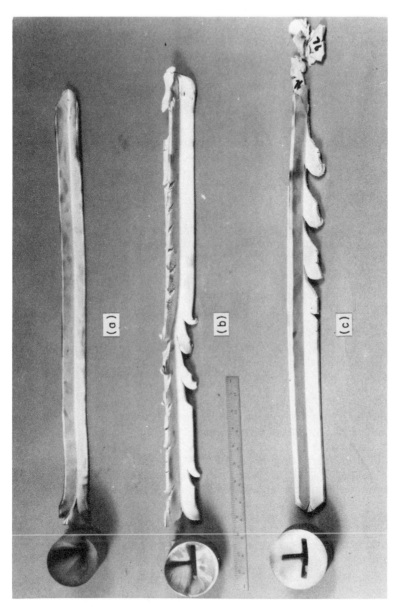

Figure 18. T-Section of 2024 Al-20 Vol.% SiC Extruded Through (a) Streamlined Die, Design Based on Metal-Flow Analysis (Using ALPID and STREAM); (b) Streamlined Die, Design Based on the Minimum-Force Criterion; (c) Conventional Die Generally Used in Industry for Extruding Conventional Alloys (Aluminum, Copper, Brass, Steel).

optimum metal-flow conditions. The material behavior has been given priority over other factors such as ram force because of the importance of obtaining products having the required mechanical properties.

The present study has shown the correlation between material flow and process variables; in some cases (for example, die length and geometry), a strong correlation was found; but in others (for example, interfacial friction), the correlation was weak.

The results of analytical modeling (i.e., velocity profiles and distribution of stress, strain, and strain rate) can be used to control the process variables (die length, die shape, temperature, ram speed, and lubrication) to achieve a desired result. This feed-back loop can be effectively used in the computer to model the process more accurately, resulting in precise process-control parameters which will, in turn, provide proper quality control of the end product.

This paper has demonstrated the importance and potential of modeling by means of the computer. The computer has also been used to design dies based on the results of modeling. The potential economic advantages of modeling and CAD/CAM of dies are clear, including reduced lead times, reduced cost, and improved quality assurance.

ACKNOWLEDGEMENT

The authors would like to thank Dr. J. S. Gunasekera, Monash University, Australia, Dr. Y. V. R. K. Prasad, Senior NRC Associate, and Mr. Jim Morgan, AFWAL/MLLM, Wright-Patterson Air Force Base, OH, for useful discussions during this study and critical comments on the manuscript. Thanks are also due to Mrs. M. Whitaker and Mrs. H. L. Henrich (Systems Research Laboratories, Inc.) for editorial assistance and preparation of the manuscript.

REFERENCES

1. R. J. Green, Int. J. Mech. Sci. 14, 215 (1972).
2. M. Oyane, S. Shima, and Y. Kono, Bull. JSNE 16, 1254 (1973).
3. S. Shima and M. Oyane, Int. J. Mech. Sci. 18, 285 (1976).
4. H. A. Kuhn and G. L. Downey, Int. J. Powder Met. 7, 15 (1971).
5. J. S. Gunasekera, H. L. Gegel, J. C. Malas, S. M. Doraivelu,
    and J. T. Morgan (submitted by T. Altan), "Computer-Aided Process
    Modeling of Hot Forging and Extrusion of Aluminum Alloys,"
    presented at the 32nd General Assembly of CIRP, Brugge, Belgium,
    Aug. 30 - Sept. 4, 1982, and published in CIRP Anals 1982
    Manufacturing Technology (Techniche Rundschau, Berne,
    Switzerland, 1982).
6. S. Kobayashi, "Thermoviscoplastic Analysis of Titanium Alloy
    Forging," AFWAL-TR-81-4130 (Air Force Wright Aeronautical
    Laboratories, Wright-Patterson AFB, OH, 1981); S. I. Oh,
    "Finite Element Analysis of Metal Forming Processes with
    Arbitrarily Shaped Dies," Int. J. Mech. Sci. 22, 538 (1980).
7. J. S. Gunasekera, H. L. Gegel, J. C. Malas, S. M. Doraivelu,
    and G. Griffin, "CAD/CAM of the Die: Computer Modeling and
    Graphics Make Complex Metal Working Dies Easier to Realize,"
    Comp. Mech. Eng. (CIME) 1, 58 (1982).
8. V. Nagpal and T. Altan, "Computer-Aided Design and Manufac-
    turing for Extrusion of Aluminum, Titanium and Steel Struc-
    tural Parts," Final Report under Contract DAAG46-75-C0054
    (Battelle Columbus Laboratories, Columbus, OH, 1976).

DEVELOPMENT OF A COMPUTER-AIDED ANALYSIS METHOD

FOR SHEET MATERIAL FORMING PROCESSES

Daeyong Lee and Catherine M. Forth

General Electric Corporate Research and Development
P. O. Box 8
Schenectady, New York   12301

ABSTRACT

   The computer-aided engineering method can be incorporated in
a number of material processing operations to reduce engineering,
manufacturing and material costs. With this objective in mind, an
interactive computer software package has recently been developed
to predict the degree of success or failure of formed sheet metal
parts at the design stage to minimize the traditional trial and
error process. For example, given the material properties, the rate
of deformation, the geometry, and loading boundary conditions, the
program computes the distriubtion of major and minor strains in the
part ,to be formed. Concurrently, the limiting strain forming limit
diagrams corresponding to the particular material are computed
based on the defect growth model. The computed strain distribution
is then directly compared with the forming limit diagrams so that
any design and processing changes can be made at the computer
terminal. Several examples of computed results are presented which,
in turn, are directly compared with the measured strain distributions obtained from the circular grid technique. Included in the
experimental work are aluminum-killed steel, enameling iron, and
high strength steel sheets.

INTRODUCTION

   With recent advances in analytical methods, graphics software,
and computing capabilities, computer-aided engineering methods are
finding numerous areas of applications in a variety of engineering

* Daeyong Lee, Department of Mechanical Engineering
  Rensselaer Polytechnic Institute, Troy, NY

disciplines. The use of computer-aided analysis methods is particularly suitable to material processing operations because of the
cost of materials and the large number of variables that are known
to control these processes. In some processing applications, e.g.,
casting, forging, stamping, drawing, and rolling, it may also be
possible to optimize the mechanical and/or metallurgical properties
of the finished product by utilizing computer-controlled processing
methods. Therefore, it if is possible to predict at the blueprint
stage whether a particular design concept could be successfully
carried out through the manufacturing process, the costs associated
with traditional trial-and-error might be reduced, thereby improving
productivity. Such an application software package, capable of
predicting the degree of success or failure of the formability of
sheet materials at the design stage, has been developed. Basic
formulations of theory, assumptions, computer software development,
and applications of the package are outlined in this paper.

Sheet metal forming, or stamping, is probably one of the oldest,
yet widely used, manufacturing methods, based on many years of
know-how and experience. But it was not until the 1960s that the
process of failure that takes place in sheet metal forming was
described quantitatively in the forming limit diagram.[1,2] Along
with further developments in plasticity theory and analytic methods,
it became possible to take an interdisciplinary engineering approach
for the sheet metal forming process. However, analyzing the deformation and failure processes in formed sheet metal parts at the
design stage is not a simple, straightforward task, because of the
complex boundary and metal-die interfacial conditions and the
difficulty of describing quantitatively the nonuniform deformation
behavior of metals. In spite of the recent developments,[3-5] computation of localized deformation and material behavior at the metal-
die interface remains a difficult problem. Among different theories
that have been proposed to account for the development of nonuniform deformation, the idea that neck grows from the local area of
initial flaw[6] has been applied by several authors[3,7] to compute the
entire forming limit diagrams. Combining such an analytical method
with a simplified constitutive description of material behavior and
a finite element analysis method, an interactive computer package
has recently been developed to predict the formability of sheet
metals at the design stage.[8]

This paper describes the manner by which analytical methods
and computer programs have been developed for use as a combined
engineering tool at the design stage. Some features of the analysis
program are explained in detail, the results of the analysis are

compared directly against experimental data using a variety of materials, and the sensitivity of the analysis program is examined.

OUTLINE OF THE GENERAL APPROACH

Material Modeling and Mechanics

In describing the material behavior, the yield function is expressed in terms of the distortion matrix, M_{ij}, which describes the variation of the flow stress with orientation,[9]

$$f = M_{ij}\sigma_i\sigma_j \tag{1}$$

where σ_i and σ_j refer to the stress vectors corresponding to the appropriate tensor counterparts, $\hat{\sigma}_{ij}$. Therefore, the equivalent stress, $\bar{\sigma}$, can be expressed as $\bar{\sigma} = (M_{ij}\sigma_i\sigma_j)\,1/2$. The well-known anisotropy parameter, R, is related to M_{33} by $R = 2/M_{33}-1$ for a planar isotropic material under the plane stress loading condition.

The strain rate hardening behavior of materials was described by expressing the equivalent strain rate, $\dot{\bar{\varepsilon}}$, in terms of the equivalent stress, $\bar{\sigma}$, or

$$\dot{\bar{\varepsilon}} = \dot{\varepsilon}_0\left[\frac{\bar{\sigma}}{k}\right]^{1/m} \tag{2}$$

where m in equation 2 is the strain rate sensitivity of flow stress. The dependence of the effective flow stress, k, on equivalent plastic strain, $\bar{\varepsilon}$, may be determined from a uniaxial tension test at a reference strain rate, $\dot{\varepsilon}_0$. It has been shown that a number of experimental data could be represented by the Swift-type equation,[10] or

$$k = k_0\,(\varepsilon_0 + \bar{\varepsilon})^n \tag{3}$$

where k_0 and ε_0 are material constants and n is the strain hardening exponent. Equations 2 and 3 can be combined to express the equivalent stress, $\bar{\sigma}$, in terms of variations in $\dot{\bar{\varepsilon}}$ and $\bar{\varepsilon}$.

Results of uniaxial tension test data are summarized in Figure 1 for the high strength steel, obtained at the reference strain rate of 2.78×10^{-4}/sec. Superimposed on the experimental data are solid lines obtained by applying equation 3. A summary of measured material parameters obtained from the tension test for all the materials are given in Table I.

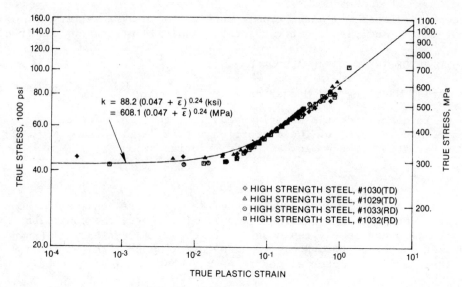

Figure 1.   True stress–true plastic strain relationship for the
high strength steel tested at room temperature at the
initial strain rate of 2.78×10^{-4}/sec, along both
rolling and transverse directions. The solid line
indicates the curve that was fitted to equation 3.

Table 1   A Summary of Mechanical Properties

	Materials		
	EIDQ Steel	AKDQ Steel	High Strength Steel
Young's modulus, E			
10^6 psi	30.0	30.0	30.0
10^3 MPa	206.9	206.9	206.9
Reference effective stress, k_o			
10^3 psi	90.0	80.0	88.2
10^3 MPa	620.6	551.6	608.3
Reference effective strain, ε_o	0.19	0.15	0.047
Strain hardening index, n	0.50	0.48	0.24
Strain rate sensitivity, m	0.0067	0.0077	0.0078
Anisotropy parameter, M_{33}	1.0	0.625	1.0
R	1.0	2.2	1.0

The method of computing the limiting strain forming limit diagrams (FLD) has been reported in a previous paper.[7] In essence, necking is assumed to develop from local regions of initial defect or nonuniformity and the equation describing the rate of thickness reduction is formulated based on the initial geometric and material conditions. The growth of initial nonuniformity was obtained by simultaneously integrating a series of equations at different but prescribed locations in the neck. In addition to the inhomogeneity index, η, the input material parameters that are required to compute the limiting FLD are m,n, ε_0, and M_{33}. The effect of initial sheet thickness on the limiting FLD was incorporated by assuming that the magnitude of measured surface flaw is independent of the sheet thickness, and the value of η is equal to $1 - (t_{min}/t_o)$ where t_{min} is the minimum thickness and t_o the maximum thickness. The results indicate that the computed FLDs are reasonably consistent with experimental observations.[7,11]

The effect of different material parameters on the computed limiting strain FLD has been examined in the previous paper.[7] The FLDs are raised with the increasing strain rate sensitivity of flow stress, m, and with the increasing strain hardening exponent, n. On the other hand, the forming limit is influenced in a complicated fashion by the variation in anisotropy parameter, M_{33}, or the R value. The calculated forming limit increases with the increasing R value in the negative region of the minor strain; the reverse is found to be true in the positive side of the minor strain.

An experience-based "shop" FLD which describes the boundary between the "safe" and "marginal" forming limit is obtained by an empirical method as described below. The plane strain intercept of the shop FLD, designated by FLD_0 (shop), was computed from the given values of sheet thickness and the strain hardening index, as outlined by Keeler and Brazier,[12] and reproduced in Figure 2. Having computed the plane strain intercept of the shop FLD, the analytically obtained limiting strain FLD curve for the given material was uniformly displaced so that the plane strain intercept coincides the value designated by FLD_0 (shop). Such an empirical method overcomes the difficulty of predicting the forming limit for the initiation of nonuniform plastic flow.

As a final analysis module, the software package utilizes a simplified finite element analysis code.[5] A two-dimensional finite element analysis code is used where the nodal point coordinates are specified along a line representing the inner surface contour of the particular formed part. Therefore, no thickness elements are present in the construction of the nodal points describing the particular geometry. Aside from using the deformation theory of time-independent plasticity, the computational method involves the determination of the initial nodal positions by the strain energy minimization process.

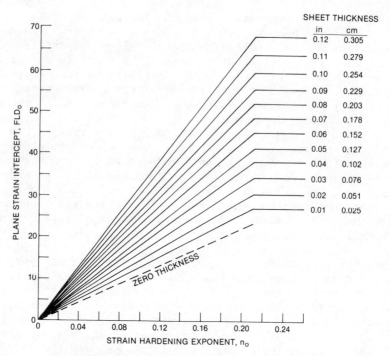

Figure 2.   An empirical relationship between the plane strain
intercept of the forming limit diagram for the "safe"
and "marginal" boundary and strain hardening exponent
for materials with different thicknesses, taken from
Keeler and Brazier[12], where $n_0 = n - \varepsilon_0$.

Interactive Computer Software Development

The Core of the on-line sheet metal formability package,
SHEETS, consists of an automated, flexible system of interactive
software which accelerates the design process through the use of
computer graphics.[13]   The package allows the user to interact with
the computer by updating, modifying and evaluating design data in
an iterative manner.

SHEETS is designed to run from a high level Digital Command
Language (DCL) program, SHEETS.com, on a Digital Equipment Corporation VAX 11/780.   The DCL program gives the user access to the six
modules which comprise the Sheet Metal Formability Package: INPUT
(for the interactive construction of data files), FORLIM (forming
limits), FEA (finite element analysis), PLOT (plot package for
analysis output), MOVIE (MOVIE.BYU, a three-dimensional graphics
program), and HELP (SHEETS and system help).   A schematic diagram
showing the control and data flow processes among all the different
modules is given in Figure 3.

SHEETS features an effective, user-friendly interface and a simple input/output control method. The flow of data control is handled internally and remains transparent to the user. For instance, after a specific module is selected by the user, the program proceeds automatically by changing the control to the appropriate software, and generates appropriately named permanent output files once the analysis is completed. Simple prompts for input are displayed at the user's terminal as required from the individual FORTRAN modules. These inputs include both data files supplied by the user and filenames of data generated by each program's modules. Additional input data, including the material properties data files, are acquired automatically by the modules when necessary. In addition, program status messages are issued to the terminal to assure the user when the program is engaged in time-consuming processes, such as the computationally intensive iteration procedure of the FEA module, file creation and data storage steps.

The SHEETS package incorporates a simple, yet efficient file management design by internally maintaining a consistent file naming scheme. Three character filename extensions are used for the designation of appropriate input and output files, making it easy for the user to remember and access the correct input and output files. Examples of this naming scheme include the following: all geometry input information files end with '.dat' (e.g., vent.dat); the files containing geometry information generated by the FEA module for input into the MOVIE module end with a '.geo' extension (e.g., vent.geo); output files from FORLIM containing information required by the PLOT module end with '.flp' (e.g., vent.flp); and those files ending with an '.out' extension (e.g., vent.out) contain a listing of calculated results including strain ratios, neck geometry, and neck strains. This file management scheme allows many of the input and output files to be accessed by the user for retrieval, updating and alteration within the VAX editor, as well as through the SHEETS package.

A brief summary of the function of each module and a list of input and output files for each module are given below.

INPUT   The INPUT module allows the user to interactively prepare an input geometry data file (.dat) using a Hewlett-Packard flatbed plotter with compacted binary instruction set and the related graphics software. The user specifies the nodal positions by placing the pen, and enters other parameters by responding to prompts. The information obtained via the INPUT module includes both geometry and boundary conditions pertaining to the part to be formed. Nodal points are entered along the final generator, keeping in mind that the parametric notation used by the program fits a smooth second degree polynomial curve through any three consecutive nodal points. Entries are made in the order of increasing node

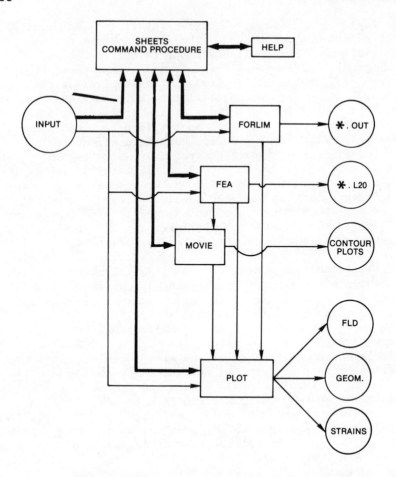

Figure 3.   SHEETS control and data flow diagram depicting the
            internal design of the package. Bold lines represent
            control flow pathways to the major modules from the
            Digital Command Language (DCL) command procedure; fine
            lines indicate intermodule data flow, accomplished by
            the creation and use of intermediate files.

values from the symmetry line. INPUT will then form elements
from three consecutive nodes.

    Input file: (a) no input files (H-P Plotter)

    Output file: (a) geometry data file for FORLIM
                        and FEA modules (.dat)

FORLIM   The FORming LIMits module provides analytical and numer-
          ical models to predict the development of nonuniform flow
          in sheet metals under various plane-stress loading

conditions. A rate dependent flow theory of plasticity was used and the computation of the detailed neck growth was repeated over a wide range of proportional loading conditions to establish the limiting strain FLDs. Input material parameters, selected from a table of available materials, include m, n, ε_0, M_{33}, and η.

Input file:	(a) material data from internal permanent files
Output files:	(a) results for use by PLOT module (.flp)
	(b) listing of results for user (.out)

FEA   The Finite Element Analysis (FEA) module contains a two-dimensional finite element analysis program which requires the nodal point coordinates specified along a line representing the inner surface contour of a particular axisymmetric part. In addition, the die speed and distance of die travel will also be entered in response to FEA module prompts. Elastic-plastic analysis of the axisymmetric body using isoparametric finite elements is made in FEA.

Several files will be created as outputs from this module, including a text file containing a summary of the results, stress/strain information formatted for the PLOT module, a geometry file formatted for MOVIE, and scalar function files describing calculated values for MOVIE contour plots of radial stress, hoop stress, effective stress, radial strain, hoop strain, sheet thickness ratio, and axial force.

Input file:	(a) geometry file created by the INPUT module (.dat)
Output files:	(a) stress/strain distribution information for PLOT module (.str)
	(b) geometry file for MOVIE module (.geo)
	(c) scalar function files containing all the computed results for MOVIE module (7 files ending with 1.sca. ...7 .sca)
	(d) results of 2-D FEA analysis (.12o)

HELP   The HELP module provides an on-line summary of the functions of each of the available modules of the SHEETS package, as well as the appropriate input and output file designations.

                    Input file: (a) none

                    Output files: (a) listing on terminal only

PLOT   The PLOT module allows the user to produce geometry and
            stress/strain plots on one of several available graphics
            devices including the Tektronix 4014, Tektronix emulators,
            the Hewlett-Packard 7221 flat-bed plotter,
            Zeta 1453 plotter or Lexidata 3400 frame buffer display.
            All graphical output in the PLOT module are produced using
            graphics compatibility system (GCS) subroutines.[14]   The
            geometry option will produce an X-Y plot of the designated
            part geometry and will identify the node by number.   The
            forming limit option will produce either true strain or
            engineering strain FLD plots with the SAFE, MARGINAL and
            FAILURE regions identified and radial and hoop strain values
            calculated by the FLD module.   The strain curve option
            prepares the strain versus the final generator plot and
            allows the user to select true or engineering values for the
            major (radial), minor (hoop), and/or thickness strains.

                    Input files: (a) geometry file from user module
                                        (.dat)
                                    (b) results of FORLIM module (.flp)
                                    (c) stress/strain information from
                                        FEA module (.str) OPTIONAL (at
                                        user's request)

                    Output: Plots on various display devices
                                    (a) geometry plot
                                    (b) true or engineering strain plots
                                    (c) forming limit diagrams

MOVIE   The MOVIE module invokes the MOVIE.BYU program developed
            at Brigham Young University.[15]   This module is used to
            display the radial and hoop strains, as well as the computed
            strain contours, for a particular geometry.   Other computed
            quantities, such as stress, axial force, etc., may also be
            displayed.

                    Input files: (a) geometry file from FEA module
                                        (.geo)
                                    (b) scalar function files from FEA
                                        (.sca)

                    Output: (a) plot of results on display
                                        terminal only

SAMPLE ANALYSIS

Several sample cases will be examined in detail to illustrate various features of the analysis program. All the axisymmetric analyses, except the sensitivity runs, represent actual forming, for which laboratory experiments have been made to verify the predictions.[16] Special features of the formed parts are summarized below:

Part	Material	Features
Range vent	EIDQ steel	Non-axisymmetric part
Dryer inner door	AKDQ steel	Anisotropic material
Test cup	High strength steel	Axisymmetric part

A brief summary of the analytical and experimental results are summarized in the following sections.

Vent Forming

A photograph of the formed range vent and the corresponding finite element representation of the circular section of the vent geometry are shown in Figures 4a and 4b, respectively. The imposed boundary conditions include the description of frictional characteristics and the edge clamping conditions. Enameling iron drawing quality (EIDQ) steel was used and the average forming strain rate was 0.00067/sec.

Given the material properties, the geometry file, and loading boundary conditions, the next step is to compute the strain distributions for the formed vent using the finite element analysis program. A parallel analysis starting from the material data base yields a computed limiting strain FLD and the shop FLD from the given set of material parameters. Plotting the calculated strain distribution on the composite limiting FLD and shop FLD diagrams gives a summary of the final result, as shown in Figure 5a. The nodal positions are identified by the nodal point numbers in the plot, and the computed radial (major) and hoop (minor) strains at each nodal point are specified in the composite FLD by the corresponding nodal point numbers. The strain levels in some of the nodal points in the vent design approach the safe/marginal boundary.

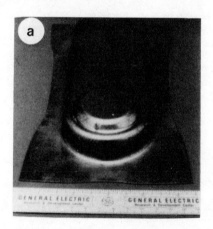
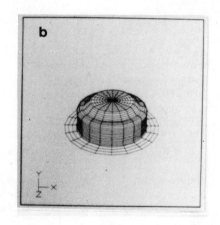

Figure 4.   (a) A photograph of the formed range vent and (b) a
            finite element model for the circular section of the
            formed part.

The computed major and minor strains are directly compared
with experimental data obtained from the circular grid method, as
shown in Figure 5B. Laboratory forming of the actual range vent
part did not show any localized necking, which is consistent with
the summary plot of Figure 5a. The numerical result shows some
discrepancy from the experimental data in Figure 5b; the major
cause for the discrepancy is believed to be the assumption that the
initial sheet geometry is circular instead of rectangular, as shown
in Figure 4a.

Vent Forming; Sensitivity Analysis

In order to examine the details of the analysis program, a
number of different analyses were made with the vent forming
problem by modifying one specific variable at a time. For example,
the role of loading boundary conditions was examined by changing
the friction coefficient from 0.14 to 0.014 in Figure 6a and by
increasing the edge clamping force from 453.5 kg (1000 lb) to
4535 kg (10,000 lb) in Figure 6c. The effect of changing the
friction coefficient was to alter the strain distribution in a
major way; no significant effect was noted from the change in the
clamping force. The speed of forming was changed from 7.1×10^{-4}
cm/sec (2.8×10^{-4} in/sec) to 71.1 cm/sec (28.8 in/sec), as shown
in Figure 6b. No major change occurred in the computed strain
distributions because the material has a very low strain rate
sensitivity of flow stress (m=0.0067). If the material had a large
value of m, the result would have been markedly different. In
the final analysis, the thickness of the sheet was changed from
0.094 cm (0.037 in) to 0.047 cm (0.0185 in) so that the initial
surface flaw would have a greater effect on the forming limit
diagram. Figure 6d shows lowering effect of the forming limits.

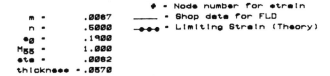

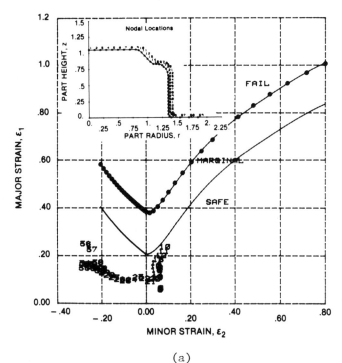

(a)

Figure 5.   (a) A geometrical description of the nodal points (inset)
            and the result of the finite element analysis super-
            imposed on the computed forming limit diagram for the
            range vent, and (b) a comparison of computed major and
            minor strains along the new generator with experimental
            data for the range vent.

(continued)

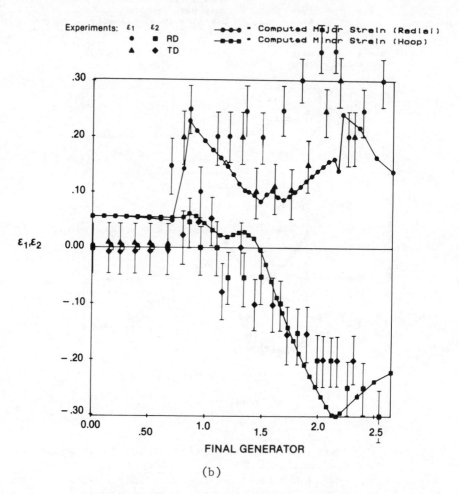

(b)

Figure 5, continued.

The detailed plot of the corresponding strain distributions shown in Figure 7 explain some of the major changes that occurred from these changes of different variables. For example, when the friction coefficient was uniformly lowered, the strain in the top of the part increased while that in the clamping area was also altered. Minor changes in other cases also illustrate complex interaction between different variables that determine the final outcome of the analysis.

Dryer Inner Door Forming

The dryer door part was formed with the aluminum killed drawing quality (AKDQ) steel which is plastically anis-tropic. The actual part was rectangular, as in the case of range vent, and only the corner was analyzed using the azisymmetric model. The friction coefficient was 0.016 and the average forming strain rate was 2.87/sec. A photograph of the part and the finite element model of the corner are shown in Figure 8.

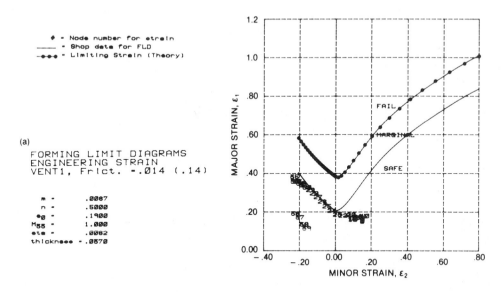

Figure 6.   Computed results for the sensitivity analysis for the range vent geometry: (a) friction coefficient was decreased, (b) forming speed was increased, (c) clamping force was increased, and (d) sheet thickness was reduced.

(continued)

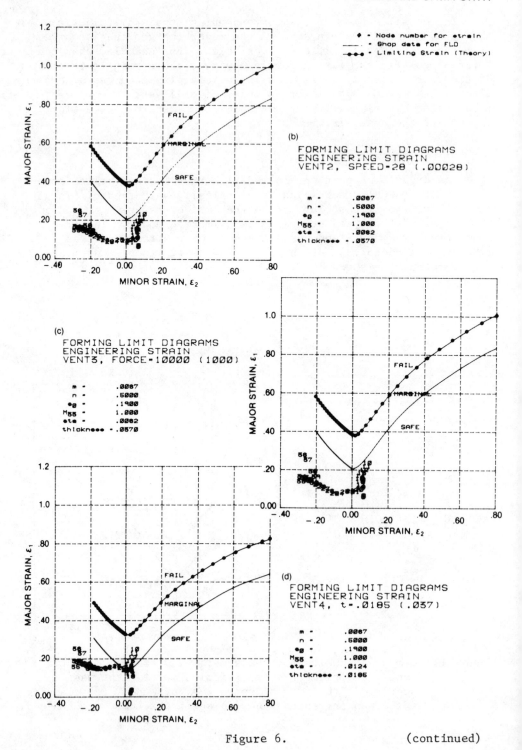

Figure 6.   (continued)

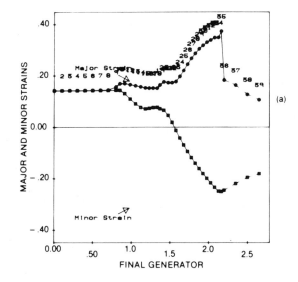

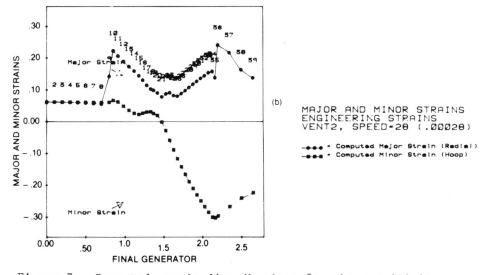

Figure 7.   Computed strain distributions for the sensitivity
            analysis for the range vent geometry: (a) friction
            coefficient was decreased, (b) forming speed was
            increased, (c) clamping force was increased, and (d)
            sheet thickness was reduced.
                                                    (continued)

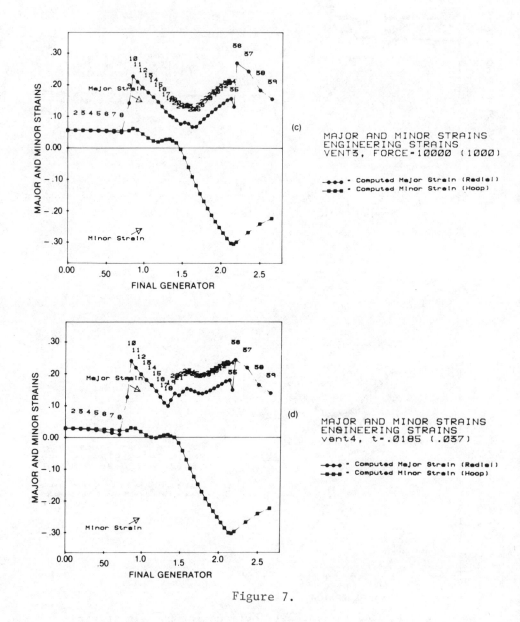

(c)

MAJOR AND MINOR STRAINS
ENGINEERING STRAINS
VENT3, FORCE=10000 (1000)

━●●● = Computed Major Strain (Radial)
━■■■ = Computed Minor Strain (Hoop)

(d)

MAJOR AND MINOR STRAINS
ENGINEERING STRAINS
vent4, t=.0185 (.037)

━●●● = Computed Major Strain (Radial)
━■■■ = Computed Minor Strain (Hoop)

Figure 7.

(continued)

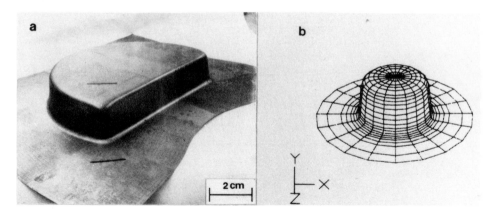

Figure 8.   (a) A photograph of the formed dryer inner door and
            (b) a finite element model for the corner section of
            the formed part.

Results of the strain distribution analysis and forming limit
calculation for the dryer inner door part are summarized in Figure
9a; the specific nodal location for the corner position is also
given in the inset. Since the location of the symmetry line could
not be predicted readily, it was obtained from the result of the
circular grid analysis.

The computed strain levels at some of the nodal locations are
near the safe/marginal boundary, indicating potential problem areas
in the actual forming process. Computed and experimental strain
distributions along the length of final generator, as shown in
Figure 9b, confirm the result of analysis. The experimental data
showed that some of the part exceeded the computed strain levels.
Aside from the necking that occurred in the selected locations,
the computed strains are in good agreement with the experimental
results.

Cup Forming

An axisymmetric cup geometry was formed using a high strength
steel sheet at an average strain rate of 0.006/sec. The material
is isotropic and the friction coefficient of 0.018 obtained from
a separate laboratory test is used in the analysis. The geometry
of the part is shown in Figure 10.

Results of the finite element analysis and forming limit
calculation are summarized in Figure 11a, showing that the particular cup could be formed readily without any major difficulty. On
the other hand, some discrepancies were noted when the computed
strain distribution was compared directly with the experimental

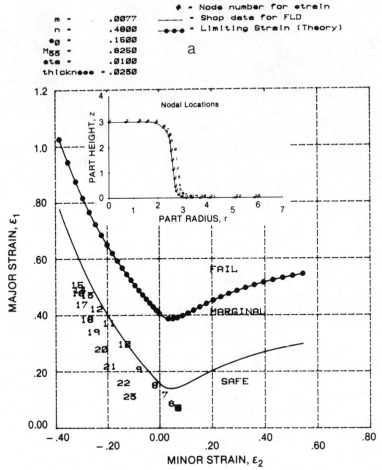

Figure 9. (a) A geometrical description of the nodal points (inset) and the result of the finite element analysis super-imposed on the computed forming limit diagram for the dryer inner door and (b) a comparison of computed major and minor strains along the new generator with experimental data for the dryer inner door.

observations, as shown in Figure 11b. For example, the measured radial strains in the vicinity of punch radius exceeded 40% engineering strain, causing the material to develop a localized neck. Except for the major strain in the punch radius region, computed strain values are in good agreement with the experimental data.

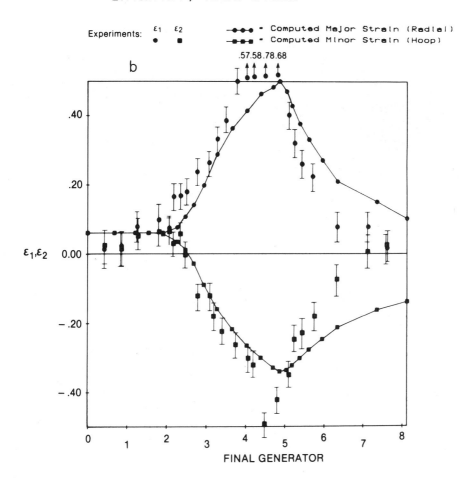

ENGINEERING STRAIN CURVES
DRYER(B), AKDQ STEEL

SUMMARY AND CONCLUSIONS

A computer-aided analysis method has been developed for a
material forming process and it has been applied to a number of
production sheet metal parts. The integrated computer code is
capable of providing an instantaneous graphic representation of
the degree of success or failure in the formability of sheet metal
parts at the design stage. Key elements of the analysis program
consist of constitutive equation, forming limit diagrams, and
finite element analysis programs. In addition, the software
package utilizes additional modules for input/output purposes and

several graphics software programs, including MOVIE.BYU.[15]

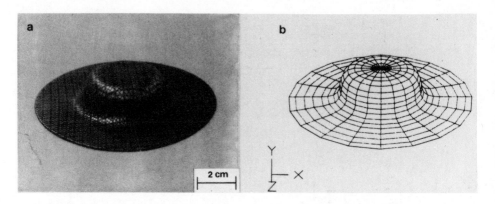

Figure 10.   (a) A photograph of the formed cup and (b) a finite
element model for the formed part.

In order to emphasize some of the key points in the analysis
program, a number of axisymmetric cases were illustrated, some of
which were used to illustrate the sensitivity of the analysis.
Results of the analysis and the preliminary experimental verification of the formed parts also demonstrated the utility of such a
computer-aided sheet metal formability analysis package. Axisymmetric analysis is indeed simpler to use as compared to a full
non-axisymmetric, or three-dimensional analysis. In addition, from
a practical point of view, a large fraction of complex parts could
be approximated as an axisymmetric problem. Such cases have been
illustrated with the examples of range vent and dryer inner door.
An effort is, however, under way to extend the capability of the
analysis to three-dimensional problems.

Although the SHEETS program is currently used in a production
environment, there are several areas of analysis that need to be
examined further in detail. Some of these areas are related to
the specification of boundary conditions, including the metal-
die frictional conditions, the description of neck growth under
the plane stress loading condition, and the ability to analyze
multipass forming processes.

ACKNOWLEDGMENTS

The authors would like to express their appreciation to Bill
Lorensen for his advice on the systems programming and the application of graphics software, to Phil Stine who provided useful information on the experimental aspect of the forming process, and
to Sam Levy for his advice on the finite element analysis method.

Craig Palmer and Paul Dupree carried out tension testing of materials. The authors are also indebted to Mark Benz, who provided support and encouragement throughout the course of this work.

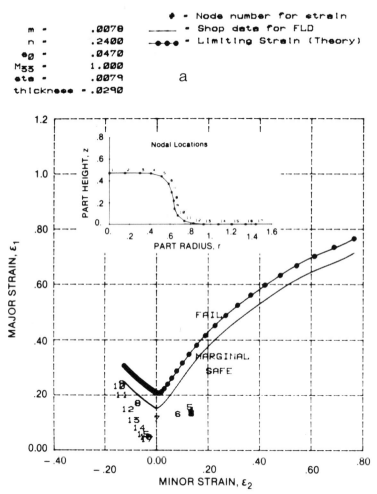

Figure 11. (a) A geometrical description of the nodal points
           (inset) and the result of the finite element analysis
           superimposed on the computed forming limit diagram for
           the cup and (b) a comparison of computed major and minor
           strains along the new generator with experimental data
           for the cup.

(continued)

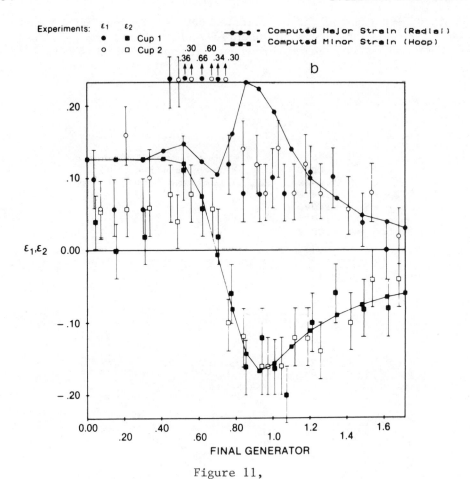

Figure 11,

REFERENCES

1. S. P. Keeler, Determination of Forming Limits in Automotive
    Stampings, Sheet Metal Industries 42, 683-691 (1965).
2. G. M. Goodwin, Application of Strain Analysis to Sheet-Metal
    Forming Problems in the Press Shop, La Metallurgica Italian
    60, 767-774 (1968).
3. J. W. Hutchinson and K. W. Neale, Sheet Necking-III. Strain
    Rate Effects, in "Mechanics of Sheet Metal Forming," D. P.
    Koistinen and N. Wang, ed., 269-285, GM Res. Lab.,
    Plenum Press (1977).

4. A. Needleman and J. R. Rice, Limits to Ductility Set by Plastic
    Flow Localization, ibid, 237-267.
5. S. Levy, C. F. Shih, J. P. D. Wilkinson, P. A. Stine and R.
    McWilson, Analysis of Sheet Metal Forming to Axisymmetric
    Shapes, in "Formability Topics-Metallic Materials," B. A.
    Niemeier, A. K. Schmieder and J. R. Newby, ed., 238-260,
    STP647, ASTM (1978), ASTM, Philadelphia.
6. Z. Marciniak and K. Kuczynski, Limit Strains in the Processes
    of Stretch-Forming Sheet Metal, Int. J. Mech. Sci. 9,
    609-620 (1967).
7. D. Lee and F. Zaverl, Jr., Neck Growth and Forming Limits in
    Sheet Metals, Int. J. Mech. Sci. 24, 157-173 (1982).
8. D. Lee, Computer-Aided Control of Sheet Metal Forming Processes,
    J. Metals 34, 20-29 (1982).
9. C. F. Shih and D. Lee, Further Developments in Anisotropic
    Plasticity, J. Eng. Mat. Tech., Trans. ASME 100, 94-302
    (1978).
10. H. W. Swift, Plastic Instability Under Plane Stress, J. Mech.
    Phy. Solids 1, 1-18 (1952).
11. S. S. Hecker, Simple Technique for Determining Forming Limit
    Curves, Sheet Metal Industries, Nov., 671-676 (1975).
12. S. P. Keeler and W. G. Brazier, Relationship Between Laboratory
    Material Characterization and Press-Shop Formability, in
    "Micro Alloying '75," J. Krane, ed., Union Carbide Corp.,
    21-31 (1977). See also Discussions.
13. C. M. Forth and D. Lee, SHEETS: An Interactive Computer Pro-
    gram to Predict Sheet Metal Formability, TIS Report
    82CRD267, General Electric Corporate Research and Develop-
    ment, Schenectady, New York (Dec., 1982).
14. R. F. Puk, Graphics Compatibility System Reference Manual, U.
    S. Army Waterways Experiment Station, Vicksburgh, Missis-
    ippi (Oct., 1977).
15. H. N. Christian and M. B. Stephenson, MOVIE-BYU-A General
    Purpose Computer Graphics Display System, in Proc.
    Symposium on Application of Computer Methods in Engineering,
    Univ. Southern Calif., Los Angeles, 2, 759-769 (1977).
16. D. Lee and P. A. Stine, Computer-Aided Prediction of Sheet
    Metal Manufacturing Processes and Its Experimental
    Verification, to be presented at the Fourth International
    Conference on Mechanical Behavior of Materials, The Royal
    Institute of Technology, Stockholm, Sweden, (August, 1983).

RECENT DEVELOPMENTS ON THE APPLICATION OF THE FINITE ELEMENT

METHOD TO METAL FORMING PROBLEMS

Shiro Kobayashi

Department of Mechanical Engineering
University of California
Berkeley, CA 94720

SUMMARY

Most recent developments on the finite element method as applied
to metal forming problems emphasize the application to three dimensional problems. There are some areas in metal forming where technological advancements can be made by full utilization of capabilities of the finite element method. An example is the development
on the use of the finite element method for preform design. In
this paper these developments are described.

Based on the rigid-viscoplastic formulation three-dimensional
finite-element analysis was performed for the block forgings, using
an 8-node hexahedral isoparametric element. The analysis of forging
of a wedge-shaped block between two flat parallel dies is presented.

The objective of the analysis of spread problems, such as
spread in rolling, in flat tool forging and spread in compression
of noncircular disks, is to predict the amount of spread. For
computational efficiency, a simplified 8-node hexahedral element
was devised and applied to the analysis of spread in rolling and in
flat tool forging.

A development in the area of sheet metal forming is the analysis
of forming of non-symmetric shapes. Based on the membrane theory,
the formulation takes into account the finite strain, normal
anisotropy, and the isotropic work hardening characteristics. New
solution is shown for punch stretching of a rectangular strip.

One of the significant industrial problems in metal forming is
the design of preforms. A new approach to this problem has been

187

introduced. The concept involved in the approach is to trace
backward the loading path in the actual forming process from a given
final configuration by the finite element method. The application
of the method to preform design in shell nosing is discussed.

INTRODUCTION

In the past a number of approximate methods for process analysis
have been developed and applied to various forming processes. They
have been useful in predicting forming loads, overall geometrical
changes of deforming workpieces, and qualitative modes of metal
flow, and determining optimum process conditions approximately.
However, accurate determination of the effects of various parameters
involved in the process on the detailed metal flow became possible
only recently when the finite element method was introduced into
the analysis of metal forming processes. Since then the method,
particularly in the area of process modeling, has assumed steadily
increased importance in metal forming technology.

A review on the subject [1] indicates that recent major accom-
plishments are the development of a user-oriented general purpose
program, namely a rigid-viscoplastic finite element code (ALPID) by
Oh [2] and the coupled analysis of transient viscoplastic deformation
and heat transfer by Rebelo [3]. It also revealed that almost all
of the applications so far have dealt with two dimensional problems,
while more complex problems were treated by approximate stress
analysis or by the upper bound method.

It is not surprising, therefore, to find that most recent developments emphasize the application of the finite element method to
three dimensional problems. There are some areas in metal forming
where technological advancements can be made by full utilization of
capabilities of the finite element method. An example is the use
of the finite element method for preform design. In this paper the
finite element formulation is first outlined and the developments in
the applications to metal forming problems are described.

FINITE ELEMENT FORMULATION

The material is assumed to be rigid-viscoplastic and obey the
Huber-Mises yield criterion and its associated flow-rule. The
deformation process of rigid-viscoplastic materials is associated
with the boundary value problem where the stress and velocity field
solutions satisfy the equilibrium equations and the constitutive
equations in the domain and the prescribed velocity and traction
conditions on the boundary. To solve this boundary value problem
we begin with a weak form of the equilibrium equation, neglecting
the body force, expressed by

$$\int_V \sigma_{ij,j} \, \delta v_i \, dV = 0 \qquad (1)$$

where σ_{ij} is the component of stress tensor, δv_i is an arbitrary variation of velocity, and a comma denotes partial differentiation.

Using the divergence theorem and the symmetry of the stress tensor and imposing the essential conditions, $\delta \underset{\sim}{v} = 0$ on S_v, where v_i is prescribed, equation (1) becomes

$$\frac{1}{2}\int_V \sigma_{ij} \, (\delta v_{i,j} + \delta v_{j,i}) dV - \int_{S_F} \sigma_{ij} \delta v_i n_j \, dS = 0. \qquad (2)$$

where n_j is the unit normal to the surface and S_F is the surface where traction is prescribed, namely $\sigma_{ij} n_j = T_i$ on S_F.

Decomposing the stress tensor into the deviatoric component σ'_{ij} and the hydrostatic component σ_m, Eq. (2) becomes

$$\int_V \sigma'_{ij} \delta \dot{\varepsilon}^0_{ij} dV + \int_V \sigma_m \delta \dot{\varepsilon}_v dV - \int_{S_F} T_i \delta v_i dS = 0. \qquad (3)$$

where $\dot{\varepsilon}_{ij}$ is the strain-rate and $\dot{\varepsilon}_v$ is the volumetric strain-rate.

By replacing the integrand of the first term with the effective stress $(\bar{\sigma})$ and the effective strain-rate $(\dot{\bar{\varepsilon}})$, and the hydrostatic part with the penalty function constraint, K, in Eq. (3), we obtain

$$\int_V \bar{\sigma} \delta \dot{\bar{\varepsilon}} dV + K \int_V \dot{\varepsilon}_v \delta \dot{\varepsilon}_v dV - \int_{S_F} T_i \delta v_i dS = 0. \qquad (4)$$

where $\bar{\sigma} = \sqrt{\frac{3}{2} \sigma'_{ij} \sigma'_{ij}}$ and $\dot{\bar{\varepsilon}} = \sqrt{\frac{2}{3} \dot{\varepsilon}_{ij} \dot{\varepsilon}_{ij}}$, and K is a large positive constant.

The effective stress for a special material is determined, by uniaxial tension or compression tests, as a function of the effective strain and the effective strain-rate.

Equation (4) is the principle of virtual work. The essential variables involved in Eq. (4) are the velocity components and their derivatives. The velocity field which satisfies Eq. (4) for arbitrary virtual motion δv_i is the solution to the original boundary value problem.

Finite-element discretization of Eq. (4) is accomplished by choosing an approximate interpolation function for the field variable in the elements, and expressing the state equation locally within each element in terms of the nodal point values. The local element equations are then assembled into the overall problem with the global constraints. Thus, Eq. (4) is approximated by the

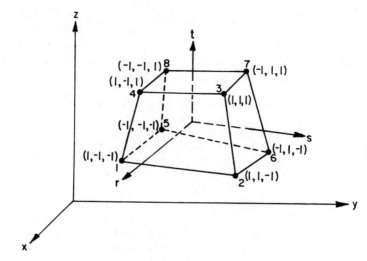

Figure 1   8-node hexahedral element and the global and the
             coordinate systems.

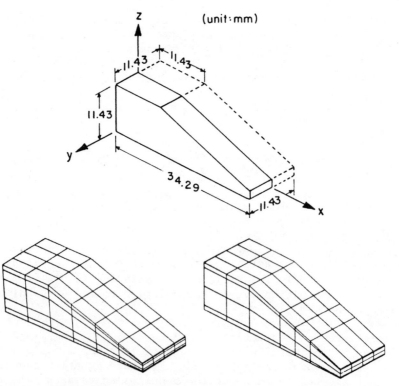

Figure 2   Workpiece dimensions and the mesh systems for wedge-
             shaped block compression.

expression in terms of global nodal point values. The condition
that the virtual motion is arbitrary, results in the stiffness
equations to be solved with appropriate boundary conditions. The
basic mathematical description of the method, as well as the solution
techniques, are given in several books [4,5,6].

THREE-DIMENSIONAL ANALYSIS OF BLOCK COMPRESSION [7].

The element used for discretization is an 8-node hexahedral
isoparametric element shown in Figure 1. The global coordinate
system (x,y,z) is transformed into the natural coordinate system
(r,s,t). The natural coordinate system is defined such that r, s
and t vary from -1 to 1 within each element. Then, an arbitrary
point (x,y,z) in an element can be expressed by

$$x = \sum_{\alpha=1}^{8} q_\alpha x_\alpha , \qquad y = \sum_{\alpha=1}^{8} q_\alpha y_\alpha , \qquad z = \sum_{\alpha=1}^{8} q_\alpha z_\alpha \qquad (5)$$

and $q_\alpha = \frac{1}{8} (1+r_\alpha r)(1+s_\alpha s)(1+t_\alpha t)$,

where x_α, y_α and z_α are coordinates at nodes in the global
coordinate system and r_α, s_α and t_α are natural coordinates at
nodes and given in Figure 1.

The velocity distribution inside an element is approximated by the
nodal velocities by

$$u = \sum_{\alpha=1}^{8} q_\alpha u_\alpha , \qquad v = \sum_{\alpha=1}^{8} q_\alpha v_\alpha , \qquad w = \sum_{\alpha=1}^{8} q_\alpha w_\alpha , \qquad (6)$$

where u_α, v_α and w_α are components of nodal velocities and u,
v and w are velocity components at any point within an element.
Discretization of Eq. (4) is accomplished with the interpolation
function given in Eq.(6) using the compatibility relationships and
the flow rules.

Introduction of the boundary conditions is straightforward,
except at the die-workpiece interface where the conditions are
mixed. The velocity is given in the direction of die movement and
the traction is specified in the form of friction on the contact
surface in the opposite direction of the relative movement of the
workpiece with respect to the die. Depending on deformation modes,
the interface may have a point or a region where relative motion
between the die and the workpiece is zero (neutral point or neutral
region), as observed in the ring compression and sheet rolling. It
has been demonstrated [8] that the velocity-dependent friction representation was effective for this type of problem. It is given by

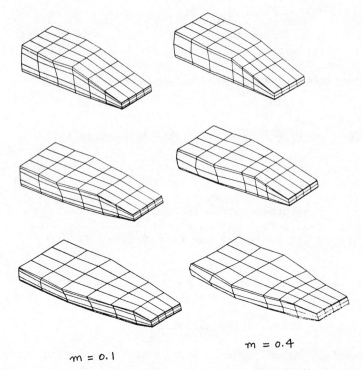

Figure 3   Grid distortions of the wedge-shaped workpiece at
             several height reductions (20, 40 and 60 percent).

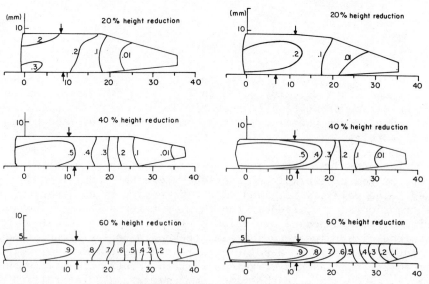

Figure 4   Effective strain distributions and locations of
             neutral zones at several height reductions.

$$\tau_f = mk_y \left\{ \left(\frac{2}{\pi}\right) \tan^{-1} \frac{|v_{DW}|}{A} \right\} \tag{7}$$

where v_{DW} is the relative velocity of workpiece material with respect to the die and A is a constant several order of magnitude smaller than the die velocity.

Besides the treatment of neutral regions, considerations must also be given to the changes of the boundary conditions during deformation. The traction-free condition changes to the contact condition at nodes which are involved in folding, and vice versa if nodes separate from the die.

An example of the analysis is given for compression of a wedge-shaped block between two flat parallel dies. The workpiece dimensions used in simulations are shown in Figure 2. Its geometry as well as the forming process are symmetric with respect to x-axis, so that a half of the workpiece is taken as the control volume for simulations. The flow stress of material Al-2024 (annealed) is

given by $\bar{\sigma} = 225.(1. + \frac{\bar{\varepsilon}}{1.6147})$ (MP$_a$). Two simulations were performed under different friction conditions: dry (m = 0.4) and lubricated (m= 0.1). Two percent height reduction was taken as one step up to 30 percent and one percent afterwards. 60 elements (126 nodes) were used for the case of lubricated condition, while 48 elements (105 nodes) for the dry condition. The mesh systems are also shown in Figure 2. The computational time for the total deformation of 60% reduction in height was 2300 seconds with 60 elements and 1700 seconds with 48 elements on the CDC 7600 computer.

Some results of computation are shown in Figures 3 and 4. Grid distortions at three stages of height reduction are shown for the two friction conditions in Figure 3. In the figure several observations on the mode of deformation can be made. The side surface of the block portion bulges outward, while the side surface of the wedge portion becomes concave. This mode of deformation is more pronounced with the dry interface condition. The contours of the die-workpiece contact area show severer distortion in the width direction for higher interface friction.

The effective strain distributions in the mid-plane (plane of symmetry) are shown in Figure 4. Also, in the figure the locations of neutral zones are indicated by the arrows. The results show that the highest concentration of the strain occurs around the lower center of the block portion and the strain decreases monotonically toward the end of the wedge. The strains are in general uniform in the height direction, particularly in the wedge portion and for low friction. The locations of the neutral zones are symmetric and approximately the mid-point of the contact surface at larger

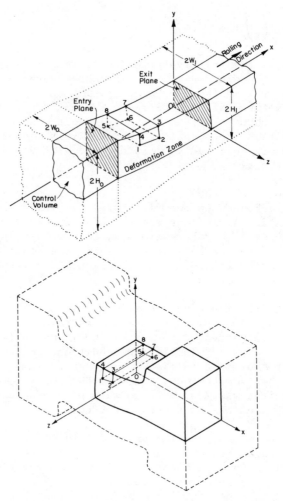

Figure 5   Element arrangements for the analysis of rolling and flat-tool forging.

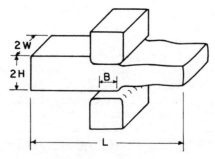

Figure 6   Process of flat-tool forging.

reductions in height. However, at 20 percent reduction the inter-face friction condition showed a significate influence on the location of the neutral zones.

In wedge compression as a workability test, usually fracture is observed on the side surface. The present analysis provides the necessary information on the occurrence of fracture, such as the strain path and the corresponding stress variations at a critical site.

SPREAD ANALYSIS IN ROLLING AND FLAT-TOOL FORGING [9,10]

For the analysis of three dimensional problems, two avenues of approach can be considered: first, the use of three-dimensional elements, and second, the use of simplified three-dimensional elements for better computational efficiency in discretization of the problem. The investigation using the second approach is worth-while, particularly in the metal-forming area, because the solution accuracy and computational efficiency must be balanced depending on the nature of the desired information.

In the analysis of spread problems, such as sideway spread in rolling, in flat tool forging, and spread in compression of non-circular disks, prediction of geometrical changes of the workpiece is of primary importance. Therefore, the spread problems are one of the problems where the use of simplified three dimensional elements is reasonable and effective.

Simplified hexahedral elements can be derived by taking one layer of elements in the direction normal to the plane of symmetry and assuming that the element sides which are initially normal to the plane remain normal during deformation. The element arrange-ments for the spread analysis in rolling and flat tool forging are shown in Figure 5. In the figure, the nodes on the plane of symmetry (XOY) will remain on the plane throughout the process and the assumption states that the velocity components u and v are independent of the z-coordinate. Then we have

$$w_5 = w_6 = w_7 = w_8 = 0$$

and

$$(u_1, v_1) = (u_5, v_5) : (u_2, v_2) = (u_6, v_6)$$

$$(u_3, v_3) = (u_7, v_7) : (u_4, v_4) = (u_8, v_8)$$

$$\tag{8a}$$

We also have

$$z_5 = z_6 = z_7 = z_8 = 0$$

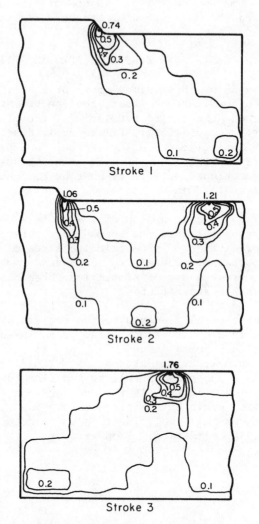

Figure 7   Strain distributions in three-bite flat-tool forging.

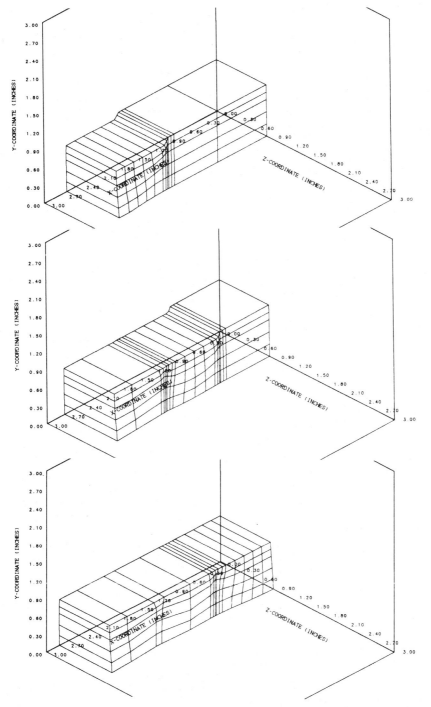

Figure 8   Spread contours in three-bite flat-tool forging.

and

$$(x_1, y_1) = (x_5, y_5) : (x_2, y_2) = (x_6, y_6)$$
$$(x_3, y_3) = (x_7, y_7) : (x_4, y_4) = (x_8, y_8)$$

(8b)

This assumption leads to a linear distribution of w in the z-direction and Eqs. (5) and (6), respectively, to

$$x = \sum_{\alpha=1}^{4} q_\alpha x_\alpha , \quad y = \sum_{\alpha=1}^{4} q_\alpha y_\alpha , \quad z = \frac{t+1}{2} \sum_{\alpha=1}^{4} q_\alpha z_\alpha$$

(9)

where

$$q_\alpha = \frac{1}{4} (1 + r_\alpha r) (1 + s_\alpha s)$$

and

$$u = \sum_{\alpha=1}^{4} q_\alpha u_\alpha, \quad v = \sum_{\alpha=1}^{4} q_\alpha v_\alpha, \quad w = \frac{t+1}{2} \sum_{\alpha=1}^{4} q_\alpha w_\alpha$$

(10)

Under this simplification, a three-dimensional 8-node element can be described by the coordinates and the velocities of only four lateral nodes instead of eight nodes. In the treatment of friction at the die-workpiece interface, the surface integration of the frictional traction is also performed according to the velocity distribution in the simplified element. The approach has been applied to prediction of sideway spread in rolling [9], and for the analysis of spread in flat tool forging [10].

During usual bar forging, the bar is moved longitudinally with respect to the dies several times between forging strokes in one pass as shown in Figure 6. Taking a three-bite flat-tool forging as a typical procedure where the first and third strokes are the beginning and the end of forging, while the second stroke represents an intermediate forging in the general case, simulation was performed with a friction factor corresponding to the dry conditions. An example of the strain distributions after each bite is shown in Figure 7 for the width-to-height ratio $(B/2W_0)$ of 0.5, and a reduction of height of 10 percent. The material is annealed Al-2024. It can be seen that the interaction of the deformation zones occur near the die edges where strain concentrations are observed. In Figure 8 three-dimensional displays of spread contours are shown for each bite with the specimen dimension $B/2W_0$ of 0.5. Figure 8 also shows the mesh system used for the computation for each bite.

It was shown that this method is capable of providing good spread predictions for the forging process. Although the problems for which simplified three-dimensional elements can be used are limited, the approach can be applied to a class of problems where

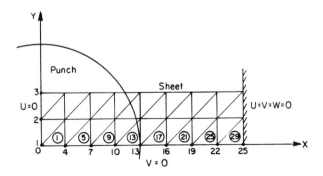

Figure 9   Rectangular strip domain used in punch stretching
             simulation.

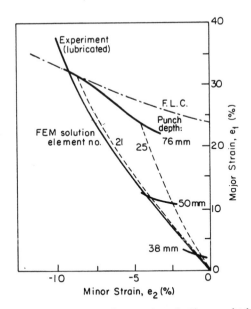

Figure 10   Comparison of strain path of the critical site predicted
             by the analysis ($\mu = 0$) with the experiment for the
             lubricated case in rectangular strip stretching.

the prediction of spread is important, such as compression of non-
circular disks. In applying simplified elements, different element
arrangements are possible for each problem, and a best arrangement
depends on the flow characteristics in the problem and on the
spread contours that are required.

SHEET METAL FORMING OF GENERAL SHAPES [11]

The variational formulation used in the previous rigid-plastic
analyses is inadequate for analyzing sheet metal forming processes.
This is due to nonuniqueness of deformation mode for the quasistatic
deformation of a rigid-plastic solid under certain types of boundary
conditions. Furthermore, out-of-plane sheet metal forming processes
involve large geometrical change during deformation. These two
points must be taken into consideration in deriving proper formulation of process modeling of sheet metal forming.

At the generic stage of deformation two configurations are of
interest: the current configuration B_1 of the body and the configuration B_2 which it will assume in the nearest future. In this
presentation B_1 is sometimes called the reference configuration and
B_2 the deformed configuration. The rectangular Cartesian coordinate
systems $\underset{\sim}{x}$ and $\underset{\sim}{x}'$ are associated with the reference and deformed
configurations, respectively. Incremental displacement u_j describes
the motion from B_1 to B_2.

According to the virtual work principle, with reference to the
current configuration,

$$\delta w = \int_S \underset{\sim}{f} \cdot \delta \underset{\sim}{u} \ ds = \int_V S_{\alpha\beta} \ \delta(dE_{\alpha\beta}) \ dV \ , \tag{11}$$

where $\underset{\sim}{f}$ is the surface traction; $\delta \underset{\sim}{u}$, virtual displacement; $S_{\alpha\beta}$,
Piola-Kirchhoff stress tensor of the second kind and $dE_{\alpha\beta}$,
incremental Lagrangian strain tensor. The Lagrangian strain tensor
is defined by

$$dE_{\alpha\beta} = \frac{1}{2}\left[U_{\alpha,\beta} + u_{\beta,\alpha} + g_{\gamma\delta} \ u^{\gamma}_{,\alpha} \ u^{\delta}_{,\beta} \right] \tag{12}$$

Where $g_{\gamma\delta}$ is the metric tensor, and a comma denotes the partial
differentiation. By assuming the flow rule according to

$$dE_{\alpha\beta} = \frac{\partial \ f(s_{\alpha\beta})}{\partial \ S_{\alpha\beta}} \ d\lambda \ , \tag{13}$$

and introducing the quantity $d\bar{E}$ defined by

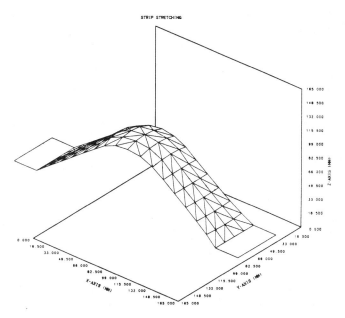

Figure 11   Geometry of the stretched rectangular strip for μ = 0.

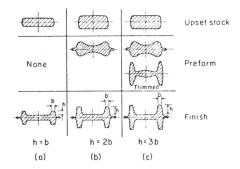

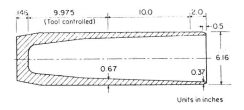

Figure 12   Preforms for finish forgings and for shell nosing.

$$d\bar{E} = \sqrt{\frac{2}{3} \, d\underset{\sim}{E}^T \, \underline{D} \, d\underset{\sim}{E}} \quad ,$$ (14)

where dE is the vector representation of Eq. (12), superscript T denoting transposé.

Equation (11) results in

$$\delta\phi = \int_V \bar{S} \, \delta(d\bar{E}) \, dV - \int_S \underset{\sim}{f} \, \delta\underset{\sim}{u} \, dS = 0$$ (15)

This is equivalent to the stationary condition of the functional ϕ. In Eq. (15), $\bar{S}$ is related to the stress-strain property of a given material through

$$\bar{S} = \sigma_0 + (h_0 - 2\sigma_0) \, d\bar{E}$$ (16)

where σ_0 is the Caushy stress and h_0 is the strain hardening rate at the current true strain ε_0. The curve of the true stress σ_0 and true strain ε_0 can be obtained, for example, by an uniaxial property test for a given material. Equation (16) is derived from the condition

that $S_{\alpha\beta} \, dE_{\alpha\beta} = \bar{S} \, d\bar{E}$, and the assumption

that $\displaystyle\int_{\varepsilon_0}^{\varepsilon} \sigma \, d\varepsilon \quad \int_{\bar{E}_0}^{\bar{E}} \bar{S} \, d\bar{E}$.

Equations (12), (14), (15) and (16) are the basis of the finite element formulation for the analysis of sheet metal forming.

The finite element formulation includes the description of the geometrical as well as tractional boundary conditions in the three-dimensional space and the transformation from the local to the global coordinate systems. The detail of formulations can be found in Reference [11].

The application is shown for the analysis of a formability test. The Nakazima method [12] is one of the test methods for determining the forming limit curves of sheet materials. It uses a hemispherical punch with rectangular strips with different blank widths to vary the straining modes of test specimens. Hecker [13] used the method to determine the forming limit curves of aluminum alloys and aluminum-killed steels.

Simulation of the test is performed using the following conditions:

material: 2036-T4 Aluminum alloy
stress-strain characteristic: $\bar{\sigma} = 592 \; (\bar{\epsilon})^{0.245}$ N/mm^2
R value: 0.78
material thickness: 1.02 mm
material width: 50.8 mm
material length: 203.2 mm
punch radius: 50.8 mm

Figure 9 illustrates the portion of sheet domain used in the computation.

The strain path for the element 21 for $\mu = 0$ was computed and
compared with that by experiment for the lubricated case in Figure
10. The forming limit curve shown in the figure was determined
from the experiments. It is seen that the experimental and theoretical results agree to each other very well. In Figure 11, an example
is shown for the geometry of the stretched rectangular strip.
These results demonstrate that the solutions were reasonable even
in detail in comparison with some experimental observations.

PREFORM DESIGN IN METAL FORMING [14]

One of the most important aspects in metal forming processes
is the design of preforms. An example of preform design for steel
finish forgings of various H-shapes is shown in Figure 12. Preform
design involves the determination of number of preforms and design
of the shapes and dimensions of each preform. A preform design
problem is also involved in shell nosing. Wall thickness increases,
but the shell may elongate or shrink in length during nosing.
Because shell specifications usually require certain wall thickness
distributions after nosing, the problem is to design the preform
with which the desired shape after the nosing operation can be
achieved. An example of such a preform is also shown in Figure 12

A new approach to preform design has been proposed. The concept involved in the approach is to trace backward the loading
path in the actual forming process from a given final configuration
by the finite-element method. A method of approach is illustrated
in Figure 13. At time $t = t_0$, the geometrical configuration x_0 of a
deforming body is represented by a point Q. The point Q is arrived
at from the configuration x_{0-1} at $t = t_{0-1}$, through the displacement field u_{0-1} during a time step Δt. Therefore, the problem is
to determine u_{0-1}, based on the information (x_0) at point Q. The
solution scheme is as follows: taking the loading solution u_0
(forward) at Q, the first estimate of P can be made according to
$P^{(1)} = x_0 - u_0$. Then, the loading solution $u_{0-1}^{(1)}$ can be calculated
on the basis of the configuration of $P(1)$ with which the configuration x_0 at Q can be compared with $P^{(1)} + u_{0-1}^{(1)} = Q(1)$. If Q and
$Q^{(1)}$ are not sufficiently close to each other, then $P^{(2)}$ can be
estimated by $P^{(2)} = x_0 - u_{0-1}^{(1)}$. The solution for loading at $P^{(2)}$
is then $u_{0-1}^{(2)}$ and the second estimate of the configuration $Q^{(2)} =$

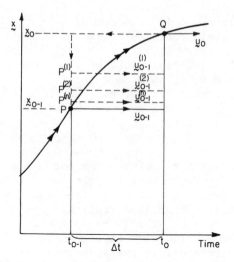

Figure 13   Concept of backward tracing scheme.

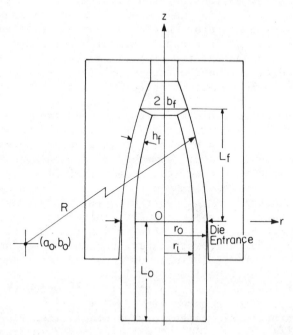

Figure 14   Die dimensions and nosed shell configuration.

$P(2) + u_0 \cdot \binom{2}{1}$ can be made. The iteration is carried out until $Q^{(n)}$ = $P(n) + u_0 \cdot \binom{n}{1}$ becomes sufficiently close to Q. The convergence is sufficient if the body is plastically deforming at all times, and if the boundary condition does not change at P and Q. We call the calculation scheme "the backward tracing scheme." It is to be noted that the capability of the finite element process analysis is the pre-requisite for applying the new approach to the problem.

Preform design in forging is complex and generally involves many problems in applying the new approach. Less complex is the preform design in shell nosing, because the number of parameters in defining preform shapes is fewer. In shell nosing, two simple preform configurations can be realized for a uniform wall thickness after nosing. The wall thickness distributions required in the preform can be accommodated by varying either the inner or the outer diameters of the problem. It is not difficult to see that different preform configurations can be arrived at by imposing different sequences of boundary conditions (such as where and when the nodal point in contact with the die should be freed from the die) during the backward process. This aspect of the problem is investigated for non-workhardening materials.

With reference to Figure 14, specifications of the final nosed shell configuration are given by (unit: nm).

$$L_1 = 48.84 \qquad r_o = 21.526 \qquad b_f = 14.47$$

$$(L_o = 47.26) \qquad r_i = 14.648 \qquad h_f = 6.88 \text{ (uniform)}$$

$$R = 259 \qquad\qquad (a_o = 237, \; b_o = 18.75).$$

Coefficient of friction at the die-workpiece interface is assumed to be $\mu = 0.05$. For the finite element calculation, 40 four node elements and 63 nodal points are used. The criterion for controlling the boundary condition during the backward tracing depends on the preform shape under consideration.

For the preform shape with constant outer diameter (Type-0 preform), $r_o^n = r_o$ for all n, where r_o^n is the radius of a nodal point located along the outer surface of the shell. The boundary condition for this type of preform is controlled in such a way that nodes which are in contact with the die ($r_o^n < r_o$) are freed from the die when the condition that $r_o^n = r_o$ is reached during the backward tracing.

For the preform shape with constant inner diameter (Type-I preform), the criterion for changing the boundary condition is that as soon as $r_i^p = r_i$ for the nodes $r_i^p < r_i$, the boundary condition of the corresponding outside node is changed from that of die-contact to the force free condition, where r_i^p is the radius of a nodal

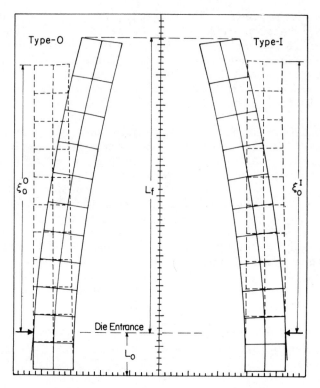

Figure 15   Nosed shell configuration and two types of preform
              shapes.

point along the inner surface of the shell.

    The two types of preforms determined by the method are compared
with the final nose configuration in Figure 15. In Figure 15, the
lengths and , for Type-0 and Type-I preforms, respectively, indicate
the portion of the preform which becomes the nosed shell length L_f
after nosing. Because the shell elongates during nosing, the
preforms are shorter than the nosed shell. It can also be observed
that the Type-0 preform is slightly shorter than the Type-I preform.
This is because the amount of deformation involved in nosing is
more for Type-0 than for Type-I, and the total elongation is larger
for the Type-0 preform.

    A new technique, the backward tracing scheme, was devised for
preform design. This is a unique application of the finite element
method to the problems in metal forming. It was demonstrated that
the technique can be applied to preform design in shell nosing.
It was also revealed that one of the critical aspects for further
development of the technique is concerned with the treatment of the
boundary conditions.

CONCLUDING REMARKS

For the analysis of three-dimensional metal flow it is concluded that for the problems of geometrically simple block compression, the finite element method provides reasonably good solutions to the detail of three dimensional plastic deformation with relatively few elements. However, it is evident that the economical constraint becomes severer for three-dimensional metal flow analysis, in comparison with two-dimensional problems. Therefore, in the development of analysis techniques for three-dimensional metal flow, emphasis must be placed on achieving the balance between the computational efficiency and the solution accuracy.

This aspect becomes even more critical when we consider the problems in warm and hot forming where coupling of temperature calculations with the metal flow analysis is required. In this connection further exploration of simplified approaches may be worthwhile.

It is seen that expanding the analysis capabilities is being made not only in the area of bulk deformation processes, but also in the sheet metal forming area. Description of the surfaces, and the treatment of the boundary conditions in the three-dimensional space in sheet metal forming could be helpful for the three-dimensional formulation for the bulk deformation problems.

Finally, an effort toward full utilization of capabilities of the finite element method in solving significant industrial problems was demonstrated in an application of the method to preform design. Further development in this area seems to be an exciting and challenging undertaking.

ACKNOWLEDGMENTS

The author wishes to thank the Army Research Office, the Battelle Columbus Laboratories-Manufacturing Technology, and the Air Force Wright Aeronautical Laboratories - Materials Laboratory, for their contracts under which the developments reported in this paper were possible. The author also wishes to thank Mrs. Carmen Marshall for typing the manuscript.

REFERENCES

1. Kobayashi, S., "A Review on the Finite-Element Method and Metal Forming Process Modeling," J. Appl Metalworking, Vol. 2, 1982, pp. 163-169.
2. Oh, S. I. "Finite Element Analysis of Metal Forming Processes with Arbitrary Shaped Dies," Int. J. Mech. Sci., 24, 1982, 479.
3. Rebelo, N., "Finite Element Modeling of Metalworking Processes

for Thermo-Viscoplastic Analysis," Ph.D. dissertation, Mechanical Engineering, University of California, Berkeley, 1980.

4. Zienkiewicz, O. C., "The Finite Element Method," 3rd. edition, McGraw-Hill, 1977.

5. Strang, G. and Fix, G. J., "An Analysis of the Finite Element Method," Prentice-Hall, Englewood Cliffs, NJ, 1973.

6. Huebner, K. H., "The Finite Element Method for Engineers," John Wiley & Sons, 1975.

7. Park, J. J., "Applications of the Finite Element Method to Metal Forming Processes," Ph.D. dissertation, Mechanical Engineering, University of California, Berkeley, 1982.

8. Chen, C. C. and Kobayashi, S., "Rigid Plastic Finite Element Analysis of Ring Compression," Applications of Numerical Methods to Forming Processes, ASME, AMD, 28, 1978, 163.

9. Li, G-J. and Kobayashi, S., "Spread Analysis in Rolling by the Rigid-Plastic Finite Element Method," Numerical Method in Industrial Forming Processes, The International Conference, 1982, p. 777.

10. Sun, J-X., Li, G-J. and Kobayashi, S., "Analysis of Spread in Flat-Tool Forging by the Finite Element Method," Proc. North American Manufacturing Research Conf., May 1983, Madison, Wisconsin, pp. 224-231.

11. Toh, C. H., "Process Modeling of Sheet Metal Forming of General Shapes by the Finite Element Method Based on Large Strain Formulation," Ph.D. dissertation, Mechanical Engineering, University of California, Berkeley, 1983.

12. Nakazima, K. Kikuma, T., Hasuka, K., "Study on the Formability of Steel Sheets," Yawata Technical Report, No. 264, 1968, 141-154.

13. Hecker, S. S., "A Simple Forming Limit Curve Technique and Results on Aluminum Alloys," General Motors Research Publication GMR-1220, 1972.

14. Park, J. J., Rebelo, N. and Kobayashi, S., "A New Approach to Preform Design in Metal Forming with the Finite Element Method," Int., J. Mach. Tool Des. Res., Vol. 23, 1983, 71-79.

RECENT MATERIAL AND PROCESS DEVELOPMENTS

IN RIM AND RRIM

W.J. Farrissey, Jr., L.M. Alberino and R.J. Lockwood

The Upjohn Company
D.S. Gilmore Research Laboratories
410 Sackett Point Road
North Haven, Connecticut 06473

INTRODUCTION

The use of reaction injection molding (RIM) and reinforced reaction injection molding (RRIM) for the production of large parts, especially in automotive applications, is currently quite commonplace.[1] In the RIM process, shown schematically in Figure 1, two liquid components are pumped at high throughputs and pressures to a self cleaning impingement type mixing head and then into a mold where the polymerization reaction occurs and the part is formed within seconds.[2] RRIM is basically the same process but short fibers or fillers are added to one or both of the components in order to obtain various special properties. Certain equipment modifications must be made in order to handle the viscous, and often abrasive slurries.

This paper will summarize current RIM and RRIM technology from a material and processing point of view and also briefly describe some of the more recent developments in these areas.

FIGURE 1. REACTION INJECTION MOLDING SCHEMATIC

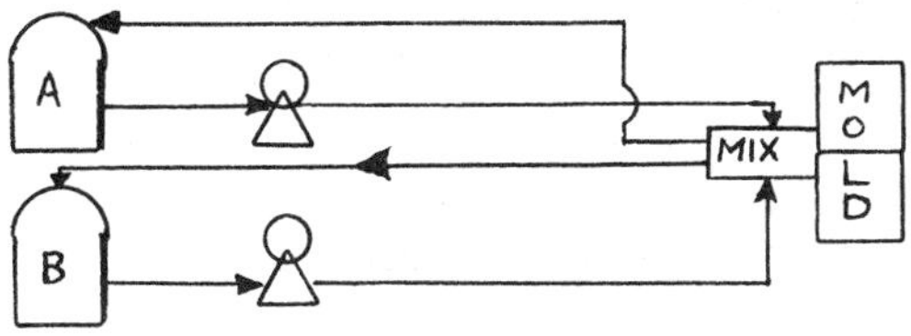

CHEMISTRY AND POLYMER STRUCTURE

The details of catalysis and mechanism of the chemistry of urethane formation may be found in standard references.[3] The overall reactions of isocyanates to form polymers in RIM AND RRIM are shown in Figure 2.

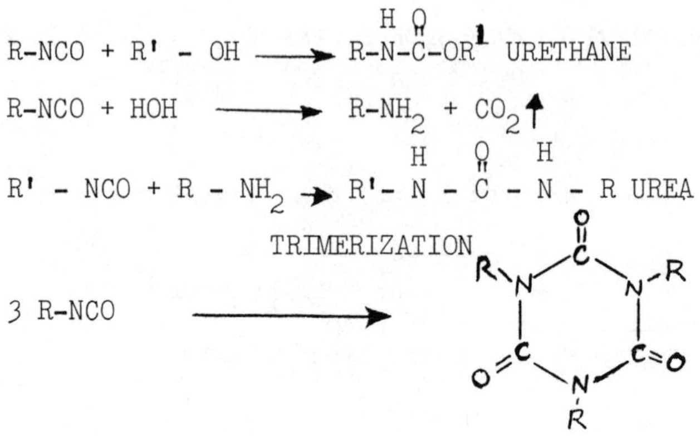

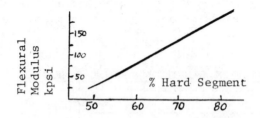

FIGURE 2. URETHANE TECHNOLOGY
GENERAL REACTIONS OF ISOCYANATES

Structurally, a typical urethane polymer consists of "hard" and "soft" segments which are interconnected in the polymer backbone. The hard segment "A" comes from the reaction of isocyanate and extender and the soft segment "B" comes from the reaction of isocyanate and polyol. Therefore, a polyurethane is a segmented block copolymer with the generalized structure $(\ldots A_n B_m A_p B_r \ldots)s$ While the usual glycol extenders have been 1,4 butane diol and ethylene glycol, systems extended with amines, such as diethyl TDA (DETDA) have become increasingly important. With either urethane hard segments, or urea hard segments, the modulus and strength values for a RIM polymer depend upon the hard segment content. This is shown in Figure 3.

FIGURE 3. FLEXURAL MODULUS VERSUS % HARD SEGMENT
FOR A URETHANE RIM SYSTEM

Thus, the wide formulating latitude in a urethane RIM polymer comes about not only from the choice of the particular chemical involved, but also from the ratio of the specific 'hard' blocks to 'soft' block. The degree of compatibility and incompatibility between the various chemical constituents often controls the ease of RIM processing and the degree of phase mixing in the final polymer strongly influences RIM polymer properties.[5] Table 1 summarizes these relationships.

TABLE 1. COMPATIBILITY RELATIONSHIPS IN A RIM POLYMER

Homopolymer		Resultant Copolymer
Hard Segment	Soft Segment	Domain
Crystalline	Compatible	Two Phase
Crystalline	Incompatible	Two Phase
Amorphous	Compatible	One Phase
Amorphous	Incompatible	Two Phase

PROPERTIES AND PROCESSING

The various considerations discussed above give rise to the range of properties for RIM polymers described in Table 2. These can range from soft highly elastic products to rigid hard plastic materials. Additionally, they can either be produced in solid form, or through the use of suitable blowing agents can be made microcellular in physical structure. Tables 3 and 4 give details of processing and properties for a 95A elastomeric RIM urethane polymer.

TABLE 2. GENERAL PROPERTIES OF
ISOCYANATE BASED RIM SYSTEMS

	Low Flex Modulus	Intermediate-High Flex Modulus	High Flex Modulus
Flexural Modulus, kpsi @ R.T.	20-75	75-200	200-400
Tensile Elongation, % @ Break	100-300	50-200	<50
Izod Impact, ft.-lb./in. Notched	10-15	5-15	<50
Impact Strenght	High	Medium-High	Low
Material Description	Elastomer	Pseudo-Plastic	Plastic
Ppplication-Automotive	Fascia	Fender	Hood or Deck Lid

TABLE 3. RIMTHANE ᵗᵐCPR® 2601–15
GENERAL PROCESSING INFORMATION

Mixing Ratio   Component "A" (Isocyanate) 40

                  Component"B"(Polyol)   60

Viscosity (cps)	"A"	1000
@ 73°F	"B"	1300
@120°F	"A"	295
	"B"	360

Processing Temperature	"A"	120°F
	"B"	120°F

Mold Temperature	150°F
Gel Time	10 seconds
Demold Time	2 seconds

TABLE 4, RIMTHANE™ CPR® 2601 –15

Property	English Units
Density	68 lbs/ft³
Hardness,Durometer	95A
Tensile Strength	2830 psi
Tensile Modulus @ 100% Elongation	1290 psi
Elongation @ Break	450 %
Flexural Modulus	15,500 psi
Tear Strength, Die C	365 pLi

Recent developments in the elastomer range of RIM products have involved the use of amine extenders. Some representative processing and mechanical property data for such a material are given in Table 5. The improvements in this technology over the glycol extended system include faster demold, better 'green ' strength and improved thermo-mechanical properties.

TABLE 5. UPJOHN RIM PREPOLYMER AMINE EXTENDED

A/B Ratio	0.88
"A" Component Temperature, ^{O}F	102
"B" Component Temperature, ^{O}F	95
Mold Temperature, ^{O}F	150
Mold Density, lbs/ft^3	62.8
Tensile Strength, psi	3500
100% Tensile Modulus, psi	2100
Elongation at Break,%	255
Die 'C' Tear, pLi	330
Flexural Modulus, psi	35,000
Heat Sag, 6" Overhang, 1 hr. @ 250°F,in.	0.66

Other recent developments in the RIM area from a materials point of view have centered on the production of higher modulus materials. At a flexural modulus level of 100,000 psi a highly toughened 'psuedo-plastic' material results with outstanding impact properties (Notched Izod ~10 ft-lb/in). As the hard segment content is further increased a truly plastic material results. Through the proper balancing of the parameters described in Table 1, a tough urethane RIM plastic can be produced. An example of the processing and properties of such a material are given in Tables 6 and 7.

TABLE 6. DSG 396 PROCESSING PARAMETERS

COMPONENT TEMPERATURE

"A" ^{O}F (^{O}C) 75(24)
"B" ^{O}F (^{O}C) 80(27)
A/B Weight Ratio 1.56

IMPINGEMENT PRESSURE

"A" atm. (psi) 150 (2200)
"B" atm. (psi) 150 (2200)
Mold Temperature,^{O}F(^{O}C) 175-180(80-82)
Demold Time,minutes 1.5
Postcure 250°F/1 hr.hr.
Gel Time, Secs. 6

TABLE 7. DSG 396 MECHANICAL PROPERTIES

HIGH MODULUS URETHANE PLASTIC
PRELIMINARY DATA

Property	English Units
Density	70 lb/ft^3
Hardness, Durometer	78D
Flexural Modulus	225,000 psi
Tensile Strength	5400 psi
Elongation @ Break	120
Notched Izod Impact	10 ft-lb/in
Heat Sag, 4" Overhang 60 min. @ 250°F	0.20 in.

By adjusting the isocyanate chemistry to form the isocyanurate structure, even higher modulus plastic materials can be RIM produced. These materials do not have good impact, but have outstanding thermo mechanical properties with heat distortion values above 150°C and a room temperature flexural modulus of 330,000 psi.

Other recent developments have involved non-conventional RIM materials such as Nylon RIM, unsaturated polyesters, and epoxy materials.[6] The most advanced of these at this time is the Nylon RIM, but its use requires specially modified RIM equipment.

The RIM process itself has been the subject of intensive research in recent years. The mixing process in the RIM mix head has been investigated.[7,8] It has been determined that Reynold's numbers of >200 are necessary in order to get adequate mixing, and the role of the striation thickness in the mixed materials has been investigated. The process of filling the mold with the mixed chemicals, and the resulting kinetic and energy equations to describe the process also have been studied.[9] The basic equations have been set forth for constant viscosity systems and for systems where viscosity is dependent upon tempreature and on chemical composition (both of which are changing due to the polymerization reaction).

Through a combination of kinetics to describe the reaction, and heat transfer equations to describe energy flow, models can be constructed which will predict temperature and concentration vs depth and time in the mold. Examples of this are shown in Figure 4 where the experimental versus predicted temperature profile for a typical urethane RIM system is shown.

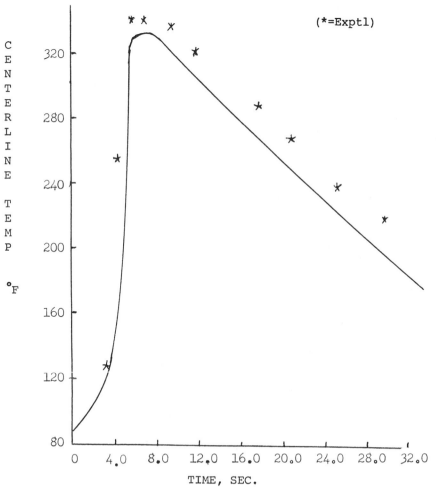

Experimental vs Predicted Temperature Profile

(*=Exptl)

FIGURE 4.

RRIM

One of the more exciting advances in the equipment areas in
the past several years has been in the design and delivery of the
equipment to handle slurries. Each component side in RRIM
equipment consists of a single large metering piston and in
addition the new heads are specially adapted to handle the
abrasive nature of the fillers. While the terminology reinforced
RIM has been adopted by the industry, the short length of the
fibers which can be handled (because of viscosity limitations)
does not allow for true reinforcement. Therefore, most of the
additives which can be handled in RRIM equipment are in reality
fillers rather than true reinforcing agents. These basic
differences are summarized in Table 8, and the effect of fiber
length is drammatically shown for a high modulus (isocyanurate)
RIM system in Table 9.

TABLE 8. REINFORCED VERSUS FILLED RIM

	REINFORCED	FILLED
Type Materials	High Aspect Ratio Fibers & Fillers Ex. $\geq$ 1/8" Chopped	Low Aspect Ratio Fibers & Fillers 1/16" Milled Glass Calcium Carbonate Hydrated Alumina
Flex Modulus	Increases	Increases
HDT	Increases	Increases
CTE	Decreases	Decreases
Impact Strength Ex. Izod Test	Increases	Decreases

However the property variations which are possible through
the incorporation of various fillers will give increased emphasis
to the RRIM process in the near future.

TABLE 9. EFFECT OF GLASS FIBER LENGTH
ON IMPACT STRENGTH

A. Milled Glass Fiber (RIM; Admiral)

	Unfilled	15% Glass
Izod	1.2	0.5 ft.-lb/in.
Gardner	0.8	0.3 ft.-lb/in.

B. 1/8" Chopped Strand (Compression Molded)

Izod	0.5	1.8 ft.-lb/in.
Gardner	0.3	0.2 ft.-lb/in.

C. Continuous Strand Mat (RIM; DSG-395-12)

Izod	0.4	9.4 ft.-lb/in.
Gardner	0.3	1.0 ft.-lb/in.

CONCLUSIONS

The new material and equipment developments which are taking place in RIM and RRIM are indications of the strength and viability of the industry. Coupled with the increased understanding of the basic RIM process through the process research, it is expected that even more sophisticated RIM applications will become a reality. Many of these new applications will be in the non-automotive area although the automotive industry will continue to dominate the technology in the immediate future.

REFERENCES

1. L. M. Alberino, "Future of RIM Development in the U.S.A.", Poly. Sci. and Technology, Vol. 18, Reaction Injection Molding and Fast Polymerization Reactions, Ed. J. E. Kresta, Plenum Press (New York).
2. R. J. Lockwood and L. M. Alberino, "Reaction Injection Molding and Reinforced RIM: A Review:, Advances in Urethane Science and Technology, Vol. 8, Ed. K. C. Frisch and D. Klempner, Technomic Press, Westport, CT. (1981).

3. J. H. Saunders and K. C. Frisch, "Polyurethanes Chemistry and Technology, Part I. Chemistry" Interscience Publishers, New York (1965).

4. G. M. Estes, S. L. Cooper, and A. V. Tobolsky, J. Macromol. Sci. - Rev. Macromol. Chemistry, C4(2), 313 (1970).

5. R. J. Lockwood and L. M. Alberino, "Phase Mixing in Urethane Polymers", Urethane Chemistry and Applications, ACS Symposium Series 172, Ed. K. N. Edwards, ACS, Washington, D. C. (1981).

6. L. M. Alberino and T. R. McClellan, "The Future of RIM in the U.S.A.", 1983 ACS Annual Meeting, Aug. 28-Sept. 12, 1983, Washington, D. C.

7. C. L. Tucker and N. P. Suh, "Mixing for Reaction Injection Molding. I. Impingement Mixing of Liquids", Polymer Eng. and Sci., 20:13, 875 (1980).

8. P. Kolodziej, C. W. Macosko, W. E. Ranz, "The Influence of Impingement Mixing on Striation Thickness Distribution and Properties in Fast Polyurethane Polymerization", Polymer Eng. and Sci., 22:6, 388 (1982).

9. J. M. Castro and C. W. Macosko, "Studies of Mold Filling and Curing in the Reaction Injection Molding Process", AIch. E Jour., 28:2, 250 (1982).

PRECISION INJECTION MOLDING OF THERMOPLASTICS

Nick R. Schott

Department of Plastics Engineering

University of Lowell

ABSTRACT

The paper reviews the state of the art of injection molding. The tolerances in terms of shrinkage and warpage are reviewed for major plastic resins. The molding process is discussed in terms of flow orientation and control of melt and machine parameters to achieve good properties and dimensions. The paper looks at automatic process control for precision parts. The use of microprocessor control is reviewed. The paper will discuss both ram velocity control and melt pressure control. The use of such a control unit in the molding of high quality optical polyurethane lenses is described. The use of computer modeling and computer aided design in the simulation of molded plastics is described from a commercial viewpoint. Also, the role of cooling of the part and the feasibility of the proposed part is described. Finally, the paper describes the need for rheological and thermodynamic data that are required for the computer modeling.

INTRODUCTION

Injection molding is the most versatile of the plastics processes for making three dimensional parts. The process is used for both thermoplastic and thermosetting resins and accounts for 32 percent of all plastics consumption.[1] Over 1200 machine models were available in 1979 of which about 40 percent have a shot capacity of less than 10 ounces. It is in this size category that most of the precision injection molding is done. Major areas that require high precision parts are the optical, medical, and electronics industry.

MOLD SHRINKAGE

The injection molding process in its present form was developed
in the early 1960's with the introduction of the reciprocating screw
plasticator/plunger.[2] Figure 1 shows a schematic of the molding
process. The resin is plasticated (melted) and accumulates in
front of the screw until sufficient melt is available to fill the
cavity. A check valve or ring prevents the melt from drooling out
of the nozzle until the proper amount of melt is accumulated. At
the proper time in the molding cycle the melt is injected into the
closed mold. Injection pressures up to 20,000 psi are standard
and up to 42,000 psi are available as custom built machines. The
melt solidifies under controlled conditions and is cooled until it
is below its heat deflection temperature (HDT test, ASTM D648).
The melt itself is compressible, which can lead to a large amount
of shrinkage as the melt cools. One can compensate for a part of
the shrinkage by pressurizing the melt in the cavity to pack in
extra molecules. This packing phase of the cycle is required until
the gate freezes off at which time no melt can escape from the cavity
and the part weight is fixed. Table 1[3] gives typical values for
the shrinkage of injection molded parts made with common thermoplastic resins. It is seen that shrinkage can be as high as 10
percent with values of 0.5 to 2% being most common. The shrinkage
is caused by the thermal contraction as the melt cools and also by
the volume change as some of the materials crystallize. The nylons,
polyethylenes, polytetrafluorine, poly-4-methyl pentene (TPX) and
polypropylene are some of the common plastics that are partially
crystalline. In these materials shrinkage is especially pronounced.

The shrinkage itself is not uniform over the length of the
part. It varies, i.e. cross-sectional area of the part, location
from the gate, rate of injection, amount (if any) of filler, cooling
channel layout, temperature of the melt, temperature of the mold,
and the amount of cushion, etc. Based on an industrial survey the
Plastics industry can predict the average amount of shrinkage
variation that will occur. Typical values for shrinkage variation
are given in Table 2. The data in Table 2 are conveniently represented in a tolerance chart as shown in Figure 2[4]. The diagram
shows that the molding can be classified for coarse, standard or
fine tolerances. Fine tolerances give the best results. The
tolerances will increase as the part dimension increases. Also,
tolerances increase as the number of cavities increase, i.e., as
one goes from a single to multi-cavity mold. An up-to-date collection together with standard custom molding practices is found in an
SPI publication[5].

WARPAGE

Warpage is a large scale deflection in a molded part due to
uneven shrinkage. The uneven shrinkage within the part sets up

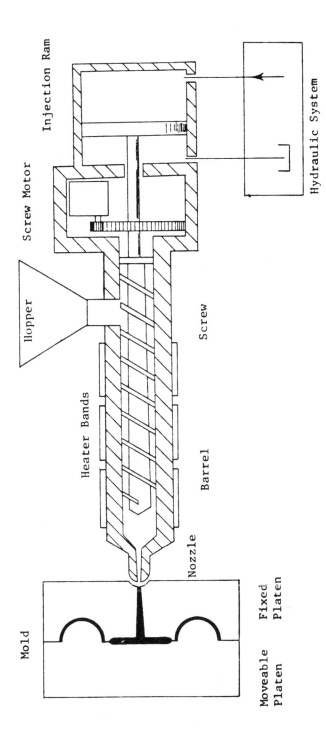

Figure 1. Injection molding – reciprocating screw process

TABLE 1
Typical Values of Mold Shrinkage in Injection Molding

Material	Mold Shrinkage (in/in or mm/mm)	Mold Shrinkage (%)
ABS		
high impact	0.005-0.007	0.5-0.7
heat resistant	0.004-0.005	0.4-0.5
medium impact	0.005	0.5
Acetal	0.020-0.035	2.0-3.5
Acrylic		
easy flow	0.002-0.007	0.2-0.7
general purpose	0.002-0.009	0.2-0.9
heat resistant	0.003-0.010	0.3-1.0
high impact	0.004-0.008	0.4-0.8
Cellulose acetate		
hard	0.002-0.005	0.2-0.5
medium	0.002-0.005	0.2-0.5
soft	0.002-0.005	0.2-0.5
high acetyl	0.002-0.005	0.2-0.5
Cellulose acetate-butyrate	0.002-0.005	0.2-0.5
Cellulose propionate	0.002-0.005	0.2-0.5
Chlorinated polyether	0.004-0.008	0.4-0.8
Ethyl cellulose	0.005-0.010	0.5-1.0
Ethylene vinyl acetate	0.010-0.030	1.0-3.0
Fluorinated ethylene-propylene copolymer	0.030-0.060	3.0-6.0
Ionomer	0.003-0.020	0.3-2.0
Nylon		
type 6-6	0.010-0.025	1.0-2.5
type 6	0.007-0.015	0.7-1.5
type 6-10	0.010-0.025	1.0-2.5
type 11	0.010-0.025	1.0-2.5
type 12	0.008-0.020	0.8-2.0
transparent (Trogamid T)	0.004-0.006	0.4-0.6
glass filled	0.005-0.010	0.5-1.0
Phenoxy	0.003-0.004	0.3-0.4
Poly 4-methyl-pentene 1	0.015-0.030	1.5-3.0
Polybutylene	0.020(molded)	2.0
	0.040(aged)	4.0
Polycarbonate	0.005-0.007	0.5-0.7
Polychlorotrifluor-ethylene	0.010-0.020	1.0-2.0

TABLE 1 (Continued)

Material	Mold Shrinkage (in/in or mm/mm)	Mold Shrinkage (%)
Polyethylene		
low density	0.015-0.035	1.5-3.5
high density	0.015-0.030	1.5-3.0
Polyphenylene oxide	0.007-0.008	0.7-0.8
modified	0.005-0.007	0.5-0.7
Polypropylene	0.010-0.030	1.0-3.0
Polystyrene		
general purpose	0.002-0.008	0.2-0.8
heat resistant	0.002-0.008	0.2-0.8
toughened	0.003-0.006	0.3-0.6
Polysulphone	0.008	0.8
Polytetrafluorethylene	0.050-0.100	5.0-10.0
Polyurethane elastomer	0.010	1.0
Polyvinyl chloride		
unplasticized	0.002-0.004	0.2-0.4
rigid	0.002-0.004	0.2-0.4
semi-rigid	0.005-0.025	0.5-2.5
flexible	0.015-0.030	1.5-3.0
Polyvinyl dichloride	0.003-0.007	0.3-0.7
Polyvinyl fluoride	0.030	3.0
Styrene-acrylonitrile	0.002-0.006	0.2-0.6
Styrene butadiene elas-tomer	0.001-0.005	0.1-0.5
Styrene methyl methacy-late	0.002-0.006	0.2-0.6
Vinylidene chloride	0.005-0.025	0.5-2.5

TABLE 2

Mold Shrinkage Variation for Injection Molding (Ref. 3)

Material	Mold Shrinkage Variation (in/in or mm/mm	(%)
ABS	0-0.004	0-0.4
Acetal	0-0.005	0-0.5
Acrylic	0-0.005	0-0.5
Nylon	0-0.010	0-1.0
Polycarbonate	0-0.003	0-0.3
Polyethylene		
low density	0-0.020	0-2.0
high density	0-0.025	0-2.5
Polypropylene	0-0.010	0-1.0
Polystyrene	0-0.003	0-0.3
Polyvinyl chloride, rigid	0-0.005	0-0.5

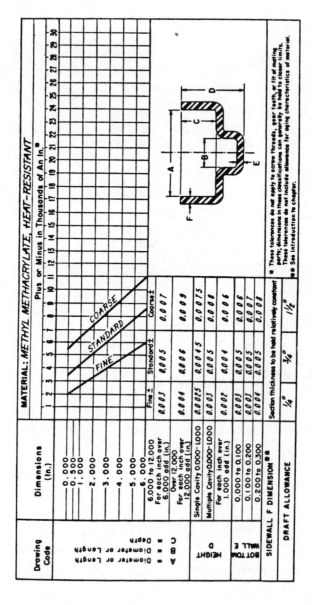

Figure 2. Typical tolerance chart for an injection molded part (Ref. 4).

frozen-in stresses that deform the part after molding. In many
cases one can predict the amount of warpage that will occur based
on the flow orientation, expected differential shrinkage and process
conditions. One can compensate for the warpage by making a correction in the tool as for example, molding an arc in a typewriter
housing which will then deform to give a flat section after warpage.

BEST EFFORT IN TOLERANCES

   In small electronic and mechanical parts, dimensional tolerances of tenths of a mil are often required. They can be achieved
by conscientious molders. They represent a best effort using
precision molds, new injection molding machines that have good
repeatability and close supervision and control of the production
process. In many cases, microprocessor based adaptive control is
used[6]. Optical parts such as lenses require the most exacting
tolerances; the use of the latest technology allows molders to
tackle tolerances once reserved for ground and polished glass
technology. In the Polaroid SX-70 camera, the viewfinder and other
"imaging" lenses have surface-contour specifications of 0.000006 in.
by interferometer measurement[6]. The tolerances are achieved via
adaptive process control that packs the acrylic or polycarbonate
part much more precisely than conventional open-loop control
systems. This assures accurate surface finish and post molding
dimensional stability. With the adaptive closed-loop control pack
and hold pressures are held within a range of $\pm$ 10 psi versus the
$\pm$ 200 psi with conventional machines.

ADAPTIVE PROCESS CONTROL

   The use of microprocessor based adaptive process control has
been the single most important development in the 70's. Although
the introduction of the adaptive control systems had been rather
slow, over 1000 microprocessor based machines out of 60,000 machines
are now in operation[6]. These machines control the fill rate, the
first and second stage pressure, the cushion, and the shot size.
The final control element in the loop is a high speed electro-
hydraulic servo valve[7]. The response time of the standard valve is
0.09 seconds while a special high speed valve has a response time
of 1/300th of a second. These control units allow one to program
the ram velocity during melt injection into the cavity. By keeping
the injection speeds within very narrow limits, thin walls are
filled without partial freeze-off and stress, while avoiding materials overheating through excessive shear at the gate. Similarly,
in optical parts, programmed injection rates eliminate flow lines
that create unseen optical imperfections[8]. In the case of electrical
connectors, the velocity control allows slow initial fill around
the core pins to prevent knit lines and incomplete fill around
stacking insert holes while at the same time preventing core-pin
displacement. After the initial fill the velocity can be increased

to prevent mold freeze-off, followed by a reduced velocity to
eliminate mold flashing before final transition to pack and hold
pressure control.

Figure 3 shows the schematic for the ram velocity control[9].
The slope of the ram position vs. time gives the velocity of the
ram. Usually at least five points are available to vary the ram
velocity. After the velocity control mode, control is transferred
to pressure control loop (Figure 4). The loop controls the pressure to which the melt is subjected. The strategy for melt pres-
urization has been mentioned earlier. An equation of state which
predicts the compressibility of the melt is:[10]

$$(P + \pi)(\nu - b) = \frac{RT}{M} \tag{1}$$

where,

P = pressure on melt
ν = specific volume of melt
R = gas constant
T = absolute temperature
π = pseudopressure
b = pseudovolume
M = molecular weight of an "interaction unit"

Equation (1) is empirical in nature and the constants π, b, and
M are determined from experimental data. The equation is thermodynamic in nature and assumes equilibrium conditions. This is not
the case in injection molding. Figure 5 shows that the specific
volume is highly dependent on the rate of heating or cooling[11].
This aspect of the molding problem again emphasizes that shrinkage
is hard to predict, since it is a function of processing history.
Shrinkage will continue as long as the polymer is above its T_g until
it reaches its equilibrium value. In precision moldings one
increases the mold temperature to get better surface finish, a
recovery of frozen-in stresses and occurrence of most of the
shrinkage while still in the mold. This leads to better tolerances
but also increases the cycle time and the molding costs.

Another variable that can be controlled is the cushion and shot
size. The cushion is the amount of melt that remains in the barrel
after the ram has travelled to its most forward position during
injection. A minimum cushion is required to prevent damage to the
screw tip or nozzle. However, any further cushion will cause more
melt material to be compressed and hence increase the shrinkage.
Figure 6 shows this effect[12].

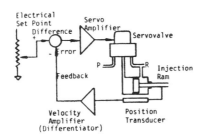

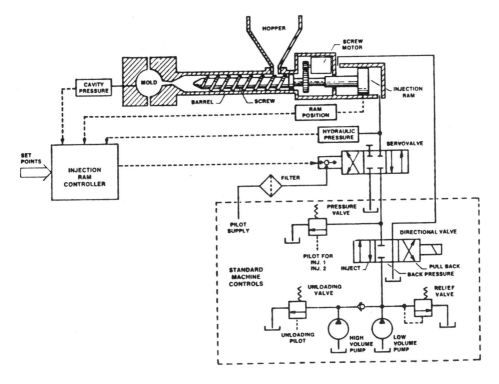

Figure 3.   Schematic for velocity control loop (9).

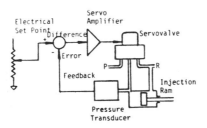

Figure 4.   Pressure control loop.

EXPANSION AND CONTRACTION

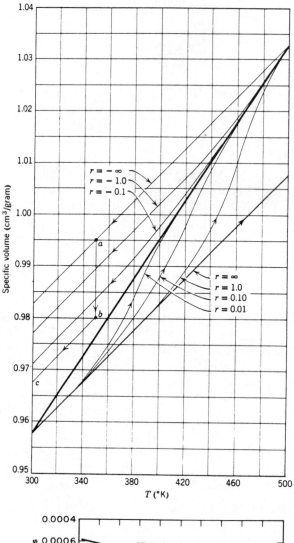

Figure 5. Calculated volume–temperature curves for the heating and cooling of polystyrene, with r (the rate of temperature change) expressed in deg C/sec (Ref. 11).

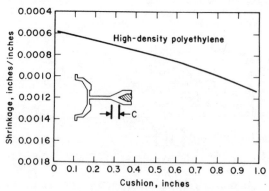

Figure 6. Effect of cushion on part shrinkage.

OTHER TECHNIQUES FOR PRECISION MOLDING

Experience in the industry has shown that one can improve moldings by adding a fraction of a percent blowing agent to prevent sink marks in thick sections. Also, one can add fillers and reinforcements which have a much lower shrinkage which thus give better dimensions. In optical moldings it is often desirable to preheat the resin; this prevents optical flow lines due to non-uniform temperatures in the melt. Additionally, motionless mixers are often placed in the nozzle to homogenize the temperatures and compositions in the melt as it goes into the cavities.

COMPUTER MODELING

Modeling of the flow in the injection molding process has been going on for the past ten years. A major advance was made with the use of finite element methods to computer model the flow and heat transfer. An ongoing effort is the work at Cornell University. This program simulates the melt flow in filling the cavity and also the flow around core pins and obstructions in the cavity. Commercially a number of programs are available[14],[15]. Mold Flow U.S.A. is a program developed in Australia that simulates the molding process either for a new mold or an existing mold. The simulation will answer the following questions:

-- Is the part, in fact, moldable?
-- Can wall thickness be reduced?
-- What part-dimension changes are needed to improve mold fill?
-- Are tooling changes necessary if a different resin is run?
-- How many gates are needed? Where should they be located?
-- Where will weld lines appear? Where should the gates be located in order to move the weld lines?
-- How should the runner system be dimensioned? Can its weight be further reduced?
-- Considering all factors, what are the optimum molding conditions?

The benefits of the computer modelling give insight into design and processing problems. Many major companies are using these commercially available programs. Their use is expected to grow at a fast rate in combination with further refinements. Currently a unified data base on polymer rheological and thermal properties is not readily available from the literature. In most cases one has to rely on the resin suppliers or generate one's own data. The models also need further refinements. In many cases, the physical properties are treated as constants whereas they should be treated as functions of temperature and shear rate.

CONCLUSIONS

Precision injection molding is making rapid strides to the point where optical quality parts are possible. The advancement of the state of the art is not due to any major breakthroughs in technology, but rather a steady growth in sophistication that is supported by improved resins, more sophisticated process control and instrumentation, process modelling, computerized part and mold design, and more reliable machines and auxiliary equipment.

REFERENCES

1. McQuiston, H., Plastics Engineering, 35, 17 (December 1979).
2. Plastics Technology, 26, 53 (December 1980).
3. Glanvill, A. B., "The Plastics Engineer's Data Book," pp. 11-12, Industrial Press, NY (1974).
4. DuBois, J., "Plastics", pp. 163-198, Reinhold Publishing, NY (1967).
5. SPI Booklet on Molding Practices, Society of the Plastics Industry, 355 Lexington Ave., NY, NY 10017.
6. Sneller, J., Modern Plastics, 58, 44 (May 1981).
7. Clark, D. C., Plastics Machinery and Equipment, 6, 29 (November 1977).
8. DeLuca, J. and S. P. Petrie, SPE ANTEC, 29, 730 (1983).
9. Davis, M. A., SPE ANTEC, 22, 618 (1976).
10. Spencer, R. S. and G. D. Gilmore, J. Appl. Phys., 21, 523 (1950).
11. McKelvey, J. M., "Polymer Processing," pp. 120-129, John Wiley and Sons, NY (1962).
12. Wang, V. W., K. K. Wang and C. A. Hieber, SPE ANTEC, 29, 663 (1983).
13. Plastics Design Forum, 5, 54 (November/December 1980).
14. Bernhardt, E. C. and G. Bertachi, SPE ANTEC, 29, 669 (1983).

OXIDE SINGLE CRYSTAL GROWTH

Dennis J. Viechnicki

Army Materials and Mechanics Research Center

Watertown, Massachusetts 02172

INTRODUCTION

Processing of ceramics from the liquid phase includes glass formation, glass-ceramics, fusion casting of refractories eutectic solidification, and single crystal growth. Ceramics include oxides, borides, carbides, and nitrides. This paper will, however, only discuss oxide single crystal growth as a form of processing. Borides, carbides, and nitrides generally sublime at very high temperatures rather than melt and, therefore, they are not processed from the liquid phase. Oxides can be processed from the liquid phase. The reasons why it is preferable to process them as single crystals if high technology applications are required will be discussed. The paper will also cover the factors that have allowed oxide single crystal growth to go from the laboratory to the factory by covering the modern crystal growth processing techniques. Materials and applications will be covered in the course of discussion.

SINGLE CRYSTAL GROWTH

Processing of crystalline ceramics from the melt must overcome two types of difficulties: i.e., those encountered in melting and solidifying, and those encountered once the materials have been solidified. Difficulties encountered in melting and solidifying ceramics include:

High melting temperatures ($1000-3000^0$C)
Large heats of fusion (28 Kcal/mol for Al_2O_3) [(1)]
Large differences in volume between liquid and solid, (22% for Al_2O_3) [(2)]
Metastability [(3)]

231

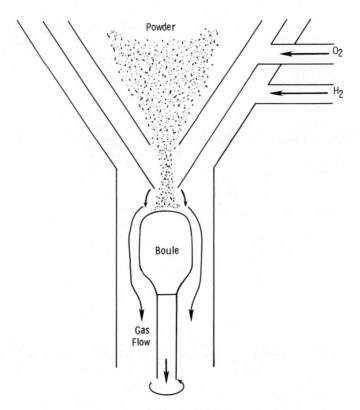

Fig. 1. Verneuil Technique

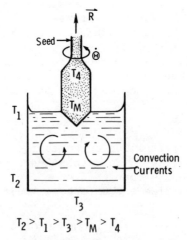

$$T_2 > T_1 > T_3 > T_M > T_4$$

CRYSTAL GROWTH BY THE CZOCHRALSKI TECHNIQUE

$\vec{R} \equiv$ pull rate $\dot{\Theta} \equiv$ rotation rate

Fig. 2. Original Czochralski Technique

It is hard to obtain these high temperatures. Once at the
melting point, a lot more energy is required to melt the material,
and then this energy must be removed to solidify the material. The
large volume differences between liquid and solid state can make a
very weak solidified material full of porosity or solidification
shrinkage. Finally, because of the existence of metastable phases
in many oxide systems, even though stoichiometry is very well
controlled, the final product may not always be the desired product
after solidification. These problems point to the necessity of
careful control of solidification to influence liquid solid interface shape and possible seeding of the melt to obtain the desired
phase. The first type of difficulty is generally equipment related,
and it has been the advances in furnace technology that have allowed
processing of crystalline ceramics from the melt and specifically
oxide single crystal growth to progress from the basic research area
to full-scale production of large ingots.

Difficulties with the solidified product include:

> Crystalline anisotrophy
> Rapid grain growth of solidified polycrystalline product
>   below the melting temperature
> Brittle nature of ceramics

Most ceramics are not cubic and will experience large thermal
stresses upon cooling because of thermal expansion differences along
different crystallographic axes. Micro and macro cracks will develop
in the material, especially at grain boundaries. If the solidified
product is polycrystalline, rapid grain growth will occur as the
material is cooled from the melting point weakening the material
further; and since ceramics are brittle, the micro cracks formed will
readily propagate through the material. The resulting material will
be quite weak indeed. The second type of difficulty dictates that
to obtain usable products for research and commercial high technology applications, crystalline ceramics must be processed or grown
from the melt as single crystals as opposed to polycrystalline, poly-
phase material made by conventional sintering and hot pressing. A
single crystal can be defined as a piece of crystalline material of
macroscopic dimensions that contains one phase, has no internal
boundaries, and has the same crystallographic orientation throughout
its macroscopic dimensions.

HISTORICAL DEVELOPMENT OF OXIDE SINGLE CRYSTAL GROWTH/FIRST GENERATION TECHNIQUES

While it is harder to make single crystal material than
polycrystalline material, over the years a variety of growth techniques have been developed. These can be separated into first
generation and second generation techniques. First generation
techniques are:

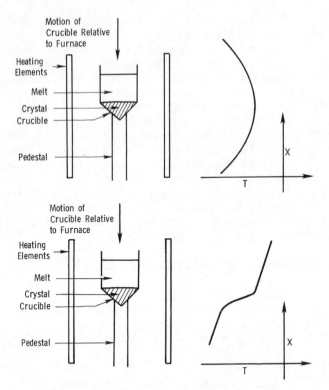

Fig. 3. Bridgman/Stockbarger Technique

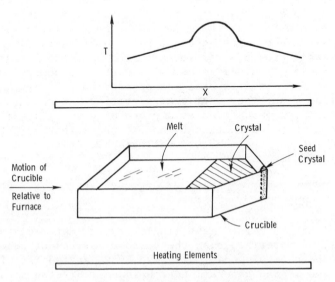

Fig. 4. Stockbarger Technique

Verneuil (4,5)
Czochralski (6)
Bridgman/Stockbarger (7,8)

They bear the names of the pioneers in crystal growth. At the time of their development, single crystal growth was viewed as a research tool required for the study of solid state physics and materials science.

The Verneuil Technique (1891-1902) was the first high temperature oxide single crystal growth mechanism (Figure 1). Oxide powder such as alumina is dropped through an oxyhydrogen torch where it is melted and then collected on a rotating pedestal. The solidified material or boule is either a low quality single crystal or a collection of columnar grains separated by low angle grain boundaries. Modern Verneuil furnaces complete with powder feed hoppers and elongated seeds, have been used to grow plate shaped rather than cylindrical boules (9). The process is single and quick; temperature is controlled by gas flow but it is not necessary to monitor temperatures closely, the single crystals are of low quality and cannot be used for high technology applications. Rather they are used for jewelry and high purity meltstock for other crystal growth techniques.

The Czochralski Technique (1917) is illustrated in Figure 2. A rotating seed is dipped into a melt and slowly withdrawn. The advantages of this technique are that impurities are left behind in the crucible and that many different kinds of materials can be grown this way. Disadvantages include control of interface shape, convection currents which affect crystal quality, and size. Many of these difficulties have been eliminated in the second generation Czochralski which will be discussed later.

The first successful high temperature crystal growth technique was the Bridgman/Stockbarger technique (1925 & 1936) where the material was melted in a crucible which was cooled by moving it through a temperature gradient in the furnace (Figure 3). Bridgman used the natural temperature gradient of the furnace (10). Stockbarger tailored the temperature gradient of the furnace to get better control of crystal nucleation and growth (11). Control of seeding was the major disadvantage to these techniques though with a major change in geometry a powerful second generation technique has developed from it.

The first generation techniques were generally used to make new materials for property measurements and were not used for production of high technology products. This was because the crystals made were small, and of low quality. Temperature measurement, seeding, and control of temperature gradients were not emphasized.

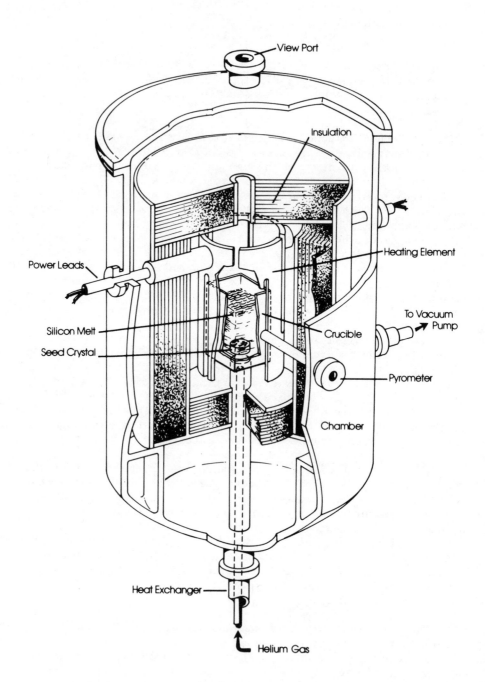

Fig. 5. Schematic of an HEM Furnace

MODERN OXIDE SINGLE CRYSTAL GROWTH/SECOND GENERATION TECHNIQUES

Second generation techniques were developed for specific applications. They are either evolutionary developments of first generation techniques or are entirely new.

They are:

> Refined Czochralski (CZ) (12)
> Horizontal Bridgman/Bagdasorov (13)
> Heat Exchanger Method (HEM)/Vertical Solidification of the
>   Melt (VSOM) (14-17)
> Skull Melting (18,19)

They are all different from the first generation techniques in the quality, quantity, and size of the products they make. These enhanced capabilities of the second generation techniques have been made possible by advanced furnace technology and other developments in processing technology. A refined or modern Czochralski furnace will have a large inductively heated iridium crucible, zirconia insulation and furnace components, a controlled inert gas atmosphere, means to rotate and pull the crystal, a mechanical or optical feedback system for crystal diameter control, and an after heater to prevent unwanted thermal stresses in the crystal during cool down. The insulation and crucible diameter control the diameter of the growing crystal, along with mechanical or optical feedback systems are employed. Interface shape is controlled by rotation and pull rate, and crucible size. Temperature is measured with an optical pyrometer. 50 mm diameter Nd: YAG and $Ga_3Gd_5 AlO_{12}$ (GGG) single crystals are grown from such furnaces.

The Horizontal Bridgman/Bagdasorov method, named after its chief Soviet inventor, is shown schematically in Figure 4. A boat shaped molybdenum crucible filled with alumina meltstock has a high temperature molybdenum heating element pass around it similar to a zone melting apparatus. A seed crystal is placed at the tapered end; after nucleation, crystal growth proceeds toward the blunt end at a rate controlled by the movement of the heating element. Temperature gradients are designed into the heating element to insure a relatively flat liquid-solid growth interface, and rejection of impurities ahead of the interface without thermal shocking the solidified crystal. Large sapphire crystals weighing 1.8 Kg measuring 217 mm X 90 mm X 25 mm were grown this way.

The Heat Exchanger Method (HEM) and the Vertical Solidification of the Melt (VSOM) are two growth techniques that employ crucibles and heat exchangers to precisely control seeding, something the other techniques are lacking. There is no movement in the furnace using either technique because the driving force for growth are the temperature gradients that are engineered into the furnace heating

Fig. 7. Schematic of a VSOM Furnace

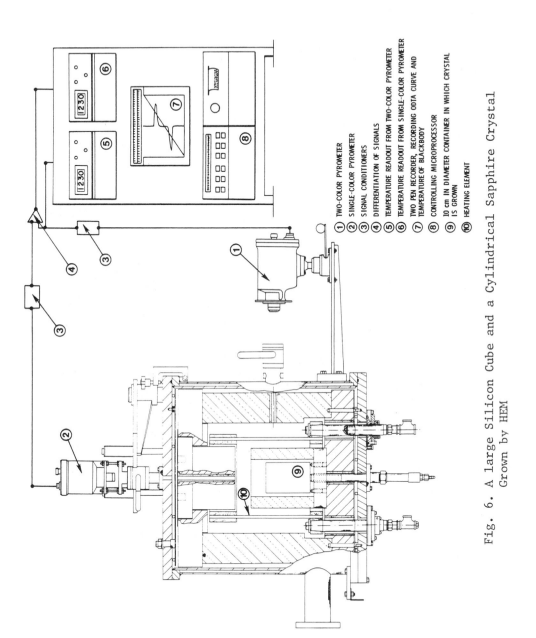

1. TWO-COLOR PYROMETER
2. SINGLE-COLOR PYROMETER
3. SIGNAL CONDITIONERS
4. DIFFERENTIATION OF SIGNALS
5. TEMPERATURE READOUT FROM TWO-COLOR PYROMETER
6. TEMPERATURE READOUT FROM SINGLE-COLOR PYROMETER
7. TWO PEN RECORDER, RECORDING ODTA CURVE AND TEMPERATURE OF BLACKBODY
8. CONTROLLING MICROPROCESSOR
9. 10 cm IN DIAMETER CONTAINER IN WHICH CRYSTAL IS GROWN
10. HEATING ELEMENT

Fig. 6. A large Silicon Cube and a Cylindrical Sapphire Crystal Grown by HEM

Fig. 8. Photograph of a VSOM Furnace

element and developed because of the heat exchangers. A detailed
discussion of crystal growth by HEM can be found in Reference 14 and
15. A schematic of an HEM furnace used to grow cube shaped silicon
ingots is shown in Figure 15. The silicon cube and a large ingot
of sapphire grown by HEM are shown in Figure 16. A detailed
description of the stages of crystal growth by the Vertical Solidification of the Melt (VSOM) technique can be found in References
16 and 17.

A schematic of VSOM furnace is seen in Figure 7, and photographs
in Figure 8. It can be noted that the power leads enter the VSOM
furnaces from the bottom helping to create a uniform temperature
gradient in the heating element. The top of the element is hotter
than the bottom. In the HEM and early VSOM furnaces the power leads
entered from the side creating a complex temperature gradient in
the element which caused different growth rates in different parts
of the furnace. Another difference between VSOM and HEM is the
shape of the liquid-solid crystal growth interface. HEM develops
a three dimensional hemispherical to conical growth interface that
allows a larger volume of material to be solidified per unit time
for a given linear growth rate. The type of material being grown
dictates the type of growth interface shape required, and the three
dimensional interface is suitable for sapphire. It is well known
in garnets, however, that to avoid faceted growth a flat interface
is required. Therefore, VSOM develops a flat growth interface and
unidirectional growth. The other important difference between
these two methods is that HEM employs a hollow helium tungsten/
molybdenum heat exchanger while VSOM employs a solid tungsten/
graphite/copper water cooled heat exchanger. The former permits
better temperature control during seeding while the latter is more
durable. VSOM has been used to grow the largest diameter Nd: YAG
ever for laser applications, weighing 2000g 75 mm in diameter and
100 mm high (17).

The fifth important second generation single crystal growth
method is the Skull Melting Technique developed in the Soviet Union
by V. V. Osiko and commercialized in the USA by Ceres Corporation
of Billerica, MA. A crucible is not used, but rather the melt is
contained by a sintered skull of the material being melted, usually
doped ZrO_2. Steep temperature gradients created by water cooled
metal fingers contain the material by preventing the skull from
melting. The material is heated inductively by direct coupling;
ZrO_2 melts at 2700^0C, and is a good conductor about 1500^0C. To get
1500^0C metallic Zr which subsequently oxidizes is added to the
batch. The solidified ingot consists of large columnar grains of
ZrO_2 which are single crystals with the sintered ZrO_2 skull at the
bottom and sides. A detailed description of the process is given
in Reference 19.

THE GENERATION GAP

The last two sections have covered first and second generation
growth techniques for oxide single crystals. While this is not
meant to be an all inclusive list, it does illustrate the generation
gap created by going from research to production. The original
crystal growth techniques were employed to synthesize new materials
for property measurements and further studies. Needless to say,
there was not a great understanding for the materials at that time.
For production purposes a thorough understanding is necessary, and
Nd: YAG is a good example. While it is relatively easy to grow
small Nd: YAG single crystals by the Czochralski technique, growth
of large crystals by any technique is difficult. The reason has
been shown to lie with the phase diagram (16,17) where equilibrium
and metastable solidification paths have been demonstrated. If
the melt is superheated or if crystal growth is attempted without a
seed, the metastable phase diagram and solidification path obeyed
and the final products of solidification are a mixture of $YAlO_3$ and
alumina but not YAG. When Nd_2O_3 is added as a dopant, similar
behavior is seen in the ternary phase diagram.

A very important reason for the generation gap is advanced
furnace technology. Early crystal growth techniques employed
expedient means of heating and little attention was paid to type and
position of insulation, and furnace atmosphere. Small crucibles of
platinum or iridium were used and heated by induction, a situation
which is difficult to control. Modern crystal growth furnaces use
controlled atmosphere and vacuum to protect graphite and refractory
metal furnace chambers and crucibles. Insulation and heat shields
are used to engineer proper temperature gradients in these furnaces
to drive crystal growth without introducing unnecessary thermal
stresses in crystal. The modern crystal growth furnaces have modern
control systems based on controlling power and/or temperature. The
latter is possible because temperature measurement is now a critical
part of crystal growth and is made possible by two color and total
radiation pyrometers and high temperature thermocouples. In the
early days of crystal growth it was important to see when the
material melted for nucleation and growth, but knowledge of actual
temperature was not necessary. In the modern furnaces temperature
measurement, rather than visual observation controls nucleation
and growth.

Finally the generation gap has been created by scale-up of
crystal growth techniques both in size and quantity of material
produced. Critical stages of crystal growth are many times more
noticeable when growing large crystals. Actual solidification
temperature and possible supercooling of the melt are evident by
the sudden increase in observed temperature while cooling as a result
of the release of the heat fusion. This knowledge leads to better
process control. Crystal quality is much better in large crystals

Table 1
Materials Grown by Second Generation Crystal Growth Techniques

APPLICATIONS

- WINDOWS
- IR DOMES
- TRANSPARENT ARMOR
- LASERS
- SUBSTRATES
- JEWELRY
- OPTICAL COMPONENTS

Table 2
Some Applications of Single Crystal Materials

MATERIALS

- SAPPHIRE

- RUBY

- Nd: $Y_3Al_5O_{12}$ (Nd:YAG)

- $Gd_3 Ga_5 O_{12}$ (GGG)

- CUBIC ZIRCONIA

just because of the decreased surface area to volume ratio, and the fact that the surface of a crystal usually contains more impurities and imperfections.

MATERIALS/APPLICATIONS

Materials grown by the second generation techniques are shown in Table I and typical applications are seen in Table II. With the exception of cubic zirconia jewelry these are all high technology applications. On the other hand, jewelry is a high volume application with outputs of 1-2 tons/month. Cubic zirconia also has high technology applications as optical components because of its high refractive index (n=2.15).

SUMMARY

This paper has covered the following:

Difficulties with liquid phase processing of oxides

1. Experimental
2. Intrinsic to material

Historical development of oxide single crystal growth/ first generation techniques.

Modern oxide single crystal growth/second generation techniques.

The generation gap

Materials/applications

CONCLUSIONS

Oxide single crystal growth is no longer a laboratory tool but an important part of materials processing because of:

Increased understanding of materials
Advanced furnace technology
Reliable methods of temperature measurement
Scale-up of crystal growth techniques

ACKNOWLEDGEMENTS

R. Belt - LAMBDA/AIRTRON
C. P. Khattak and F. Schmid - Crystal Systems, Incorporated
J. Wenckus - Ceres Corporation

REFERENCES

1. Janef Thermomechanical Tables, The Dow Chemical Co., Midland,
    MI, USA March 1964.
2. A. D. Kirshenbaum, and, J. A. Cahill, The Density of Liquid
    Aluminum Oxide, J. Inorg. Nucl. Chem., 14:283. (1960)
3. J. L. Caslavsky, and D. J. Viechnicki, Melting Behavior and
    Metastability of Yttrium Aluminum Garnet (YAG) and YAlO$_3$
    Determined by Optical Differential Thermal Analysis, J. Matls
    Sci., 15:1709. (1980)
4. A Verneuil, Production Artificielle du Rubis par Fusion, Paris
    Acad, Sci, Compt. Recd., 135:791.(1902)
5. K. Nassau, Dr. A. V. L. Verneuil: The Man and the Method,
    J. Cryst. Growth, 13/14: 12.(1972)
6. J. Czochralski, Ein reues Verfahrer zar Messurg der Kristal-
    listionsgeschurinchigkeit der Metalle., Phys. Chem., 92:219
    (1917)
7. P. W. Bridgman, Certain Physical Properties of Single Crystals
    of Tungsten, Antimony, Bismaith, Tellurium, Cadimium, Zinc, and
    Tin, Proc. Am. Acad.Arts Sci. 60:305.(1925)
8. D. C. Stockbarger, Large Single Crystals of Lithium Flouride,
    Rev. Sci.Inst., 7:133.(1936)
9. H. Djevaherdjian, Procede de Fabrication d'ein Corps en Peirre
    Synthetique, et Installation Pour la Mise en Oeuvre de ce
    Procede, Swiss Patent No. 354428, July 1961.
10. W. D. Lawsen, and S. Nielson, Semiconducting Compounds in
    "The Art and Science of Growing Crystals", J. J. Gilman, ed.
    John Wiley and Sons, Inc., New York, p. 372 (1963).
11. Ibid, p. 371.
12. R. Uhrin, and R. F. Belt, "Growth of Large Diameter Nd: YAG
    Laser Crystals", Interim Report, 1 April 1980 to 31 March 1981,
    Contract No. DAAB-07-77-C-0375, Modification P00003, US Army
    Electronics R&D Command Night Vision and Electro-Optics Labora-
    tory, Ft. Belvoir, VA 22060.
13. K. H. S. Bagdosarov, The Synthesis of Large Monocrystals of
    Corundum, in "Ruby and Sapphire", L. M. Belyayev, ed. Moscow
    Nauka 1974. English translation by US Army Foreign Science and
    Technology Center, Charlottesville, VA, 22091, (1975) p. 28.
14. F. Schmid and D. Viechnicki, Growth of Sapphire Disks from the
    Melt by a Gradient Furnace Technique, J. American Ceramic
    Society, 53:528.
15. D. Viechnicki and F. Schmid, Crystal Growth Using the Heat
    Exchanger Method (HEM), J. Crystal Growth, 26:162. (1974)
16. J. L. Caslavsky and D. Viechnicki, Melt Growth of Nd: Y$_3$ Al$_5$
    O$_{12}$ (Nd:YAG) Using the Heat Exchanger Method (HEM), J. Crystal
    Growth, 46:601. (1979)
17. J. L. Caslavsky and D. J. Viechnicki, Resolution of Factors
    Responsible for Difficulty in Growing Single Crystals of YAG,
    AMMRC TR 82-34, US Army Materials and Mechanics Research Center,
    Watertown, MA 02172, June 1982.

18. V. I. Aleksandrov, V. V. Osiko, A. M. Prokhorov, and V. M. Tatarintsev, Production of Refractory Single Crystals and Molten Ceramics by a New Method, Vestrik Akad. Nauk. USSR. 12:29.(1973)
19. K. Nassau, Cubic Zirconia: an update, Lapidary J. 35:1194.(1981)

RECENT ADVANCES IN SOLIDIFICATION PROCESSING

D. Apelian

Department of Materials Engineering
Drexel University
Philadelphia, PA   19104

INTRODUCTION

Control of the cast structure is the underlying object of
solidification processing. Recent advances and developments in
solidification processing are making possible the production of
high purity castings, fine grained superalloy and titanium ingots,
rapidly solidified structural components and castings having unique
microstructures. In addition, recent advances in processing technology have developed which now allow us to have better producibility
and reliability in aluminum castings. These developments have all
stemmed from a good understanding of the science of solidification
processing as well as an appreciation of the merits of structural
control via processing. Specifically, the following recent developments will be reviewed and discussed: (i) Vacuum Arc Double Electrode Remelting (VADER); (ii) Rapid Solidification by Plasma Deposition (RSPD); (iii) rapid cycle casting of ferrous components by
Diffusion Solidification; (iv) refining of the melt via filtration
prior to casting; and (v) thermal analysis of aluminum castings.

VADER MELTING PROCESS

The VADER melting process allows one to produce equiaxed
dendritic superalloy castings having a fine grain size, and castings
which subsequently can be forged. This is a tremendous achievement;
in the past, to produce complex superalloy components which could
subsequently be forged, one had to resort to powder metallurgy and
sophisticated forging technologies. The VADER (Vacuum Arc Double
Electrode Remelting) process was developed and patented by Special

247

Metals Corporation of New Hartford, N.Y., 13413, USA (1).

A schematic of the VADER melting process is shown in Figure 1.
Two consumable electrodes are arced against each other, and the
molten metal is allowed to drip into a stationary, rotating or a
withdrawal mold located under the electrode. The VADER melting
process enables one to manufacture fine grained superalloy ingots
which subsequently can be forged to gas turbine components. Comparable size VIM-VAR (vacuum induction melted and vacuum arc remelted,
respectively) products have large grains or irregular shape and
internal cracking is prevalent. The contrasting structural features
between VIM-VAR and VADER melted ingots are shown in Figure 2. The
VADER melting process produces superalloy ingots having equiaxed
grains (4-8 ASTM).

In the VAR process, droplets from the electrode are suspended
to form a molten pool which solidifies in a columnar dendritic mode.
Rejection of phases along solidification fronts and other mechanisms
may generate areas of macrosegregation. Vapor which deposits on the
mold wall can fall into the molten pool and thus cause surface or
internal defects in the ingot. In addition large amounts of energy
are required to offset the losses to the water cooled mold. In
contrast, in the VADER melting process droplets form as a result of
arc melting between the two horizontal electrodes. The droplets
fall with very little, if any, superheat. Drop temperature measurements on IN-718 alloy were found to be 20°F below the liquidus as
measured by DTA techniques. Figure 3 schematically illustrates the
contrasting features between VAR and the VADER melting processes.
In the VAR process, a mushy zone exists between the liquidus and
solidus isotherms and the control of this mushy zone is of paramount
importance in controlling the resultant cast structure. Specifically,
the width of the mushy zone and the time spent to solidify in this
two phase region have pronounced effects on the resultant cast
structure. These two parameters are further illustrated in Figure
4 which shows the progress of the liquidus and solidus isotherms
(T_L and T_S) during unidirectional solidification... solidification
progressing from the bottom to the top of the ingot. The vertical
distance between the liquidus and solidus curves represents the
region of the ingot over which solid and liquid phases coexist at a
given time during solidification. This vertical distance is the
width of the mushy zone; casting characteristics such as feeding,
hot tearing and macrosegregation are strongly influenced by the
width of this zone. The horizontal distance between the solidus
and liquidus curves, shown in Figure 4, is a measure of the local
solidification time, t_f. The latter is inversely proportional to
the average cooling rate at a given location during solidification
and has a dominant effect on the scale of microsegregation throughout
the structure.

There are additional advantages to the VADER melting process.

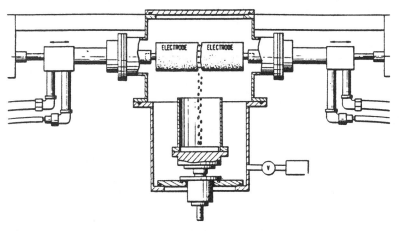

Figure 1.   Schematic diagram of the VADER melting process.

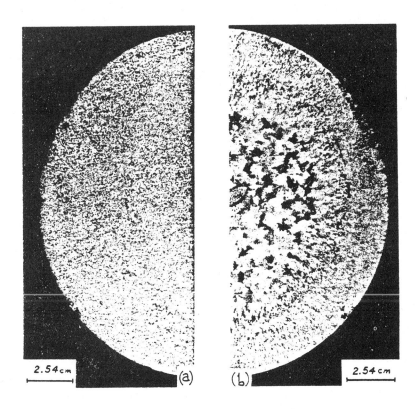

Figure 2.   Macrostructure of (a) VADER processed IN-718 ingot and
            (b) VIM-VAR IN-718 ingot.

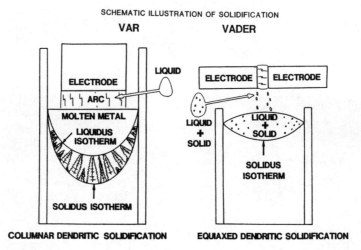

Figure 3.   Model of columnar dendritic solidification in the VAR
            and the VADER melting processes respectively.

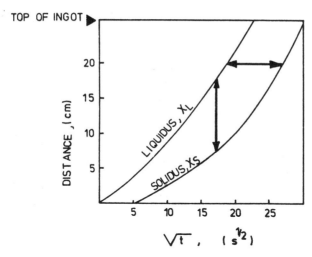

Figure 4. Progress of liquidus and solidus isotherms during
           unidirectional solidification.

About 40% less energy is required in the VADER melting process for
a comparable melt rate with the VAR process. Moreover, melt rates
of at least 3 times that of VAR are feasible. Another subtle feature
of the VADER melting process is that is offers flexibility in manufacturing in terms of electrode preparation and conditioning.

    Tensile properties (at room temperature and 650°C) and stress-
rupture property data of U718 alloy VADER processed are shown in
Figure 5. The observed increase in reduction in area and elongation
are impressive.

RAPID SOLIDIFICATION BY PLASMA DEPOSITION (RSPD)

    Resource scarcity, high costs of manufacturing and energy,
together with the quest for novel microstructures have caused a fast
paced evolution in material processing. During the last six years
various rapid solidification technologies (RST) have been developed
because of the resultant unique structure/property relationships in
metallic materials. Specifically, RST allows the production of
refined and homogeneous materials, as well as new alloy compositions,
microstructures and phases by extending solid solubility, and
allowing new phase reaction sequences.

    Even though the structural advantages rapid solidification

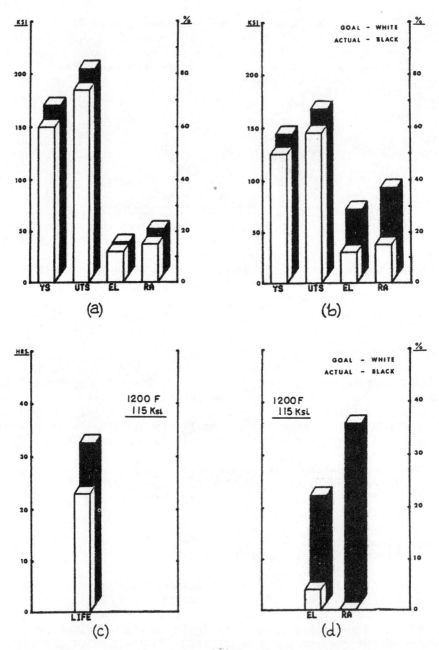

Figure 5.   Properties of U718 VADER processed material; white regions
            indicate the goal and black shaded regions indicate actual
            measured values.   (a) Room temperature ingot-roll down
            tensile properties; (b) 1200°F ingot-roll down tensile
            properties; (c) and (d) ingot-roll down stress rupture
            property data.

offers have been known since Duwez's pioneering work (3) in 1960 and that of Grant (4) and Flemings (5), it was the emergence of novel RST processes which was the catalyst for commercial development. The commercial RST processes include (i) centrifugal helium atomization whereby cooling rates on the order of $10^6 - 10^7$ K/s have been achieved (6); (ii) melt spinning for the production of metallic glassy ribbons (7); (iii) processes where a high energy source such as a laser or an electron beam is utilized to modify the metallic substrate's surface such as layer glazing (8); (iv) plasma deposition where powder material is injected into a high temperature inert gas plasma whereupon the metal particulates melt and subsequently are resolidified at a substrate (9-14).

Rapidly quenched atomized powders coupled with traditional powder metallurgy (P/M) processing have resulted in components having high temperature performance; additionally, the capability of generating "near net shape" components help conserve the use of scarce raw materials. Here one produces homogeneous microstructures through the use of atomized metal powders where each particle becomes a micro-casting of uniform composition. These powders can then be hot extruded, hot closed-die forged, hot isostatically pressed, or liquid-phase sintered into a near net shape which requires minimal machining thus conserving materials. Because of the potential of P/M technology, a great deal of research has gone into the identification of one optimum technique of consolidating powders to produce high-density high-performance components. A paradox exists since the product of most RST processes is either in powder or flake form which mandates further processing to make useful shapes. Unfortunately, subsequent processing steps such as consolidation/sintering can significantly alter the desirable as-quenched structure.

RSPD possesses the potential for circumvention of the paradox because it combines melting, quenching, and consolidation in one single operation. The process involves the injection of powder particles into a plasma gas stream of high temperature created by heating of an inert gas via an electric arc in a confined water cooled nozzle. The powder particles injected into the plasma jet (temperatures approximately 10,000 K and higher) are melted and subsequently resolidify when they impact a substrate. The cooling rates are typically 10^5 to 10^6 K/s and resulting grain sizes are ¼ to ½ μm in size (14).

Plasma spraying in air typically results in deposits which exhibit low densities due to oxidation of the deposited material. Developed about 25 years ago (15) the arc plasma spray gun has been used extensively to apply protective coatings to metal parts (16). Coatings applied in room atmosphere are typically about 75 to 85% dense and have an oxygen content of up to 2%. The problem of oxidation can be reduced by the use of an inert atmosphere or a shielded plasma jet. Densities in the range of 85 to 95% can be

attained while the oxygen content is reduced to less than 1/3 that
of air-deposited coatings. However, these deposit densities are
still unacceptable for high performance structural components.
Conventionally plasma arc sprayed deposits have inherent defects
such as unmelted particles and oxide inclusions which preclude
their use as high performance structural parts.

Recent developments in the field of plasma arc spraying, namely
the introduction of the low pressure plasma deposition (LPPD)
process, has resulted in a renewed interest in the capabilities of
plasma processing. In conventional processes the deposition is
carried out at atmospheric pressure. Whereas in low pressure plasma
deposition (LPPD), the process is carried out in an evacuated chamber
thus permitting higher pressure ratios which yield much higher
velocities in the range of Mach 2 to 3. In contrast to conventional
plasma spraying, LPPD provides major benefits such as (i) higher
particulate velocities which create greater than 98% dense droplets,
(ii) broad spray patterns which produce large deposit areas, and
(iii) transfer arc heating of the substrate which improves deposit
density. These characteristics provide LPPD with the potential of
becoming a viable method of consolidating powders for high performance applications.

When thicker than usual coating structures are deposited in a
low-pressure chamber and separated from the substrate, the oxygen
contents of the structures are typically found to be 300-500 ppm,
while densities are measured to be 97% or more of the theoretical
density. As a result of the inherent rapid solidification, an
ultrafine cellular microstructure is formed, see Figure 6; moreover,
mechanical behavior of the RSPD material is superior to cast or
wrought materials (17). However, isolated partially melted or
unmelted particles were also included in the microstructure, Figure
7. At Drexel University's Solidification Processing Laboratory work
is underway to study the melting and subsequent solidification of
injected particulates during RSPD. The models developed for particle
melting during RSPD is presented elsewhere (18). In addition, the
RSPD process can be automatically regulated to make controlled
deposits on complex geometries at reasonably high deposition rates
(up to 50 kg/hr). In sum, RSPD offers unique manufacturing flexi-
bilities not attainable by the other alternative RST processes.

The RSPD Process

A plasma is formed within the interior of the plasma gun by
ionizing the gases with an electric arc, as shown in Figure 8.
Commonly used gases are Ar or N_2 with additions of He or H_2. Plasma
temperatures of approximately 10,000 K cause a rapid volume increase
within the gun, so that the plasma gases exiting through the gun
nozzle are accelerated to velocities as high as Mach 3 into the

Figure 6. As-deposited cellular structure in region of fully
melted material shown by transmission electron microscopy.

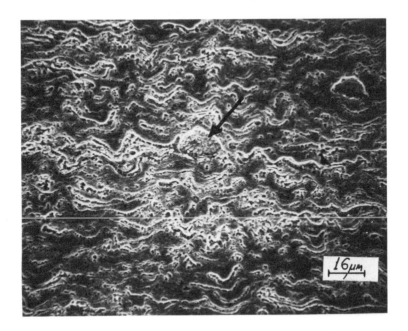

Figure 7. Unmelted particle (arrowed) observed in plasma sprayed
Fe-Mn 16 μm particulates under Ma 3 conditions.

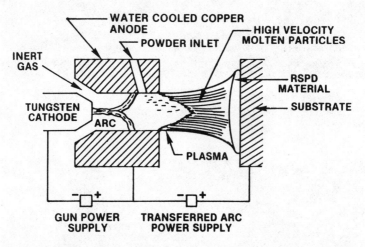

Figure 8.   Schematic of a plasma spray system. For reversed
            transferred arc cleaning, the polarity of the transferred
            arc power supply is reversed.

low-pressure chamber. At low pressure, the collisions of the plasma with the surrounding atmosphere, which cools and slows the plasma gases, are minimized. Even at distances of half a meter from the gun nozzle, the plasma temperature is several thousand degrees Kelvin and the plasma has a velocity of several thousand meters per second. The exact conditions will depend on chamber pressure, plasma gas composition and mass flow, arc power, and gun design.

The powder to be deposited is injected into the plasma stream, either within the throat of the gun or just beyond the end of the nozzle, external to the gun. A gas stream, usually Ar, carries and accelerates the powder particles so that they can be injected into the plasma gases. Because of the acceleration in the plasma, time-of-flight to the substrate is about a millisecond. If particles are too large, they will be heated insufficiently to melt and will not reach a great enough velocity to be deformed and attached on impact with the substrate. If particles are too small, they may be vaporized in the plasma or be so dominated by momentum of the plasma gas that they follow the gas stream around the substrate rather than impact on the substrate. Common practice is to use powder screened to a particular mesh size, where most of the coarser particles will melt, and ultrafine particles will be evaporated or swept away. A typical screened powder is -400 mesh, where the largest spherical particles will be 37 μm in size, and where a substantial fraction of the particles will be smaller than 5 μm. The temperature and velocity at impact for the powder particles will depend on particle size and material physical properties, as well as the plasma characteristics.

When proper conditions are maintained, it is difficult to determine where particle boundaries are located for the fully melted particles. However, it is possible to vary from optimum conditions so that layering is observed. In this condition, the greatest particle thickness observed has been on the order of 1.5 μm. If a 37 μm powder particle is considered, its volume would produce a 1.5 μm thick disc approximately 150 μm in diameter. For a powder feed rate of 20 kg/h, more than 10^6 particles/s are injected into the plasma. With every particle forming a disc overlapping those that arrived earlier, each point on the original surface would be covered 350 times per second. If particles solidified just before the next incoming droplet impacted the surface, a solidification rate of 0.05 cm/s would be achieved. The cell size of 0.25 to 0.5 μm indicates a solidification rate in excess of 1 cm/s, so there is ample time between impacts for mass quenching of the liquid droplets (14).

DIFFUSION SOLIDIFICATION

In diffusion solidification casting of steel, high-carbon

liquid iron is brought into contact with a low-carbon solid iron isothermally, and the liquid solidifies by rejection carbon to the surrounding solid iron. The mold is first filled with uniform-sized low-carbon steel shot, then heated and subsequently quickly infiltrated with liquid cast iron (2-4% C) under moderate pressure.

Consider the liquid and solid region of the phase diagram shown in Figure 9, which compares conventional casting, rheocasting and casting by diffusion solidification (SD). In the rheocasting process, solidification occurs by manipulation of temperature on an isocomposition line, whereas the diffusion solidification process is carried out isothermally by manipulation of composition, i.e., by solute rejection from the liquid phase. The general process steps for diffusion solidification are:

- The liquid is initially held at its liquidus temperature, which is the process temperature. In Figure 9, this corresponds to an f_L' amount of liquid of composition C_L'... at T_1.

- The solid particles (f_S' amount of composition C_S') are heated and held at the same process temperature, T_1.

- The liquid is rapidly infiltrated into the solid particles, filling the mold before significant SD takes place and thus blocking further infiltration.

In practice, a refractory mold is filled with low-solute shot, a barrier of coarse refractory particles is placed on top, and then the melt charge is added as shown in Figure 10. The particle valve (19) between shot and melt prevents premature infiltration between the two until sufficient pressure is exerted above the melt to break the liquid's surface tension; the particle valve and shot are then infiltrated and SD proceeds. The necessary sequence of operations for the entire process (using a wax pattern/investment mold logic) is:

- Make the core.

- Place the core in an injection molding machine and form the wax pattern around it.

- Invest the pattern and core with a refractory slurry to make the mold.

- Dewax and bond the mold.

- Emplace the low solute shot, particle valve, and the high solute melt charge.

- Heat the total assembly to the casting temperature in vacuum.

- Pressurize the casting vessel to infiltrate the casting.

- Homogenize.

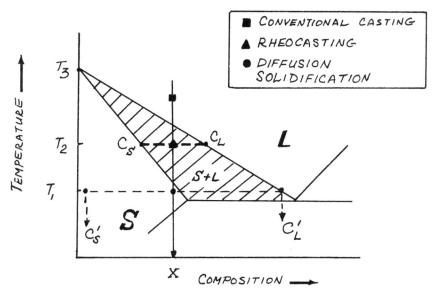

Figure 9.   Phase diagram showing conventional casting at T_3, rheocasting at T_2 and diffusion solidification at T_1.

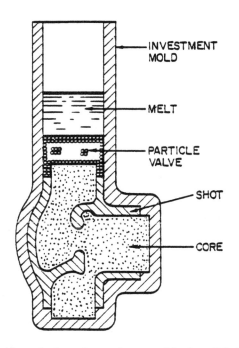

Figure 10.   Sectional drawing of assembled mold for SD casting.

• Cool to room temperature and remove mold, core and particle
valve.

The general casting logics which utilize the SD concept and can
thus lend themselves to rapid cycle casting of metallic components
are the permanent mold logic and the investment mold logic. These
are shown in Figures 11 and 12, respectively.

Process Advantages

In conventional casting solidification occurs via heat transport over a temperature range, and the final structure is dendritic.
In SD casting, solidification occurs via mass transport, the process
is isothermal and the liquid-solid front is planar. These fundamental differences result in totally unique microstructures; the source
of conventional macrosegregation is no longer the interdendritic
fluid and solute mass dictates the resultant structure.

Advantages of the SD process are that casting takes place at
temperatures reduced by 150-200°C and the process can be carried out
isothermally to cause 100% solidification and to obtain complete
homogenization of the resulting casting, all without rejecting any
heat to the mold.

Solidification time in conventional casting processes depends
on casting dimensions and mold characteristics. In SD casting, the
solidification time is essentially independent of casting dimensions
and is controlled by the infiltrable shot size. Furthermore, the
mold characteristics do not control the solidification time in SD
casting. In brief, solidification time and mold-filling time in
the SD process are shorter than in conventional casting.

In conventional casting, t_s, time for solidification, is proportional to ℓ^2, where ℓ is the dimensional term, length (Chvorinov's
Rule). In SD casting, on the other hand, t_{sd}, time for solidification, is proportional to $(\ell^{6/5}/P^{2/5})$, where P is the infiltration
pressure available (20).

The weaker dependence of solidification time on workpiece
dimension can be advantageous when considering the SD process for
automation. Cycle times comparable to die casting ought to be
achievable with the SD process. There is much less of a problem of
thermal shock to the "die" (mold), and the microstructure of the
resultant casting is homogeneous with respect to carbon because of
its rapid diffusion over these short distances.

In SD, the preheated solid particles occupy approximately 5/8
of the final volume of the casting prior to infiltration of the
liquid phase; therefore, solidification shrinkage and heat of solidi-

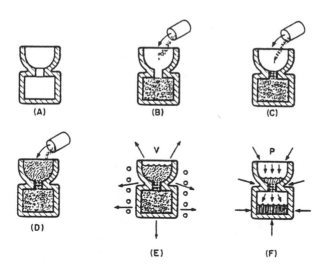

Figure 11. SD permanent mold logic; (A) - (E) indicate the sequence
of events.

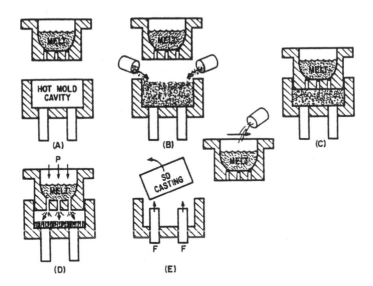

Figure 12. SD investment mold logic where (A) - (F) indicate the
sequence of events. In (B) the mold is filled with the
shot; in (C) the particle valve is inserted; in (D) the
infiltrant is emplaced; (E) the setup is evacuated and
heated to process temperature; (F) infiltration takes
place.

fication to be accommodated are at least proportionately reduced. Although in conventional castings a riser is needed, in SD the solidification shrinkage is not isolated, uniformly distributed, and smaller in amount. If the workpiece will subsequently be worked or hot isostatically pressed, or if it must be free of connected porosity, this better control of the shape and distribution of casting porosity is an advantage.

Diffusion solidification as a casting technique produces unique properties and microstructures without relying on extreme pressures, and it may therefore compete favorably with hot isostatic pressing as a way of consolidating atomized metal powder (shot). The atomized shot has the benefits of rapid solidification and of the controlled melting and solidification environments; the infiltrating liquid has the benefits of vacuum melting and of filtration by the particle valve. The distribution and magnitude of shrinkage cavitation are controlled by the particle-size distribution of the shot and by the content of inclusions (such as aluminum oxide or silicates, in steel). Steel castings have been made by SD with more that 99% of theoretical density, for example (21,22).

Since solidification proceeds simultaneously throughout the casting during SD, hot tearing and macrosegregation of impurities and alloying elements are decreased. The microstructure of diffusion-solidified steels is more like that of wrought steels than of cast steels - there is no columnar zone (20,22). The grain size and grain orientation of the diffusionally solidified casting is instead controlled by the grains in the initially solid portion of the charge.

It is possible to produce completely homogeneous microstructures free of microsegregation by SD since the initial shot particle size is chosen so as to minimize the freezing time by minimizing the diffusion distance consistent with successful forced infiltration under a reasonable external pressure. Most alloy systems to which SD can be applied are nearly completely homogenized soon after the completion of freezing. Design nomograms giving the infiltration depth and time at casting temperature to achieve a level of casting macrosegregation and homogeneity have been published (21,22,23).

There are other advantages for SD castings. Pieces with small surface-to-volume ratios of those with drastic changes in cross section can easily be produced because the casting solidifies without rejecting heat to the surroundings. Also, the casting will be free of laps and cold shuts because the mold is heated prior to infiltration.

The major limitations of SD castings are: (i) the initial shot should be of high metallurgical quality for the assurance of good mechanical properties; (ii) in addition, not all alloy systems can

be SD cast. There are certain criteria which must be met. One
needs a distribution ratio (c_S/c_L) above 0.4 and preferably close to
1.0 so that solidification can occur isothermally or even adiabati-
cally. Also, a steep solidus line is desirable so that the shot does
not have to be heated too close to its melting point. Lastly, rapid
diffusion of the SD alloying element in the solid compared to the
self-diffusion of the principal element is needed so that the shot
will not sinter excessively.

Applications

The iron-carbon system is ideal for SD because of the large
interstitial solubility of carbon in face-centered-cubic austenite.
Therefore, the particle size of the low-carbon shot and the low
process temperature relative to the melting point of the shot combine
to eliminate any chance of sintering shrinkage by bulk diffusion of
the iron from the shot/shot grain boundaries to the free surfaces
of the shot. Diffusion solidification casting and welding have been
applied to plain carbon steels from 0.1 of over 1% C; current research
at Drexel University continues to emphasize this system.

Excellent ductility of SD steel castings can be obtained if
they are adequately infiltrated and if metallurgical bonding is
achieved at the original solid-liquid interface. Surface and sub-
surface scales such as silicates are especially harmful and lead to
dotted-line fractures along the particles outlining the original
shot surface. All that is necessary to correct this fault is to
leave sufficient carbon in the original shot to reduce the silicates
during heating to the process temperature. On the other hand, if
one is interested in making castings/structures having a specifically
"designed" fracture path, then the SD process is one attractive
alternative.

MELT PURIFICATION BY FILTRATION

Metal filtration using various types of ceramic filter media
has been very effective in controlling the level and particle size
of inclusions in the lower-melting temperature alloys. In fact, the
filtration of aluminum has been an effective commercial process for
over 15 years (24). Some newer developments in high-temperature
ceramic filters have resulted in: 1) extruded, continuous, fixed-
pore geometry ceramic monoliths and, 2) reticulated (foam), open-
pore ceramic foams (25,26). The ceramic foams are currently
available in a variety of refractory compositions and pore sizes,
and generally contain 75 to 90 percent of volume of open pores.
Most of the emphasis has been placed on the use of the ceramic foams
because of their greater open porosity and because the pores are
interconnected in such a manner that the melt must proceed through

the filter by a tortuous path (which favors greater particle contact with the filter wall or web). These new ceramic foams and the development of advanced filters enables us to purify superalloy and steel melts prior to casting.

Ceramic foams are characterized by the number of pores per linear inch (ppi), i.e., 30 ppi foams have 30 pores per linear inch. There is a variation in the pore diameter throughout the foam; for a 10 ppi foam the average diameter is about 0.070" (1778 µm) with a range of 0.023" – 0.146" (584 – 3708 µm); for a 30 ppi foam, the average diameter is 0.28" (711 µm), with a range of 0.009" – 0.056" (229 – 1422 µm). Figure 13 compares the average pore diameters of 10, 20, and 30 ppi ceramic foams with particles having diameters ranging between 0.001" and 0.010" (2.5 µm and 254 µm).

Two mechanisms predominate in the removal of particles when melts pass through ceramic filters, as shown in Figure 14. The first is a simple mechanical screening or blockage of coarse particles at the surface of the filter. The second involves the entrapment of smaller particles on the interior surfaces of the filter. The ability of a filter to entrap particles depends on the relative interfacial surface energies between the melt and the particles. If the interfacial energy is high (non-wetting), the particles will be "pushed out" of the melt when they contact another solid surface, such as that of other particles, crucible walls, of the surface of filter materials. Once the particles come in contact with a solid surface, they become attached and bond to that surface. The thermal energy of the melt is sufficient for the particles to "sinter" to the walls of the filter, or "sinter" to the other particles already bonded to the filter.

The ability of a filter to capture (remove) particles from a melt depends on many factors, such as: the concentration and type of particles in the melt; the melt characteristics (composition, viscosity, surface tension); the temperature; and the filter characteristics (composition, structure, porosity, permeability, etc.). A discussion of these factors is given in detail elsewhere (27). However, for a given filter and cross-sectional area, the flow rate of the metal passing through the filter and the thickness of the filter are the two major factors affecting particle removal.

Apelian and Ali (28) have filtered steel melts containing carefully prepared "synthetic" Al_2O_3 inclusions achieving high removal efficiencies, see Figures 15 and 16. Sutton and Apelian (29) have also successfully applied this technology to superalloy melts.

THERMAL ANALYSIS OF ALUMINUM CASTINGS

Aluminum castings such as the A-356 alloy is a high strength

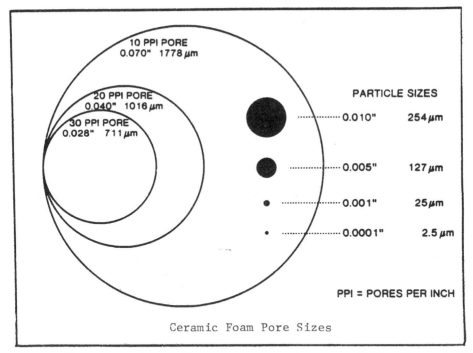

Ceramic Foam Pore Sizes

Figure 13.   Comparison of average pore diameter of 10, 20, and 30
ppi foam materials with particles of various diameters.

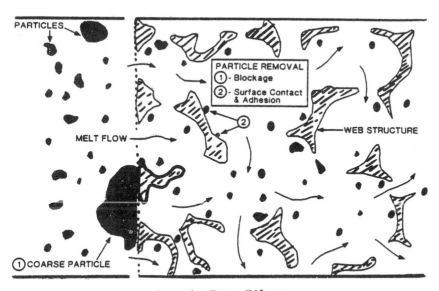

Ceramic Foam Filter

Figure 14.   Cross-sectional view of metal filtration and mechanisms
for particle removal by: (1) blockage, and (2) particle
contact/adhesion to web surfaces of ceramic foam.

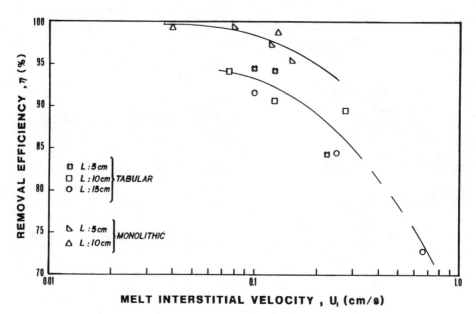

Figure 15.   Effect of melt interstitial velocity on weight based
             inclusion removal efficiency (synthetic Al_2O_3 inclusions).
             Data show effectiveness of tabular alumina and extruded
             monolithic filters of various lengths.

aluminum alloy which is widely used for both commercial and military
applications. In these castings the structure is controlled via
several processing variables such as: the extent of grain refining;
the level of modification which pertains to the morphology of the
eutectic phase distributed within the interdendritic interstices;
the solidification rate, and finally the post casting heat treatment
given. The first two of these processing variables have traditionally
been difficult to control.

     In fact, in most castings test coupons have been intentionally
added into the design of the pattern, which can then be sectioned
and metallographically evaluated to check the structure's grain size
and level of modification. If the evaluation of the test coupon is
negative, then the casting must be scrapped which is an unacceptable
situation.

     To have a high degree of quality assurance in the product, one
needs a control situation which assesses the melt's "quality" prior
to it being cast. If the melt's quality index is unacceptable,
then changes to the melt can be made... action can be taken to alter
the situation... prior to the final operation, or prior to casting.
This is essentially accomplished through thermal analysis of the

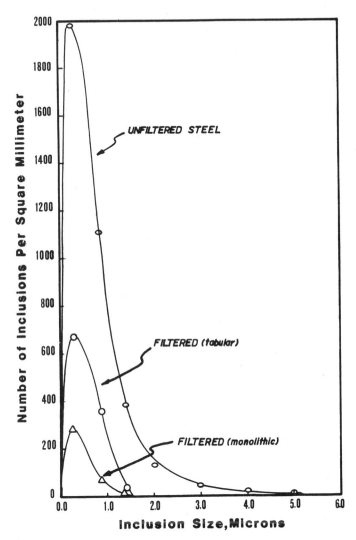

Figure 16. Al$_2$O$_3$ size distribution in unfiltered and filtered
            steels. The steel melt was filtered through tabular and
            extruded monolithic filters; filter length 5 cm and melt
            flowrate 0.085 cm/s.

melt via microprocessors. One such product was recently introduced
to the market; it is called the Alu–Delta G.S. and is marketed by
Metallurgical Products and Technologies Inc. of Willow Grove, PA.

The advantages and benefits of such a process control equipment
will allow one to accurately assess the extent of grain refining as
well as the level of modification. Prior to casting, a sample of
the melt is poured into a sand cup which has emplaced at its bottom
a thermocouple connected to the microprocessor of the Alu–Delta G.S.
Figure 17 is typical of the cooling curve obtained when the molten
sample is solidified in the sand cup. The Alu–Delta analyzes the
magnitude and difference between the undercooled temperature, T_u, and
the liquidus temperature, T_L, at the onset of solidification. The
microprocessor analyzes the data and relates this to a grain size
value which is then shown digitally as well as printed out on tape.
Furthermore, a green light flashes if the grain size value obtained
is within the specifications desired indicating to the operator that
all is well and to go ahead with pouring/casting. A yellow light
indicates that recheck or further interpretation is needed, and a
red light flashes when the melt's "quality" is out of the desired
specifications.

In addition, the Alu–Delta shows and prints out the actual
eutectic transformation temperature, T_{eut}, read from the sand cup
test. The equipment is manually set for the eutectic temperature of
the specific alloy being used in the unmodified case, and the microprocessor compares the actual T_{eut} (which is that for the modified
melt) to that for the unmodified melt. If the ΔT obtained via
modification is not satisfactory, then further modifiers such as Sr
or Na are added. Figure 18 shows the set-up and the three digital
screens for grain size, T_{eut}, and ΔT due to modification.

Such instruments are now (and will be in the near future)
revolutionizing the manner in which decision analysis and assurance
of casting quality from the point of view of solidification structural
control are conducted and maintained.

SUMMARY

Recent advances in Solidification Processing have been reviewed.

The Vacuum Double Electrode Remelting process offers advantages
not achieved by conventional VIM-VAR melting of superalloys by manipulating the fluid flow in the mushy zone as well as controlling the
nucleation of fine grains in the solidifying ingot.

Rapid Solidification by Plasma Deposition is a process whereby
one combines both atomization and consolidation in one step giving
rise to rapidly solidified microstructures. This process must not

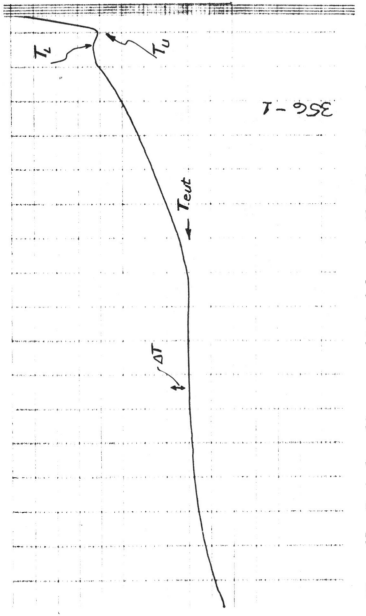

Figure 17. Typical cooling curve obtained from the Alu-Delta G.S. sand cup test.

Figure 18. The Alu-Delta G.S. equipment. The digital readouts and
            the printer for the grain size level; the eutectic
            temperature, T_{eut}; and the ΔT - level of modification
            are shown.

be confused with one which is used for coating applications only but
rather it is a process for rapidly solidified structural buildup.

Another recent development is the <u>Diffusion Solidification</u>
process where solidification proceeds via mass transport rather than
heat transport. The process has the advantage that diffusion solidification time is not controlled by the size of the casting but rather
by the size of the original solid shot particulates. It has applications in rapid cycle casting of ferrous materials.

Melt purification technology of aluminum, steel and superalloy melts by _filtration_ methods has become an acceptable process - commercial for both aluminums and superalloys, and being developed for steels. This acceptance has come by better understanding the effect of the process variables on the filter performance and by the emergence of high performance ceramic filters.

With the advent and progress made in microprocessor technology, we are now able to accurately determine a priori the aluminum melt's level of modification (% Sr or Na added) as well as the extent of grain refinement (% TiB_2 added) before casting. The operation and advantages offered by the Alu-Delta G.S. have been reviewed. Such an equipment offers the manufacturer and the user of the casting a level of assurance which could only be had previously via destructive testing of the end product, the casting.

REFERENCES

1. F.H. Soykan, J.S. Huntington: U.S. Patent 4,261,412, April 14, 1981.
2. J.W. Pridgeon, F.N. Darmara, J.S. Huntington, W.H. Sutton: "Principles and Practices of Vacuum Induction Melting and Vacuum Arc Remelting", in Metallurgical Treatises, J.K. Tien and J.F. Elliott: editors, Published by AIME-TMS, November 1981, Warrendale, PA.
3. W. Klement, R. Willens and P. Duwez: _Nature,_ 1960, 187, p. 869.
4. M. Lebo and N.J. Grant: _Met. Trans.,_ 1974, 5, p. 1547.
5. M.C. Flemings: Report of the ARPA Materials Research Council, 1976.
6. P.R. Holiday, R. Cox and R.J. Patterson: in "Rapid Solidification Processing Principles and Technologies", Eds. R. Mehrabian, B.H. Kear and M. Cohen, p. 246; 1978, Baton Rouge, LA, Claitors Publishing Division.
7. H.H. Liebermann: "Coaxial Jet Melt-Spinning of Glassy Alloy Ribbons:, Technical Information Series, Report No. 80 CRD 117, June 1980.
8. E.M. Brienan and B.H. Kear: in "Rapid Solidification Processing Principles and Technologies", Eds. R. Mehrabian, B.H. Kear and M. Cohen, p. 87; 1978, Baton Rouge, LA, Claitors Publishing Div.
9. D.A. Gerdeman and N.L. Hecht: "Arc Plasma Technology in Materials Science", 1972, New York, Wien, Springer-Verlag.
10. P.R. Dennis et al.: "Plasma Jet Technology", 1965, NASA Report No. SP-5033.
11. B. Gross, B. Grycz and K. Miklossy: "Plasma Technology", English translation edited by R.C.G. Leckey, 1969, American Elsevier Publishing Company, New York.
12. K.D. Krishnanand and R.W. Chan: "Rapidly Quenched Metals", Eds. N.J. Grant and B.C. Giessen, p. 67, 1976, M.I.T., Cambridge.
13. S. Shankar, D.E. Koenig and L.E. Dardi: _J. of Metals,_ Oct. 1981,

p. 13.

14. M.R. Jackson, J.R. Rairden, J.S. Smith and R.W. Smith: _J. of Metals_, Nov. 1981, p. 23.

15. G.M. Giannini, "The Plasma Jet", _Sci. Amer._, p. 2, Aug. 1967.

16. T.A. Roseberry and F.W. Boulger, "A Plasma Flame Spray Handbook, Rpt. No. MT-043, Naval Sea Systems Command, Dept. of the Navy, March 1977.

17. "RSPD for Fabrication of Advanced Aircraft Gas Turbine Components", P.A. Siemers, General Electric Company, Report to Air Force Systems Command, January 1982.

18. D. Apelian et al.: "Melting of Powder Particles in a Plasma Jet", Conference Proceedings of the 6th International Symposium on Plasma Chemistry, Montreal, Canada, July 1983.

19. M. Paliwal, D. Apelian, G. Langford; _Met. Trans._, 1980, Vol. 11B, pp. 39-50.

20. D. Apelian, G. Langford: Drexel University, unpublished work.

21. G. Langford, D. Apelian: _Journal of Metals_, Vol. 32, September 1980, pp. 28-33.

22. G. Langford, R.E. Cunningham: _Met. Trans._, 1978, Vol. 9B, pp. 5-19.

23. C. Lall, D. Apelian, G. Langford: "Aluminum-Lithium Castings Made by Diffusion Solidification", 109th AIME Annual Meeting, Las Vegas, Nevada, February 1980.

24. L.C. Blayden and K.J. Brondyke: "In-Line Treatment of Molten Aluminum", _Light Metals_, AIME, pp. 473-503, 1973.

25. D. Apelian and R. Mutharasan, "Filtration: A Melt Refining Method", _J. Metals_, _32_ (9), pp. 14-19, 1980.

26. R. Mutharasan, D. Apelian and C. Romanowski, "A Laboratory Investigation of Aluminum Filtration Through Deep-Bed and Ceramic Open-Pore Filters", _J. Metals_, _33_ (12), p. 12-17, 1981.

27. D. Apelian, et. al., "Commercially Available Porous Media for Molten Metal Treatment: A Property Evaluation", _Light Metals_, Ed., J. Andersen, Pub. AIME-TMS, 1982.

28. S. Ali, D. Apelian and R. Mutharasan, "Filtration of Steel Melts - A Refining Process", Submitted to Met. Trans, June 1983.

29. W. Sutton, D. Apelian, "Filtration of High-Temperature Vacuum Melted Alloys", to be published.

PROCESSING OF STEEL INGOTS FOR

THE PRODUCTION OF HIGH QUALITY FORGINGS

Vito Colangelo and Steve Tauscher

AMCCOM, Benet Weapons Laboratory
Watervliet, N. Y. 12189

INTRODUCTION

The subject paper describes recent advances in the production
of alloy steels of a 4335 + V alloy used for the manufacture of
high quality forgings. The framework of the discussion will be
the impact of new steel production methods upon the producibility/
reliability/affordability of the final end item.

In accord with the focus of the conference, the presentation
will concentrate on recent steel making practices which either:

1. take advantage of better steel production control
    methods,
2. are based on a more thorough understanding of the
    role of critical process parameters, or
3. demonstrate better integration of materials science,
    engineering design and advanced weapon component
    manufacturing methods.

While the properties obtained from an alloy are also
controlled by heat treatment and forging practice, the current
discussion will focus on the melting and degassing methods used
in the production of steel.

The properties of a specific alloy steel will be compared
as the melting procedures have evolved over a 20-year period.

The presentation will include properties obtained from a
single steel alloy produced by the following processes: electric

furnace melting, electric furnace plus vacuum degassing, electric furnace plus vacuum arc remelt, electric furnace plus argon-oxygen decarburization, electric furnace plus argon-oxygen decarburization plus vacuum arc remelting, electric furnace plus electroslag refining, electric furnace plus vacuum degassing plus calcium treatment.

The forgings produced from these steels were evaluated on the basis of mechanical properties including strength, ductility, impact strength at -40°F and in some cases, fracture toughness.

PROCEDURE

Steels manufactured by the following melting processes were examined:

- ° A. Electric furnace
- ° B. Electric furnace + Vacuum Degassing
- ° C. Electric furnace + Vacuum Arc-Remelt (VAR)
- ° D. Electric furnace + Argon Oxygen DeCarb (AOD)
- ° E. Electric furnace + AOD + VAR
- ° F. Electric furnace + Electroslag Refined (ESR)
- ° G. Electric furnace + Vacuum Degassing + Calcium treatment

The composition of the steels examined were in accordance with the following specification:

$\underline{C}$	$\underline{Mn}$	$\underline{Si}$	$\underline{Cr}$	$\underline{Ni}$
.32/.36	.55/.70	.25 max	.90/1.15	2.0/3.5

$\underline{Mo}$	$\underline{V}$	$\underline{P}$	$\underline{S}$
.38/.60	.08/.12	.010 max	.010 max

Chemical analyses were taken on the top and bottom of each forging and averaged to obtain a mean chemical composition.

The chemical analyses for each process combination is shown in Table I.

As stated, the steels were manufactured by various combinations of processes but all were forged to the same final configuration. The forging produced by process combinations - B, C, D, E, F, and G - were forged from the same size preform; i.e., 13" O.D. x 4-1/2" I.D. x 76" long and, therefore, received the same degree of forging from the preform to the final configuration. This was approximately 2:1 at the maximum diameter and 4-1/2:1 at the minimum diameter as shown in Figure 1.

Preform (13" O.D. x 4.5" I.D.)

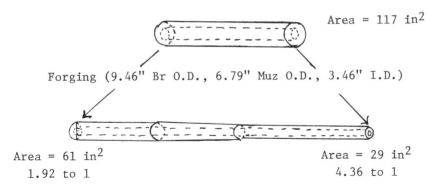

Area = 117 in^2

Forging (9.46" Br O.D., 6.79" Muz O.D., 3.46" I.D.)

Area = 61 in^2 Area = 29 in^2

1.92 to 1   4.36 to 1

FIGURE 1.   Forging reduction, preform to final
            configuration

There were undoubtedly differences in the overall amount of
forging being received from the ingot to the final configuration.
However, the question of forging reduction has not been addressed
here nor was it intended to be. This question has been adequately
evaluated in other studies.

The steels were heat treated in the same general manner with
only slight variations in procedure occurring because of equipment
limitations.

The steels were heat treated to attain a yield strength of
160-180 ksi.   The desired microstructure was 100% tempered
martensite.

Heat treatment consists of the following procedure:

> Forging was air cooled to room temperature,
> austenitized @ 1550°F - 1580°F, water quenched,
> tempered at 1020° - 1100°, depending upon the
> producer.

Mechanical properties were taken from discs cut from each
end of the fully heat treated forging representing the minimum
and maximum forging reduction. Test specimens representing the
tangential direction were then machined. The results obtained are
presented in Table II.

Table I

Mean Chemical Composition of Forgings

		C	Mn	P	S	Si	Cr	Ni	Mo	V
A.	(EF)	.34	.70	.009	.010	.2	1.14	2.68	.58	.12
B.	(EF + VD)	.34	.63	.003	.008	.05	1.04	2.74	.44	.08
C.	(EF + VAR)	.35	.66	.006	.006	.20	1.13	2.57	.51	.10
D.	(EF + AOD)	.36	.67	.003	.002	.18	1.10	2.86	.39	.08
E.	(EF + AOD + VAR)	.36	.56	.003	.001	.17	1.08	2.85	.38	.08
F.	(EF + ESR)	.34	.62	.008	.007	.10	1.05	2.40	.49	.09
G.	(EF + VD + CT)	.33	.64	.008	.005	.12	1.01	2.18	.48	.11

Table II

Summary of Mechanical Properties for 4337 + V Steel Forgings

METHOD * OF MANUFACTURE	0.1% Y.S. (KSI)			REDUCTION IN AREA (%)		
	MAX. DIAM.	MIN DIAM.	ALL DATA	MAX. DIAM.	MIN. DIA.	ALL DATA
	MEAN (STD.D)	MEAN (STD)	MEAN (STD DEV)	MEAN (STD DEV)	MEAN (STD DEV)	MEAN (STD DEV)
(A) Electric Furnace	173.4 (3.53)	170.0 (3.47)	171.7 (3.50)	34.3 (4.37)	31.7 (3.92)	33.0 (4.15)
(B) Vacuum Degas	166.0 (1.66)	164.9 (2.16)	165.5 (1.91)	38.0 (4.04)	38.8 (3.92)	38.4 (3.98)
(C) VAR			173.4 (0.69)			53.0 (1.10)
(D) AOD			164.1 (2.75)			41.5 (9.88)
(E) AOD + VAR			162.8 (1.50)			59.5 (0.60)
(F) ESR	172.8 (3.09)	165.8 (3.09)	169.3 (3.09)	44.6 (4.03)	47.2 (3.15)	45.6 (3.60)
(G) Vacuum Degas + Calcium Treatment	169.7 (3.09)	167.1 (3.00)	168.4 (3.05)	45.2 (3.54)	44.9 (4.03)	45.0 (3.79)

Table II

Summary of Mechanical Properties for 4337+V Steel Forgings (Continued)

METHOD OF* MANUFACTURE	FRACTURE TOUGHNESS K_{Ic} (KSI √IN)			CHARPY V (-40°F) (FT-LBS)		
	MAX. DIAM. MEAN (STD DEV)	MIN. DIAM. MEAN (STD DEV)	ALL DATA MEAN (STD DEV)	MAX. DIAM MEAN (STD DEV)	MIN. DIAM MEAN (STD DEV)	ALL DATA MEAN (STD DEV)
(A) Electric Furnace			101.0 (6.55)*	15.5 (2.67)	15.7 (3.47)	15.6 (3.07)
(B) Vacuum Degas	126.3 (7.27)	96.0 (5.85)	111.5 (6.56)	19.1 (2.22)	16.5 (1.60)	17.8 (1.91)
(C) VAR			146.3 (5.69)			21.0 (0.50)
(D) AOD			155.3 (7.89)			35.1 (2.40)
(E) AOD + VAR			195.5 (0.71)			49.4 (4.21)
(F) ESR	130.23 (10.25)	116.8 (10.38)	123.5 (10.32)	24.5 (3.12)	26.65 (3.61)	25.6 (3.37)
(G) Vacuum Degas + Calcium Treatment	135.8 (6.16)	127.9 (7.81)	131.9 (6.99)	24.6 (2.60)	27.7 (2.73)	26.2 (2.67)

RESULTS

The results of the chemical analyses for the steel forgings produced are presented in Table I. As can be seen, the compositions were within the range required by the specification. The compositions obtained with the exception of sulphur were unremarkable. Nickel tended to be on the low side of the range and molybdenum and vanadium were on the high side. These were the results of the aims of the producers and were based upon the producers experience with the alloy. These trends were not considered significant. Sulphur on the other hand ranged from the maximum compositional limit of 0.010% to a low of 0.001%, a ten fold decrease.

The mechanical properties which included 0.1% yield strength, ductility, impact strength and fracture toughness data, where available, were presented in Table II. These data represented the mean and standard deviation obtained on test specimens taken from the forgings.

DISCUSSION

In reviewing the data presented, several facts should be kept in mind:

1. the data for each category represents a minimum of ten forgings manufactured in a full-scale production facility,
2. the data for each manufacturing process is valid only for the particular vendor and process. It is the author's experience that considerable variation in properties occurs between vendors even with the same process,
3. finally, these data are useful only as an indication of what may be attainable from a particular process. The data should not be construed as guarantees that these levels will be met if a process is specified.

Sulphur Level

The chemical composition, with the exception of sulphur and possibly phosphorus, is not particularly notable. The sulphur level, however, ranges from a maximum of 0.010% to a low of .001%. A comparison of the sulphur levels to the impact strength obtained, as shown in Figure 2, reveals a distinct relationship between the sulphur level obtained and the impact strength. As the sulphur level decreases below 0.010%, there is a marked increase in the impact strength. These data verify other observations made by

the authors on similar material using individual heat populations.
This is no great surprise since sulphur has long been known to
exert a detrimental influence[2] upon the mechanical properties of
metals. However, this study shows that an improvement can still
be obtained even at very low levels of sulphur.

Inclusion Morphology

Sulphur level alone is not the only reason for improved
mechanical properties. The inclusion morphology must also be
considered. The addition of calcium to steel of this composition
results in a change in the shape of the inclusion from an elongated,
axially oriented form to globular sulphides. Examples of the
inclusion morphology evident on the fracture surface of test
specimens are shown in Figures 3 and 4. Treatment with calcium
is not new and its use has been discussed extensively[3,4,5].

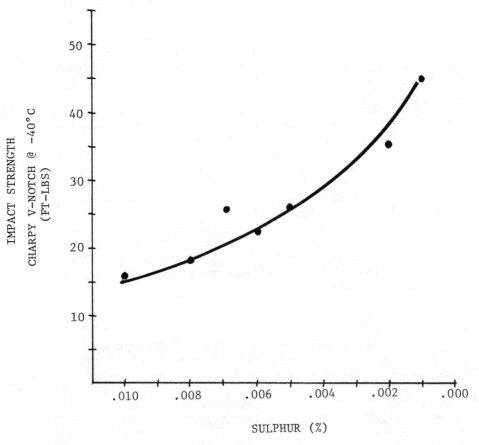

SULPHUR (%)

FIGURE 2.

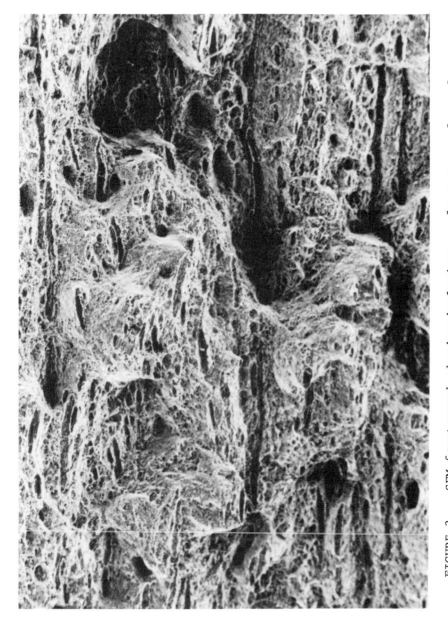

FIGURE 3.   SEM fractograph showing inclusions on fracture surface of a Charpy bar (vacuum degassed steel)

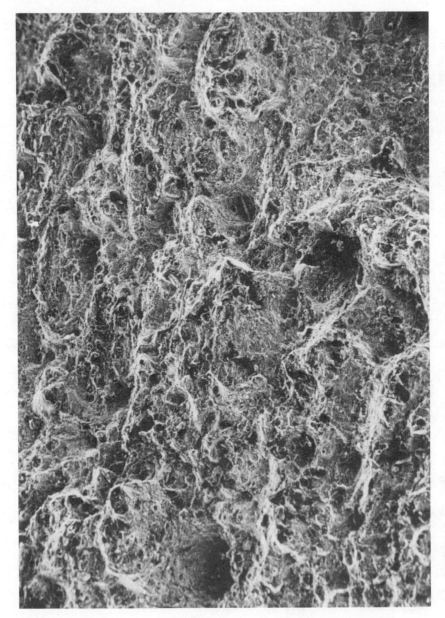

FIGURE 4.   SE: Fractograph showing inclusions on fracture surface of a
            Charpy bar (calcium treated steel)

In general, there has been reasonable agreement that calcium additions will change the morphology of the sulphides and improve transverse mechanical properties. What has not been clarified is the role of oxygen when calcium additions are made or the level of calcium required. Levicek, et al[6] indicate that in the carbon steels studied, 100 ppm calcium would give optimum results and that this level could be achieved only with extremely low oxygen levels.

These data support another study made here on alloy steels identical to those presented herein. The study, as yet unpublished, indicated that, when calcium additions were made to a heat where residual oxygen concentrations were in excess of 130 ppm, no improvement in mechanical properties resulted whereas, when the residual oxygen concentration was less than 90 ppm, a threefold increase in impact strength was obtained along with the expected change in inclusion morphology.

The explanation appears to lie in the fact that calcium is an extremely effective deoxidizer. When high levels of oxygen are present in the heat, the calcium additions act to deoxidize the steel rather than desulphurize it.

Electroslag Refined Vs. Vacuum Arc Remelt

An examination of the data presented in Table 2 indicates that the ESR and VAR material were essentially equivalent in most respects with ESR being slightly superior in impact strength and VAR being superior in % R.A. and fracture toughness. Other studies have also shown comparable properties between ESR and VAR products. Witek et al[7] compared a bearing steel melted using the ESR and VAR processes and concluded that both reduced the volume and size of non-metallic inclusions with somwhat greater purity being obtained with VAR. In order to determine which process might have a competitive advantage, one would have to examine other criteria such as surface finish of preform. For our purposes, the superior surface of an ESR preform (as cast) has permitted its direct conversion into a forging whereas a VAR requires extensive conditioning.

Argon-Oxygen Decarburization

An examination of the data relative to AOD processing reveals some interesting trends. The properties attained with AOD material were exceptionally high, however, the reduction in area data displays some scatter as evidenced by a relatively large standard deviation. When this material was vacuum arc remelted, the net values for % R.A., Charpy and fracture toughness increased, and the scatter decreased, as might be expected. This data verifies

other experimental data gathered here on AOD. Production forgings
have shown considerable variations within heats on Charpy and RA.
In all cases this variation was traced to either cleanliness
problems or high gas content. In one case, the entire AOD heat
of eighteen forgings was scrapped because of hydrogen flakes.
These scrapped preforms were manufactured by the same producer
that made the preforms cited in Table II. These experiences
indicate that, even though the AOD process is capable of yielding
exceptional mechanical properties, it is subject to great variability and gas content and inclusion count must be controlled to
yield maximum results. In order to be maximally effective, AOD
should be preceded by a degassing operation.

SUMMARY

As can be seen from an examination of the data, there has
been a progressive increase in the ductility and impact strength
obtained over the last 20 years. The current manufacturing
processes are capable of producing alloy steels of the type
examined with 2 - 3 times the impact strength at the same yield
strength level than those of 20 years ago.

Not stated, but implicit in the study, is the fact that by
means of recent advances in melting technology such as calcium
injection, these properties can be obtained at lower costs than
possible before.

REFERENCES

1. Tauscher, S. G., The Correlation of Fracture Toughness
    with Charpy V-Notch Impact Test Data, ARRADCOM,
    Watervliet (1981)
2. Agricola, G., Translation by Hoover, H. C., and
    Hoover, L., DeRe Metallica, The Mining Magazine. (1912)
3. Wilson, A.D., Effect of Calcium Treatment on Inclusions
    in Constructional Steels, Metal Prog. 121:41 (1982)
4. Wilson, A.D., Calcium Treatment of Plate Steels and Its
    Effect on Fatigue and Fracture Toughness Properties,
    Proceedings Eleventh Annual OffShore Technology Conference.
    11:4 (1979)
5. Takenouchi, Tand Suzuki, K., The Influence of Ca-Al
    Deoxidizer on the Morphology of Inclusions, Trans.
    ISIJ. 18:344 (1978)
6. Levicek P., Stransky, K., et al, The Silicocalcium
    Deoxidation of Carbon Steels, Slevarenski. 28:331 (1980)
7. Witek, C., Kalisyewski, E., et al Hutnik. 2:54 (1979)

METAL ALKOXY-DERIVED POWDERS

K. S. Mazdiyasni

Air Force Wright Aeronautical Laboratories

Wright-Patterson Air Force Base, OH 45433

ABSTRACT

The use of metal alkoxides to prepare high purity oxide powders is emphasized. Thermal and hydrolytic decomposition of metal alkoxides, $M(OR)_n$, have been employed to obtain submicron size <50nm single and mixed oxides powders. The alkoxy technique in powder synthesis allow the mixing of residual concentration of an alloying element to be done at something approaching the molecular level. The high surface activity associated with the alkoxy-derived powders make possible relatively low temperature processing of the powder compact to near theoretical density and uniform fine grain size bodies. Transmission electron microscopy, X-ray, ir, DTA and BET surface area measurement are used to show nucleation, crystallite growth and morphology as well as polymorphs of the oxide powder synthesized and microstructural features observed.

INTRODUCTION

The sophistication of org nometallic research in the last quarter of a century has offered the ceramist, and the solid-state chemist an almost unlimited approach to the preparation of tailor-made high-temperature refractory borides, carbides, nitrides, oxides, and silicides. Organic compounds containing oxygen-metal bonds, nitrogen-metal bonds, and carbon-metal bonds have found prominent positions in many phases of modern materials research. These organometallic compounds have served a major role in the synthesis of very high purity ceramics, fibers, and films. Moreover, because of the diversity of their molecular structures, organometallic compounds are now the test subjects of many physicochemical studies. These investigations continue to furnish an ever-deepening under-

standing of the nature of chemical bonding-knowledge directly
applicable to the solution of the basic problems inherent to high-
temperature nonmetallic materials.

Emphasis in this Chapter has been placed on the use of metal
alkoxides to prepare high purity oxides. Alkoxides are $M(OR)_n$
compounds where n is the valence of the metal, M, and R is an organic
group. Selection of the appropriate organic group is the key to
the method and necessitates considerable research in metal-organic
reactions. The organic group must impart sufficient stability and
volatility to the alkoxide so that a clean break of the M-OR bonds
and MO-R is achieved to produce the desired oxides free of organic
contaminants.

In work on the synthesis and the properties of metal alkoxides,
Bradley [1] systematically studies the various alkoxides, especially
the Group IV B transition-metal alkoxides. Hunter [2], Hunter and
Kimberline [3], and Kimberline [4] reported applications of the
formation of oxides from alkoxides by aqueous hydrolysis. The
alkoxide in an alcoholic solution is intimately mixed with water
and the finely divided oxide precipitates. This suggested the
alkoxides as a promising class of compounds for the preparation of
high purity oxides as powders, sols, gels, fibers and films.

The usual methods of making a fine-particle oxide have been to
make salts such as the oxalates, acetates, and carbonates and to
thermally decompose these to oxides. Hydroxides have also been used.
Other methods are the hydrolysis of materials such as chlorides, and
the vapor-phase reaction of metal halides with steam or steam plus
CO_2 to prepare the oxides in the vapor phase. Relatively novel
methods such as the plasma arc with induction coupling and the
Vitro electric arc process and freeze drying of inorganic salt
solutions do not achieve purities better than the old-fashioned
thermal decomposition and precipitation techniques, although particle
size may consistently be 100nm or less.

Thermal and hydrolytic decomposition of metal alkoxide, $M(OR)_n$,
can be employed to obtain submicron size <50nm single and mixed
oxides simply by employing an alkoxide or mixed alkoxides as the
starting materials.

The high surface activity associated with these powders make
possible relatively low temperature processing of the powder compact
to near theoretical density and uniform fine grain size bodies.
This approach to high-purity fine particulates, 3-50nm, offers
almost unlimited flexibility because of the wide variety of organic
compounds that have become available in recent years.

The alkoxy technique describes briefly the synthesis of some Group IV B, yttrium and lanthanide, aluminum and alkali earth metal alkoxides, the application of metal alkoxides decomposition to refractory, ferroelectric, piezoelectric and electrooptic oxides bodies with improved thermophysical properties over conventional ceramics.

SYNTHESIS OF GROUP IV B METAL ALKOXIDES

The chemistry of Si, Ti, Zr and Hf alkoxides has been a subject of extensive research, [1,5-8] because of their industrial importance as precursor materials for various applications. Among these are the preparation of ultrahigh purity submicron refractory oxide powders, [9-11] the vapor deposition of oxide thin films, [12-13] and the preparation of single and mixed phase ferroelectric materials [14-17]. The methods described here are most useful for laboratory-scale preparation of Group IV B transition metal alkoxides and with minor changes in the preparative procedures may be well suited, for other groups in the periodic table, and for industrial production.

Most metal alkoxides are extremely sensitive to moisture, heat, and light. Therefore, the main emphasis must be directed toward the utilization of very clean glass or stainless steel reaction vessels in order to obtain a high purity product in ~90% theoretical yield. The specific experimental conditions for particular preparations have been adequately described in the literature [18].

GENERAL PREPARATIONS

Ammonia Method

The most economic and simple method for the large scale production of alkoxides involves the addition of commercial grade anhydrous metal halide to a mixture of 10% anhydrous alcohol in a diluent (benzene or toluene) in the presence of anhydrous ammonia [19-20].

$$MCl_4 + 4ROH + 4NH_3 \xrightarrow[5\ C^0]{C_6H_6} M(OR)_4 + 4NH_4Cl \tag{1}$$

where M = Ti, Zr, Hf and R = i-C_3H_7.

The Group IV B metal tetrakis (isopropoxide) is readily purified by fractional distillation or recrystallization. Because the removal of NH_4Cl by filtration is always cumbersome and time-consuming, the ammonia method may be carried out in the presence of an amide or nitrile. In this particular method the metal alkoxide separates out as the upper layer, while the ammonium chloride remains in solution in the amide or nitrile in the lower layer, thus the filtration step is eliminated [20-21].

An improved method of preparation of titanium alkoxides based on a recent report for the preparation of Zr and Hf analogs has been suggested by Anand [22]. This procedure consists of treating the anhydrous metal halide saturated in benzene with HCl with various esters, such as ethyl formate, ethyl acetate, diethly oxalate, n-propyl acetate, isopropyl acetate, n-butyl acetate, and isobutyl acetate, in the presence of anhydrous ammonia. The alcohol formed in the reaction is itself consumed in the formation of metal alkoxides.

$$MCl_4 + 4HCl + 4CH_3COOR + 4NH_3 \longrightarrow$$

$$M(OR)_4 + 4CH_3COCl + 4NH_3Cl \tag{2}$$

ESTER EXCHANGE REACTION

A valuable method for converting one alkoxide to another is an ester exchange reaction [23]. The method is particularly suited for the preparation of the tertiary butoxide from the isopropoxide and t-butyl acetate. The reaction is as follows:

$$M [O-CH(CH_3)_2]_4 + 4CH_3COOC(CH_3)_3$$

$$\longrightarrow M [O-C(CH_3)_3]_4 + 4CH_3COOCH(CH_3)_2 \tag{3}$$

Since there is a large difference between the boiling points of the esters, the fractionation is simple and quite rapid. Another advantage appears to be the lower rate of oxidation of the esters as compared to that of the alcohol.

Alcoholysis Reaction

Substitution of other branched R groups with the lower straight chain alcohols has been carried out in an alcohol interchange reaction as shown in Eqn (4).

$$M(OR)_4 + 4R'OH \xrightarrow{C_6H_6} Ti(OR')_4 + 4ROH \tag{4}$$

The distillation of an azeotrope drives the reaction to completion. The interchange becomes slow in the final stages with highly branched alcohols, probably owing to steric hindrance. The rise in temperature to the higher boiling alcohol, however, reliably signals the end of the reaction. In this method of preparation, speed, reversibility and absence of side reactions play major roles.

YTTRIUM AND LANTHANIDE ALKOXIDES

The electronegativity of yttrium and the lanthanides places these elements between metals such as aluminum, which form covalent alkoxides, and sodium, which form ionic alkoxides. The degree of

ionic character of the M-O bond, which is dependent on the size and
electronegativity of the metal atom, is important in determining the
character of the alkoxide. This significant factor is made apparent
by comparing the electronegativity of the metal atom with that of
the oxygen atom using the Pauling Scale [24].

Disregarding the contribution of the strongly covalent alkyl
C-O bond where Xo-Xm = 3.5 - 2.5, whenever Xo-Xm is >2.4 more of
the ionic alkoxides are found; where Xo-Xm is <2.3, the alkoxides
are predominantly covalent in nature. Some metals, notably the
rare earths, fall between these differences and only highly branched
organic radicals impart sufficient covalent character to the
alkoxides. It should be noted, however, that all the alkoxides would
be classified as ionic according to the original Pauling concept in
which the electronegativity difference of 1.7 and above means the
bonds are 50% ionic. Hannay and Smith, [25] have placed this
difference at 2.1 which is consistent with the covalent character
observed in this simple model.

The reactions within the lanthanide series to form alkoxides
are all slow and follow no significant trend. The size, shape, and
oxygen content of the metal used vary the surface area thus
influencing both the reaction rate and % yield. When mixed rare
earth (misch-metal) is used, a synergistic effect is found. The
reaction rate is increased twofold or more over that for a single-
phase rare earth, Brown and Mazdiyasni [26].

The study of alkoxides of yttrium and the lanthandes has been
limited by expensive starting materials and by the difficulties in
preparation and handling. Successful characterization of most of
the alkoxides is complicated by their extreme sensitivity to
moisture, heat, light, and atmospheric conditions.

SYNTHESIS

The yttrium and lanthanide tris isopropoxides are prepared by
the reaction of metal turnings with excess isopropyl alcohol and a
small amount of $HgCl_2$ (10^{-4} mol per mol of metal) as a catalyst,
Mazdiyasni, et al [17].

$$Ln + 3C_3H_7OH \xrightarrow{HgCl_2} Ln(OC_3H_7)_3 + 3/2\ H_2 \tag{5}$$

After filtration, the crude product is purified by recrystallization from hot isopropyl alcohol or vacuum sublimation. Yields of
75% or better are realized with this method. For some of the
larger metals ions (lanthanum through neodymium) the reaction rate
and % yield are increased by using for the catalyst a mixture of
$HgCl_2$ and $Hg(C_2H_3O_2)$ or HgI_2, Brown and Mazdiyasni [26].

Substitution of other R groups for the isopropoxy groups is done by the alcohol interchange technique.

$$Ln(Oi-C_3H_7)_3 + 3ROH \longrightarrow Ln(OR)_3 + 3i-C_3H_7OH \tag{6}$$

where R is C_4H_9, $s-C_4H_9$, $t-C_4H_9$, $n-C_5H_{11}$,

$i-C_5H_{11}$, $s-C_5H_{11}$, $t-C_5H_{11}$, etc.

The equilibrium shown in Eqn (6) is an overall reaction and is assumed to include the intermediate stages of mixed alkoxides such as $Ln(OR)_x(OC_3H_7)_{y-x}$ as well. It is best to use a small excess (∿10%) of higher boiling ROH; otherwise, the rate of exchange, especially the last stages of the interchange, will be very slow. The isopropyl alcohol liberated during the course of the alcoholysis reaction is fractionated out azeotropically. A wide variety of solvents (benzene, toluene, carbon tetrachloride, cyclohexane, etc.) may be used to act as an inert diluent. The i.r. spectra of the yttrium and lanthanide tris isopropoxides are quite similar. The spectra have proved useful in controlling the preparative results for completeness of reaction and identification of products.

ALUMINUM ALKOXIDE

Similarly, aluminum tris-isopropoxide is synthesized by the method of Adkins and Cox [27].

$$Al + 3C_3H_7OH \xrightarrow[\text{exothermic}]{\text{HgCl}_2} Al(OC_3H_7)_3 + Hg + 2HCl + 1/2H_2 \tag{7}$$

The alkali or alkali earth metal alkoxides are prepared by the reaction of metal with alcohol. The reaction is highly exothermic with evolution of excess hydrogen.

$$M + ROH \xrightarrow[\text{exothermic}]{} M(OR)_n + H_2 \tag{8}$$

where M is the metal and R is the isopropyl radical and n= 1 or 2

PYROLYSIS MECHANISM

When zirconium tertiary butoxide is pyrolyzed the mechanism of reaction is one in which olefin and alcohol are split in successive steps

$$Zr(OR)_4 \longrightarrow ZrO_2 + 2ROH + Olefin \tag{9}$$

and the suggested mechanism is as follows:

$$Zr(OC_4H_9)_4 \longrightarrow Zr(OC_4H_9)_3OH + CH_3-\underset{\underset{CH_3}{|}}{C}= CH_2 \tag{10}$$

$$Zr(OC_4H_9)_3OH \longrightarrow ZrO_2 + 2C_4H_9OH + CH_3-\underset{\underset{CH_3}{|}}{C}= CH_2 \tag{11}$$

After the initial decomposition steps, the reaction is very fast,
so no intermediates found are isolated in the decomposition. The
products found are olefin, alcohol and oxide. The rate of reaction
is dependent on the concentration of the starting material, and the
rate equation is given as a function of the concentration

$$V = kK_1 \, [Zr(OR)_4] \tag{12}$$

However, if isopropoxide or, for this matter, t-butoxide are used
in a closed system, one can have the following mechanism. In this
instance the rate controlling step might be the dehydration of
alcohol. A molecule of H_2O produced from one molecule of alcohol
would then regenerate two molecules of alcohol by hydrolysis of
zirconium alkoxide, and hence the chain reaction would then be set
up in accordance with following equation

$$Zr(OR)_4 \xrightarrow{K_1} ZrO_2 + 2ROH + R = R \tag{13}$$

$$ROH \xrightarrow{K_2} R = R + H_2O \tag{14}$$

$$2 \overset{\diagup}{\underset{\diagdown}{-}} Zr - OR + HOH \longrightarrow \overset{\diagup}{\underset{\diagdown}{-}} Zr - O-Zr \overset{\diagup}{\diagdown} + 2ROH \tag{15}$$

and in this case an intermediate product of hydrolysis, zirconium
oxide alkoxide will go under rapid disproportionation

$$2ZrO(OR)_2 \longrightarrow ZrO_2 + Zr(OR)_4 \tag{16}$$

This mechanism is used to produce very large quantities of oxide by
hydrolytic decomposition technique. The basic requirement for the
alkoxides is that the hydrolysis reaction be rapid and quantitative.

Where the alkoxy (-OR) group may be quite stable to hydrolysis,
a small amount of a mineral or organic acid may be added to catalyze
the reaction. The oxides are at this point in a finely divided state
of submicron (maximum agglomerate size 0.1mμ) particle size and of
extremely high purity.

VAPOR-PHASE DECOMPOSITION

Mazdiyasni, et al [28] have shown that the thermal decomposition of metal alkoxides can be employed to produce oxide coatings on high temperature substrates and for preparation of high-purity submicron refractory oxide powders. The method has been employed to prepare zirconia and hafnia as ultra-high purity submicron powders on a continuous basis. Also improved synthesis of yttrium and rare-earth compounds have considerably extended the usefulness of this approach to synthesis of new ceramic materials.

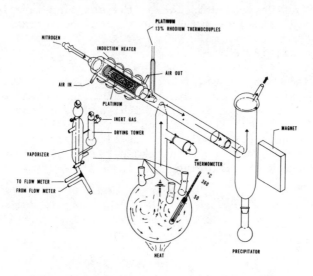

Figure 1. Flow Diagram for Preparation of Mixed Oxides.

The apparatus employed in this method is shown in Fig. 1. The alkoxide is vaporized in the round-bottom flask and carried by a dry inert-gas stream into the decomposition chamber where a preheated (platinum induction furnace heater) inert gas impinges on the alkoxide vapor providing substrateless insitu vapor-phase decomposition of the alkoxide to the oxide. The fine particulate powder is collected with the assistance of an electrostatic precip- itator. The magnet is used as appropriate for magnetic materials such as γ Fe_2O_3 using $FeCo_5$ in a partial pressure of oxygen, Lynch, et al [29]. However, a representative reaction is the decomposition of zirconium tetra-tertiary butoxide to zirconium oxide. Vaporization of the alkoxide is done at 190-210 C^0 at 760 mm Hg and decomposition at 325-500 C^0:

$$Zr(OC_4H_9)_4 \longrightarrow ZrO_2 + 2C_4H_9OH + 2C_4H_8 \tag{17}$$

The particle size of this powder was determined by Mazdiyasni, et al [12] from an alcohol dispersion of the powder sprayed onto a carbon substrate on a copper mesh screen and viewed in an electron microscope. The results are shown in Figure 2.

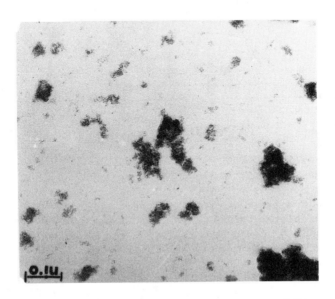

Figure 2.   Electron Micrograph of as Prepared Dispersed Zirconia Powder.

The average particle size was found to be 3-5nm with evidence that the larger particles were agglomerates of individual particles of 2nm or less. Particle size range was generally ∿ 2nm with only an occasional large particle of 0.1μ to 0.5μ (100 to 500nm). It is likely that these particles are also agglomerates. The powder as formed was amorphous to X-rays, but electron diffraction patterns showed strong lines for cubic zirconia. The high purity (> 99.95%) of the powder is indicated in Table I. The table lists purities from the starting tetrachloride and its subsequent conversion to the isopropoxide $Zr(OC_3H_7)_4$, to the butoxide $Zr(OC_4H_9)_4$, and finally to the oxide. Three batches are given in separate analyses for the oxide demonstrating the reliability of high purity levels on a continuing basis. Even the hafnium content is sharply decreased as an impurity in this process.

Table I Emission Spectrographic Analyses of $ZrCl_4$ to $Zr(OR)_4$ to $Zr(OR')_4$ to ZrO_2

Elements	$ZrCl_4$ p.p.m.	$Zr(OR)_4$ p.p.m.	$Zr(OR')_4$ p.p.m.	ZrO_2 p.p.m.	ZrO_2 p.p.m.	ZrO_2 p.p.m.
Al	500	180	5	ND<5	ND<5	ND<5
Ca	80	70	ND<5	ND<5	ND<5	ND<5
Fe	500	160	20	ND<10	ND<10	ND<10
Hf	<10,000	5000	1000	500	<200	<200
Mg	25	200	ND<10	ND<10	ND<10	10
Mo	50	ND<10	ND<10	ND<5	ND<5	ND<5
Na	500	500	50	<100	<100	<100
Si	1900	700	500	300	300	300
Sn	100	ND<10	ND<10	ND<10	ND<10	ND<10
Ti	100	20	5	5	5	5
W	<50	<50	<50	<50	<50	<50
Cu	200	20	<10	<10	<10	<10

ND < = Not detected, less than.

PURITY

The values shown in Table I are taken from results of emission spectroscopic analysis of the starting material to the final oxide product. The increased purity obtained through the successive steps of preparation of the isopropoxide, the ester exchange reaction to form tertiary butoxide, and the distillation and decomposition of this product to the oxide is remarkable.

Very high purity metal chloride as a starting material should result in an even purer final product. Economic considerations, however, would make it desirable to use a technical-grade metal chloride rather than a higher purity starting material in any large scale process. As long as the alcohols and solvents are recovered at each step of the reaction, the costs of such a purification method should not be excessive. The results produce a zirconia which is far purer than any commercially available high purity product.

PARTICLE SIZE

The most interesting aspect of oxide preparation in this manner, however, is not the very high purity but the unusually small particle size.

Fig. 3 shows an electron micrograph of particles of dispersed zirconia powder. A particle size distribution curve for particles greater than 2nm in diameter as measured from electron micrographs is given in Fig. 4. This was the smallest size which could be measured. The resolution of the electron microscope was consistently better than 1nm. The largest particles observed were less than 30nm in diameter. The mean average particle size was approximately 5nm and 80 percent of the particles were less than 10nm. Samples were prepared by dispersal in ethyl alcohol using ultrasonic vibration, followed by spraying the dispersion on thin carbon substrates. Other techniques such as dispersal in distilled water or xylene resulted in agglomerated samples and unsatisfactory determinations.

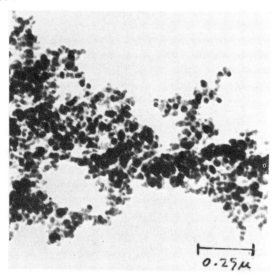

Figure 3. Electron Micrograph of Uncalcined Alkoxy-Derived Zirconia Found Growing While Observed in Electron Microscope.

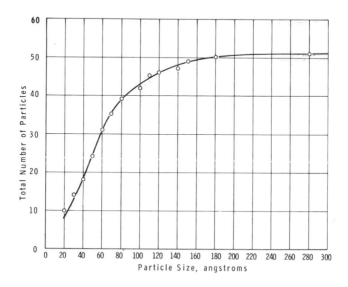

Figure 4. Particle Size Distribution Curve of Ultrafine Zirconia.

METASTABLE TRANSITIONS OF ZIRCONIA

AMORPHOUS ZrO_2

Mazdiyasni, et al [16] have shown that an ultra-high purity
submicron zirconium oxide powders prepared by thermal decomposition
of alkoxides was amorphous to X-ray, but electron diffraction
patterns showed strong lines for cubic zirconia. After the powder
was aged for several months at room temperature there was essentially
no change. The oxide powder was then heat-treated from 140 to
1000 C^0 and re-examined. Up to 300 C^0 it remained in the amorphous
metastable cubic state, from 300 to 400 C^0 it was a mixture of
tetragonal and monoclinic phases, and at higher temperatures only
the monoclinic phase remained. To determine the temperatures at
which the transformations occurred, and whether they occurred
simultaneously or concurrently, powder samples were isothermally
aged at 5 to 10 C^0 intervals from 150 to 500 C^0. In the aging
studies a thermal balance was used to continuously monitor and
hold temperatures constant to ±0.50 C^0 or better over long periods
of time. The transformation temperature range was reproducible to
within the precision of the measurements (±.5 C^0 or better). X-
ray diffraction patterns indicated that a sharp transition occurs
from the metastable cubic to a metastable tetragonal phase at 300
to 305 C^0. The metastable tetragonal phase was found without the
monoclinic phase only at 305 C^0. At 310 C^0 monoclinic phase was
clearly evident. At 335 C^0 the diffraction pattern showed that
most of the oxide was a monoclinic with only weak tetragonal peaks
to indicate that the metastable teteragonal phase was still present
at low concentrations. At 400 C^0 the pattern indicated that only
the monoclinic phase was present.

Table II shows the results obtained from many X-ray patterns.
In the X-ray diffraction powder data for samples aged at 150 to 300
C^0 for 150 hrs. (150 hrs. was determined to be adequate to attain
equilibrium), four strong peaks were found. The powder is in
transition from the amorphous (to X-rays) to the crystalline state
at this point and definitive evidence of the phases present was
obtained from the electron diffraction patterns which indicated that
only the cubic phase was present. X-ray diffraction results after
treatment at 350 C^0 for 150 hrs. show that the powder is well
crystallized in the tetragonal phase.

The X-ray diffraction powder data of this material heat-treated
at 335 C^0 for 150 hrs. show that most of the intense tetragonal peaks
have virtually disappeared whereas the most prominent monoclinic
peaks appear at high intensity. In the sample heat-treated to 400 C^0
for 24 hrs. most of the tetragonal phase had disappeared and the
powder consisted principally of the monoclinic phase. Considering
the reported sluggishness of zirconia transformations, it is remarkable that the transformation from a metastable cubic phase to a fully

crystalline tetragonal phase occurs within a 5 C^0 interval.

Table II X-Ray Diffraction Powder Data for Submicron ZrO$_2$
(CuKα Radiation)

150°-300°C, 150 hr		305°C, 150 hr		335°C, 150 hr		400°C, 24 hr		Phase
I	d(A)	I	d(A)	I	d(A)	I	d(A)	
						vw	5.096	M
						w	3.670	M
				vs	3.195	vs	3.170	M
vs	2.952	vs	2.951	vs	2.962			T C
				s	2.865	vs	2.844	M
		s	2.591	w	2.602	w	2.614	M T
s	2.533	w	2.533	w	2.548	w	2.546	M T C
						vw	2.335	M
						vw	2.219	M
						vw	2.186	M
						vw	2.025	M
						w	1.996	M
						w	1.848	M
				w	1.820			T
vs	1.807	s	1.812	vw	1.799			T C
		vs	1.796			vw	1.787	M T
						w	1.697	M
						w	1.651	M
		s	1.555					T
vs	1.537	vs	1.534	w	1.541			T C
						vw	1.518	M
		s	1.479	w	1.481	w	1.481	M T
		s	1.292					T
		s	1.272	vw	1.275			T
		s	1.178	vw	1.178			T
		s	1.169	vw	1.170			T
		s	1.154	vw	1.154			T
				vw	1.152			T
		s	1.139					T
		s	1.041	vw	1.044			T

NOTE: vs = very strong; s = strong; w = weak; vw = very weak; M = monoclinic; T = tetragonal; and C = cubic.

Since it is difficult to obtain quantitative measurements of particle sizes in the 1 to 20nm region. Consequently, the variation in particle size from batch to batch can be obtained semiquantitatively by random counting from enlarged electron micrographs. (In the region of 10 to 100nm, for example, an absolute accuracy by line-broadening X-ray diffraction is about ±25%). Therefore the relation of temperature of transformation of the metastable cubic to the metastable tetragonal phase may be determined only empirically. The diffraction results were obtained on powders of 1 to 20nm particle size with a mean average size of 5nm. Powder samples from other batches of material with average particle sizes of approximately 8 to 10nm were also investigated. Transformation temperatures for the metastable cubic to metastable tetragonal phase were 250 to 270 C^0 (for fractions of larger average particle size). Fractions with an average size below 5nm were not examined.

It is concluded that:

(1) Ultra-high purity submicron zirconium oxide prepared by alkoxide vapor decomposition is metastable cubic as formed. (2) Transition of metastable cubic to metastable tetragonal zirconia is relatively sharp and occurs at 300 to 305 C^0 in a powder with average particle size of 5nm. (3) Powders with average particle size of 8 to 10nm transformed at 250 to 270 C^0 to the metastable tetragonal form. (4) Transformation of the metastable tetragonal to the monoclinic form is more gradual and occurs between 305 and

400 C^0 in a powder with average particle size of 5nm.

HIGH TEMPERATURE X-RAY STUDIES

    The X-ray diffraction pattern of hydrolytically decomposed
submicron particulate ZrO_2 was observed by Mazdiyasni [10] over
the temperature range from 250 to 1200 C^0 on heating using CuKα
radiation (Fig. 5). The following sequence of phases was observed

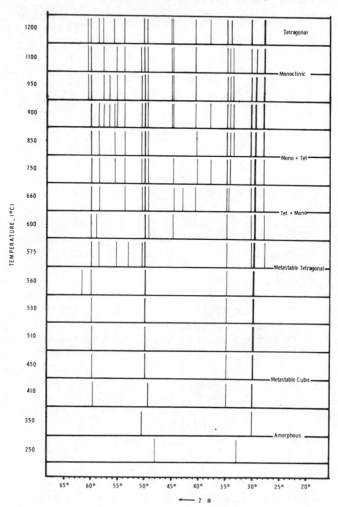

Figure 5. High Temperature X-Ray Variations of Metastable ZrO_2.

in the alkoxy-based powder as the temperature increased to 1200 C^0:
amorphous, to metastable cubic, to metastable tetragonal, to
tetragonal + monoclinic, to monoclinic + tetragonal, and finally to
tetragonal. Calcination studies were also undertaken Fig. 6 to

observe the growth of primary crystallites into larger particles
and the influence of tumbling in different atmospheres, such as air,
oxygen, hydrogen, and vacuum, on growth. No measureable change in
size and morphology of crystallites was observed as a result of
varying the atmospheric conditions during calcination. Figs. 7 and
8 are two typical electron micrographs showing,

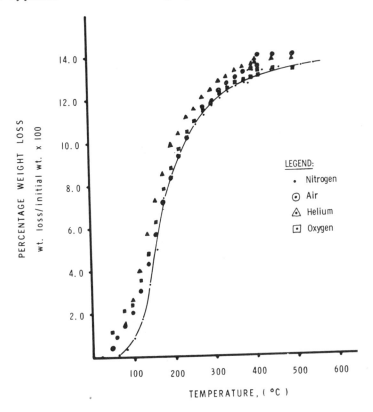

Figure 6. Effect of Calcination Atmospheres on Alkaxy-Derived
ZrO_2.

respectively, the effect of calcination at 500 C^0 in air for 24 hrs.
without tumbling, and calcination at 750 C^0 in air for 24 hrs. with
tumbling. Tumbling is seen to be an effective method of controlling
crystallite growth while atmospheric conditions have a negligible
effect during the pre-sintering stage of processing these powders.

INFRARED AND RAMAN SPECTRA OF ZIRCONIA POLYMORPHS

Infrared and Raman spectroscopy may be used for the characterization of the ZrO_2 phases and for a method of non-destructive
analysis. McDevitt and Baun [30] clearly distinguished the ir
spectra of the monoclinic and stabilized cubic phases of ZrO_2; at

the same time, they showed that monoclinic samples with the same
X-ray diffraction pattern could have considerable different ir
spectra. These workers suggested that ir spectra might be more
sensitive than X-ray diffraction patterns to

Figure 7. TEM of Submicron ZrO_2 Showing the Effect of Calcination
at 850 C^0 in Air For 24 hrs. With Tumbling.

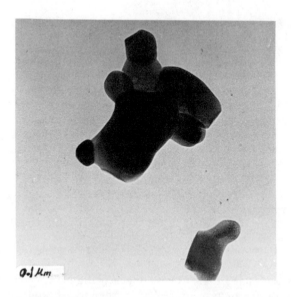

Figure 8. TEM of Submicron ZrO_2 Showing the Effect of Calcination
at 500 C^0 in Air for 24 hrs. Without Tumbling.

small changes in the crystal lattice of the phases of ZrO_2. This suggestion implies that ir and Raman spectroscopy may yield practical new information on destabilization phenomena because both methods can be used non-destructively.

Phillipi and Mazdiyasni [31] used high-purity (>99.99%) submicron alkoxy-derived metastable ZrO_2 powder as a starting material to follow the phase transformation sequence in unstabilized ZrO_2 at the lower temperatures. Commercially available monoclinic and CaO- and MgO-stabilized ZrO_2 were also investigated, as was a 6.5 mol% alkoxy-derived Y_2O_3-stabilized ZrO_2 "Zyttrite".

Powder transmission was used for ir spectra. Phillipi and Mazdiyasni's [31] investigation focused on the fundamental vibration frequencies in the 900 to 200 cm^{-1} region, rather than on the weak bands found in some samples at higher frequencies.

INFRARED POWDER TRANSMISSION SPECTRA

The spectrum of the as-prepared metastable cubic submicron ZrO_2 without a stabilizing additive indicates that the fundamental ir frequency is near the strong 480 cm^{-1} band. The incipient broad weak shoulder on its high frequency side was better developed in stabilized cubic and stabilized tetragonal spectra. The broad band in the 3500 cm^{-1} region may result from atmospheric moisture retained by the CsI pellet or from OH^- groups from the starting material. The two bands in the 1300 to 1700 cm^{-1} region were found in some, but not all, of the samples investigated. Their frequencies are variable; the bands occurred at 1630 and 1350 cm^{-1}, the maximum separation, in some samples, and at 1540 and 1400 cm^{-1}, the minimum separation, in others. Their intensities were also variable; they were affected by thermal history stabilizing additives. The bands generally were present in the spectra of powders which were ground before they were blended with CsI, suggesting absorption of atmospheric constituents. Broad bands in this region are commonly observed in spectra of metal oxides with absorbed CO_2, Hair [32], and are assigned to surface carbonate groups which are variable with respect to temperature and crystal structural of the oxide. Therefore, similar bands in the ZrO_2 spectra are attributed to impurities formed from absorbed atmospheric CO_2.

The effects of grinding on the ZrO_2 polymorphs were also studied by Phillipi and Mazdiyasni [31]. The ir spectra of the as-received samples blended directly with the CsI were compared with those of the same powders dry-ground in B_4C. In each case there was a general improvement in contrast of the spectral structure with grinding for equal concentrations of sample. The main transmission minimum in the 500 cm^{-1} region shifted to frequencies higher by 10 to 50 cm^{-1}, and there were small shifts of the sharper bands. The monoclinic 270 and 235 cm^{-1} bands each changed in contour on the low-

frequency sides, suggesting that broad unresolved bands disappeared
with grinding. A poorly defined shoulder in the 320 cm^{-1} region of
both strained and unstrained monoclinic ZrO_2 also disappeared on
grinding, and no new bands appeared.

The ir spectra of cubic ZrO_2 stabilized with Y_2O_3, and CaO,
and MgO were examined. No significant differences attributable to
the type of concentration of stabilizing agent were detected.

To determine the effects of thermal history, portions of
alkoxy-derived pure ZrO_2 were heated for 24 hrs. each in air at
approximately 100 C^0 intervals between 384 and 1078 C^0 in a resistance furnace. Each sample in it's Pt boat was withdrawn, cooled
quickly to room temperature, and blended with CsI without grinding.
Infrared spectra of these samples are presented in Fig. 9. An
orderly progression of spectral changes with increasing temperature

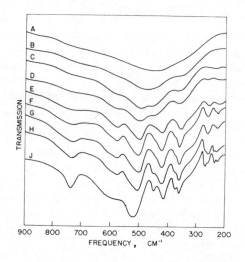

Figure 9. Powder Tranmission Spectra of Metastable Cubic ZrO_2
          Heated for 24 hrs. at Temperatures Indicated: a.   as
          received; b.   384 C^0; c.   478 C^0;d.   573 C^0; e.   668 C^0;
          f.   766 C^0; g.   870 C^0;h.   974 C^0; and i.   1078 C^0.
          Ordinate Scales Displaced for Clarity

was observed which corresponds to the low-temperature metastable
ZrO_2 polymorphic transitions. Changes in the basic character of
the spectra occurred near 600 and 1000 C^0, in agreement with previous
X-ray diffraction work described by Mazdiyasni [10]. The X-ray
diffraction pattern of the sample calcined at 478 C^0 indicated
tetragonal ZrO_2 with a small amount of monoclinic, whereas the sample
calcined at 766 C^0 was pure monoclinic. Both patterns had diffuse
lines in the back-reflection region which indicated lattice strain.

The spectrum of the sample calcined at 1078 C^0 closely resembled that of monoclinic ZrO$_2$.

RAMAN SPECTRA

In contrast with ir spectra, in which the overall band contours provide as much information as the exact band frequencies, Raman spectra ordinarily consist of discrete lines. Accordingly, the spectral data are summarized in Table III. The Raman spectrum of a

Table III   Raman Frequencies of Zirconia Polymorphs (Cm^{-1})

Metastable cubic	Stabilized cubic	Tetragonal	Monoclinic
490 w,b	625 m,b	640 s	638 m
	480 w,b	615 ?	617 m
	360 w,b	561 w	559 w
	250 w,b	536 w	538 w
	150 w,b	473 s	502 w
		380 m	476 s
		332 m	382 m
		263 s	348 m
		223 w	337 m
		189 s	307 w
		179 s	223 w
		148 s	192 s
			180 s
			104 m

NOTE: w = weak, m = medium, s = strong, and b = broad.

pure alkoxy-derived submicron metastable cubic ZrO$_2$ showed only a weak broad line at 490±20 cm^{-1} which was perhaps 80 cm^{-1} wide. A well-annealed Y$_2$O$_3$-stabilized cubic ZrO$_2$ had a distinct a symmetrical line at 625 cm^{-1} and several other weak lines, whereas, for a highly strained cubic sample, several of the weaker lines were better developed. Tetragonal and monoclinic zirconias had many more lines. The poorly resolved tetragonal pair around 640 and 615 cm^{-1} were well resolved in the monoclinic phase. The tetragonal and monoclinic phases have many lines in common, but two strong lines at 263 and 148 cm^{-1} are peculiar to the tetragonal phase, whereas the strong line at 348 cm^{-1} and two weaker ones are peculiar to the monoclinic phase.

The number of ir bands in the powder spectra increases from the cubic phase to the tetragonal and monoclinic phases. This observation is consistent with the predictions of group theory applied to single-crystal specta. The ir spectrum of material depends on its crystal structure; if it is inhomogeneous a different spectrum is associated with each local structure. There is evidence that additional ir-active modes may be activated in inorganic spectra by mechanical damage to the surfaces of single crystals and particles. In particular, the work of Barker [33] showed that forbidden modes could be made to appear in the reflection spectrum of single-crystal corundum as a result of relaxation of certain of the selection rules by strain.

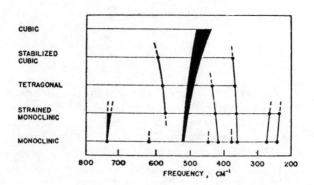

Figure 10. Frequencies and Range of Existence of Powder
          Transmission Bands for Phases of ZrO$_2$. Broad Zones
          Denote Variability with Sample History and Mode of
          Preparation.

The association of the powder bands is shown in Fig. 10.
Considering the data from many spectra collected, it is possible to
accommodate the somewhat different spectra of stabilized cubic and
strained monoclinic zirconias by regarding them as intermediates
between those of three classical crystal structures. The curved
lines connecting several frequencies show the continuity of existence
of a given band over several phases; they do not necessarily imply
a frequency shift of a resonance.

Instead of the existence of three discrete cubic, tetragonal,
and monoclinic polymorphs, the ir spectra suggest a more continuous
transformation as crystal symmetry is altered. Strain, whether
induced by thermal history or mechanical energy of grinding, develops
localized changes of structure which change selection rules and
cause shifts and splitting of vibrational modes. Similarly, the
introduction of stabilizing ions produces a cubic structure with
certain sites distorted sufficiently that the "tetragonal" modes
appear around 580 and 360 cm^{-1}. Thus ir spectroscopy is a sensitive
indicator of site symmetry in zirconias.

DIFFERENTIAL THERMAL ANALYSIS

Differential thermal analysis (heating and cooling rate of 3
C^0/ min) of the ultra high purity, 1-5nm, ZrO$_2$ particulates was
carried out by Mazdiyasni [10]. The differential analysis was made
with an electrically shielded apparatus, using an alumina cell and
a HfO$_2$ standard, as described by Ruh, et al [34]. Voltage differences between the standard and sample thermocouples were amplified
with a microvolt amplifier and plotted as a function of temperature
on a high sensitivity recorder. DTA results were in very good
agreement with the high temperature X-ray diffraction line data and

the infrared analyses, time-temperature dependent phenomenon playing
a major role in the mechanism of phase transformation of the 1-5nm
particle size ZrO_2, Mazdiyasnim et al [16]. An examination of the
DTA heating curve Fig. 11 for an

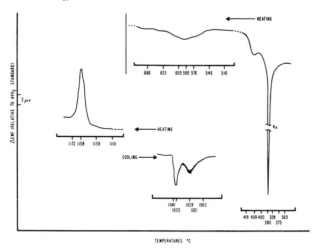

Figure 11. Thermogram of Metastable Submicron ZrO_2.

amorphous submicron ZrO_2 powder shows a transition from metastable
cubic to metastable tetragonal in the relatively narrow temperature
range of 365 to 400 C^0, marked by a very large (60-80mv) exothermic
peak followed by a gradual transformation of the room temperature
stable monoclinic phase. This result has been thermodynamically
related to a function of surface energy (specific surface area) as
a determinant of the most stable phase of ZrO_2 by Gravie [35]. The
high temperature X-ray diffraction results shown in Fig. 5 and the
medium infrared analysis, Fig. 9 are in good agreement with the DTA
studies with respect to the identification and transition temperatures of these two phases. There is, therefore, independent evidence
of the cubic or tetragonal phase being the most stable low temperature phase for ZrO_2 particulates below 30nm in size.

MECHANISM OF CRYSTALLITE GROWTH

     On heating, crystallite growth occurs (agglomeration of small
particles, followed by rapid growth) leading to conversion to the
monoclinic phase. The DTA curve of the first heating cycle (Fig.
12, upper left) shows the monoclinic-tetragonal inversion beginning
at about 1139 C^0. Maximum deflection occurs at 1159 C^0, a value
somewhat lower than the 1170-1190 C^0 reported in the literature,
and temperature equilibrium is again attained at approximately
1172 C^0. Differential temperatures were also recorded on cooling.
In this case the tetragonal to monoclinic inversion peak (Fig. 12,
lower right) is shifted to a slightly higher temperature, 1034 C^0,

with appearance of a second exothermic peak at approximately 1020 C^0. The 1020 C^0 peak has not been previously observed during differential thermal analysis of a zirconium dioxide system. It is interesting to note that the second peak always develops as discontinuous "noise" peaks super-imposed on a main peak, in a manner reminiscent of the

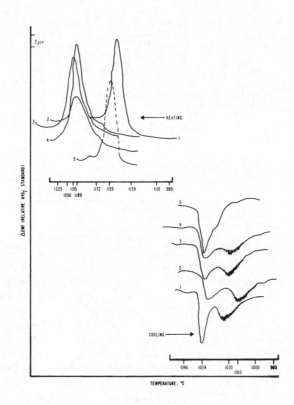

Figure 12. Thermogram of Submicron ZrO_2 During Monoclinic Tetragonal Inversion.

"Barkhausen effect". The Barkhausen effect is commonly observed in ferromagnetic materials during magnetization by an applied magnetic field. The discontinuities are associated with irregular motions of the magnetic domain walls, William and Shookly [36], and Dekker [37]. The occurrence of small exothermic peaks on a larger exothermic peak may be interpreted as spontaneous energy releases, either by motion of ferroelectric domain walls, or by abrupt changes in crystallite size.

The fact that an extra peak was observed during the first heating cycle suggested the occurrence of some fundamental change in the system. To establish this point, the effect of annealing upon the tetragonal-monoclinic inversion was further investigated by

Mazdiyasni [10] as shown in Fig. 12. It was noted that the
monoclinic-tetragonal peak was shifted to the higher temperature of
approximately 1180 C⁰, during heating cycles 2, 3, and 4, and that
it became much smaller and broader in shape. The respective cooling
curves, however, continued to exhibit distinct peaks at 1034 C⁰ and
approximately 1020 C⁰, with a gradual disappearance of the 1020 C⁰
peak on cycling. The 1034 C⁰ peak was predominant throughout and
became more pronounced on each annealing cycle. It is reasonable
to conclude, therefore, that the 1034 C⁰ peak, although higher than
the 980 C⁰ usually reported in the literature, is normal for the
tetragonal—monoclinic inversion on cooling and that the 1020 C⁰ peak
may be associated with growth and nucleation processes.

     fter prolonged heating, at about 1300 C⁰, the second peak
(Fig. 12, Cycle 5), together with the "noise effect," were eliminated.
A subsequent run gave only the DTA result normally observed for the
tetragonal to monoclinic phase transformation. At this point the
powder apparently had lost all its activity.

     An electron micrograph of two or more crystallites while in
growing stage is shown in Fig. 13. The solid state reaction appears
to proceed by a process of nucleation and growth at an interface,
where the grain boundaries are formed. The crystallite growth
mechanism, in the experiments carried out may be involved first, an
atomic rearrangement of powder particles (or grain), and then the
generation of nucleus at the interface, which may result in the
formation of a larger crystallite.

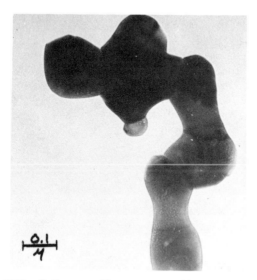

Figure 13. TEM of Two or More Crystallite of ZrO₂ While in
           Growing Stage and Sintering.

DYNAMIC CALCINATION STUDIES

Table IV shows that the heat treatment of the as-formed powder (for BET studies) in a tumbling furnace resulted in marked differences in specific surface area of the powder. The great reduction in surface area with increased temperature is, to a large extent, a result of the crystallite growth, the particles having grown from 1-5nm to approximately 100-150nm in size.

Table IV Specific Surface Area of Submicron ZrO_2

No.	A Single Point BET		B Nelson and Eggersten[15] BET	
	T-outgassing 1/2 hr °C	Specific Surface m^2/g	T-outgassing 24 hrs °C	Specific Surface m^2/g
1	60	222	66	217
2	90	230	100	200
3	120	210	-	-
4	150	198	150	195
5	180	186	-	-
6	210	157	200	165
7	-	-	300	135.7
8	320°C/152 hrs	50	480	48.1
9	-	-	940	0.96

The solid state reactions of ceramic powders depend strongly on the characteristics of the initial powder. Nucleation and crystallite growth of 1-5nm particulates of alkoxy-based zirconium oxide are particularly illustrative of this strong dependence as shown previously by Mazdiyasni, et al [16]. Literature references indicate that there are at least three major possible mechanisms by which nucleation and crystal growth may take place, Burke and Turnbull [38], Coble [39] and Anderson and Morgan [40]. Since a definitive choice of mechanism in the alkoxy-based system is not yet feasible on the basis of available data, the experimental observations only are presented.

SYNTHESIS FOR BULK CERAMIC POWDERS

Hydrolytic Decomposition: The alkoxides react rapidly with traces of water to form alkyl alkoxides.

$$M(OR)_n + H_2O \longrightarrow MO(OR)_{n-2} + 2ROH \tag{18}$$

M is Ti, Zr, Hf, Th, R is an alkyl group, and n is the valence of the metal. The alkyl alkoxides then decompose to the oxides.

$$2MO(OR)_{n-2} \longrightarrow MO_2 + M(OR)_4 \qquad (19)$$

This hydrolysis of alkoxides has been used to prepare high-purity of a wide variety of mixed oxides such as mullite, $3Al_2O_3.2SiO_2$ and "Zyttrite," 6 mol% Y_2O_3. ZrO_2, and 7 mol% Y_2O_3. HfO_2, $BaTiO_3$, and PLZT. etc.

POWDER PREPARATION AND ANALYSIS

MULLITE

The mixed oxide containing 71.8 wt% Al_2O_3 and 28.2 W% SiO_2 was prepared by Mazdiyasni and Brown [41] via hydrolytic decomposition of solution of the corresponding metal alkoxides. Aluminum tris isopropoxide was added to a stoichiometric amount of silicon tetrakis isopropoxide. The mixed alkoxides were refluxed in excess isopropyl alcohol for 16 h before hydrolysis to ensure thorough mixing. The hydroxyaluminosilicate was prepared by slowly adding the alkoxide solution to ammoniated triply distilled deionized water according to the reaction:

$$6Al(OC_3H_7)_3 + 2Si(OC_3H_7)_4 + xH_2O \xrightarrow[\text{or trace of } HNO_3]{H_2O + NH_3}$$

$$2Al_3Si(OH)_{13}. xH_2O + 26C_3H_7 OH \qquad (20)$$

The resulting hydroxyaluminosilicate was repeatedly washed with dry isopropyl alcohol and dried in a vacuum at 60 C^0 for 16 hrs:

$$2Al_3Si(OH)_{13} \xrightarrow[60 \ C^0]{vacuum} 3Al_2O_3. 2SiO_2 + 13 H_2O \qquad (21)$$

At this stage in the preparation, the mixed oxide was amorphous to X-ray diffraction.

REFRACTORY OXIDES

In case of Y_2O_3 stabilized ZrO_2 and HfO_2, high purity spectrographic grade tetrakis tertiary amyloxide of zirconium or hafnium was added to a stoichiometric amount of yttrium tris-isopropoxide. The alkoxides were refluxed in a solution of benzene and tertiary amyl alcohol for 16 hrs. before the hydrolysis reaction to ensure complete mixing. After the solution was cooled to near room temperature, it was slowly added to a slight excess of deionized triply distilled water and stirred vigorously for 1/2 hr. The metal hydroxides were precipitated quantitatively from the solution by simultaneous hydrolytic decomposition of the alkoxides according to the reaction:

$$M(OC_5H_{11})_4 + 2Y(OC_3H_7)_3 + 10 H_2O \longrightarrow$$

$$M(OH)_4 \cdot 2Y(OH)_3 + 4C_5H_{11}OH + 6C_3H_7OH \tag{22}$$

The double decomposition reaction is general for most of the complex oxides, e.g. $SrTiO_3$, and $SrZrO_3$ have been prepared at 99.95% purity in the 5 to 50nm particle size range in this manner by Smith, et al [42].

The intimately mixed hydroxides were repeatedly washed with tertiary amyl alcohol; they were then refluxed with vigorous stirring in dry tertiary amyl alcohol for 2 hrs. to remove the excess adsorbed water as an azeotrope. The mixture was filtered and vacuum-dried at 60 C^0 for 12 hrs. to the corresponding mixed oxide.

$$Zr(OH)_4 \cdot 2Y(OH)_3 \qquad\qquad ZrO_2 \cdot Y_2O_3 + 5H_2O$$

$$\xrightarrow[\text{vacuum}]{60\ C}$$

$$Hf(OH)_4 \cdot 2Y(OH)_3 \qquad\qquad HfO_2 \cdot Y_2O_3 + 5H_2O \tag{23}$$

At this stage in the preparation, the mixed oxide (with maximum agglomerate size 0.5μ) was found to be amorphous by X-ray diffraction analysis. The purity after calcination at 550 C^0 for 1/2 hr. was >99.95%. Atomic absorption analysis showed that nominal compositions of 6 and 7 mol% Y_2O_3 containing ZrO_2 and HfO_2 was maintained respectively.

The aforementioned new concept in powder synthesis may be extended to other oxides of interest such as the rare earth oxides and rare earth oxide doped and undoped ferroelectric and electro-optic materials. Simultaneous hydrolytic decomposition of the mixed metal alkoxide allow the mixing of a residual concentration of a dopant material to be done at something approaching the molecular level.

For example, the barium bis isopropoxide was prepared by the following method:

$$Ba + 2C_3H_7OH \xrightarrow{82\ C^0} Ba(OC_3H_7)_2 + H_2 \tag{24}$$

Titanium tetrakis tertiary amyloxide was formed by the method of Bradley and Wardlaw [43], Mehrotra [23], and Brown and Mazdiyasni [26].

$$TiCl_4 + 4C_3H_7OH + 4NH_3 \xrightarrow[5\ C^0]{C_6H_6} Ti(OC_3H_7)_4 + 4NH_4Cl \tag{25}$$

$$Ti(OC_3H_7)_4 + 4C_5H_{11}OH \xrightarrow[24\ hrs]{reflux} Ti(OC_5H_{11})_4$$

$$+4C_3H_7OH \tag{26}$$

BaTiO$_3$ Powder Preparation. Typically, 1.1 g of recrystallized high-purity barium bis-isopropoxide, Ba(OC$_3$H$_7$)$_2$, and 1.87 ml of frac-tionated titanium tetrakis tertiary amyloxide, Ti(OC$_5$H$_{11}$)$_4$, were dissolved in a mutual solvent such as isopropyl alcohol (C$_3$H$_7$OH) or benzene (C$_6$H$_6$). The solution was refluxed for 2 hrs with vigorous stirring before the hydrolysis reaction. While the stirring was continued, drops of deionized triply distilled water were slowly added. The barium titanate precipitated quantitatively from the solution as a result of simultaneous hydrolytic decomposition of the alkoxides according to the reaction:

$$Ba(OC_3H_7)_2 + Ti(OC_5H_{11})_4 + 3H_2O \xrightarrow{\text{exothermic}}$$

$$BaTiO_3 + 2C_3H_7OH + 4C_5H_{11}OH \tag{27}$$

The reaction was carried out in a CO$_2$-free atmosphere. The hydrated oxide was dried in a vacuum or in a dry helium atmosphere at 50 C^0 for 12 hrs. At this stage the oxide was finely divided, stoichiometric titanate with 50 to 150nm (maximum agglomerate size 1μ) particles and was more than 99.98% pure.

Barium titanate powders doped with 0.1 to 0.34 mol% Ln$_2$O$_3$ or Sc$_2$O$_3$ were also prepared by Mazdiyasni and Brown [44] via hydrolytic decomposition of mixtures of their respective metal alkoxides. Typically, high-purity crystalline Ba(OC$_3$H$_7$)$_2$, colorless mobile Ti(OC$_5$H$_{11}$)$_4$, and a crystalline lanthanide tris-isopropoxide, Ln (OC$_3$H$_7$)$_3$, or scandium isopropoxide, Sc(OC$_3$H$_7$)$_3$, were dissolved in a mutual solvent such as isopropanol. The alkoxide mixture was hydrolyzed by adding it to a water-alcohol solution and further processed as described above.

$$Ti(OC_5H_{11})_4 + Ba(OC_3H_7)_2 + 2M(OC_3H_7)_3 + 6H_2O \longrightarrow$$

$$BaTiO_3 \cdot M_2O_3 + 4C_5H_{11}OH + 8C_3H_7OH \tag{28}$$

Where M is La, Nd, or Sc.

PLZT

Another example of perovskite that may be cited is the synthesis of a more complex quaternary mixed oxide with nominal lead zirconate-lead titanate molar ratio of 65/35 containing 10 at % lanthana, electrooptics, PLZT by Brown and Mazdiyasni [45] and Heartling and Land [46] and Heartling [47].

$$0.9Pb(OR'')_2 + 0.1La(OR)_3 + 0.63375Zr(OR')_4 + 0.34125Ti(OR')$$

$$+3H_2O \xrightarrow{xsH_2O} Pb_{0.9}La_{0.1}Zr_{0.63375}Ti_{0.34125}(OH)_6 \tag{29}$$

$$+0.3ROH+3.9R'OH+1.8R''OH+xsH_2O$$

$$Pb_{0.9}La_{0.1}(Zr_{0.63375}Ti_{0.34125})(OH)_6 \xrightarrow[\text{vacuum}]{60 \ C^0}$$

$$Pb_{0.9}La_{0.1}(Zr_{0.63375}Ti_{0.34125})O_3+3H_2O \tag{30}$$

The crystallite morphology observed in these ferroelectric materials are rectangular or cubical in symmetry with particle diameters in the range of 5-35nm.

Fundamental understanding of the relation of microstructural effects to the behavior of refractory, ferroelectric, or piezoelectric ceramic materials is still quite limited when compared, for example, with that for metals and metal alloys. Several problems experienced with pure oxides and mixed oxides may be inherent in the material, but other problems, just as serious and limiting, can be cited. Of particular concern is the lack of control over the many variables during processing from powder synthesis to final compact firing. Of major importance are impurities present in the starting powders or inadvertently introduced during processing and their influence on the properties of refractory or electronic materials. Although the presence of such impurities may not be a priori adverse, it is important to understand their influence on the final properties or to eliminate them completely, if necessary.

Table V Emission Spectrographic Analysis of Fine Particulate Mixed Oxides

(ppm)

Elements	(Mullite)	$HfO_2 \cdot Y_2O_3$	$BaTiO_3$	PLZT
Na	<300	nd< 3	nd< 10	500
Si	Major	<10	100	20
Ba	nd< 1	nd< 1	Major	nd< 5
Mg	1	nd< 3	nd< 30	nd< 30
Mn	nd< 1	nd<10	nd< 30	30
B	1	nd<10	nd< 10	nd< 10
Sn	nd< 3	nd<10	30	nd< 30
Pb	nd< 1	nd<10	nd< 10	Major
Al	Major	nd< 3	nd< 10	nd< 30
Fe	nd< 5	nd<10	nd<100	nd<100
Ni	nd< 5	nd<10	nd< 5	nd< 30
Co	nd<30	nd< 3	nd< 5	nd< 10
Cr	nd<10	nd<10	30	nd< 30
W	nd< 5	nd<10	10	nd< 10
Ca	nd<10	nd<10	10	30
Ti	nd<10	nd< 5	Major	Major
Zr	nd< 5	100	nd< 5	Major

NOTE: nd< = not detected less than

Mazdiyasni [48] has demonstrated with relative ease the elimination of major impurities in various oxides and mixed oxides. Spectrographic analysis for impurities in the mixed oxides (Table V) of the calcined powders at 500-800 C^0 for 1/2 hr. showed no measurable impurities present in two or more representative batches of mixed alkoxides decomposed to the mixed oxides. Separate analyses for the oxides synthesized therefter demonstrated that high purity levels were maintained on a continuing basis. None of the unlisted impurities were higher in concentration than those shown in Table V.

SURFACE AREA, CRYSTALLITE SIZE AND MORPHOLOGY

The mullite powder as-prepared and calcined at 600 C^0 statically for 1 and 24 hrs. and dynamically for 24 hrs. examined by electron microscopy. The powders were dispersed ultrasonically in a solution of absolute ethyl alcohol; a small amount of the dispersion was then placed in a Freon nebulizer and sprayed onto a carbon substrate on a Cu mesh screen.

In Fig. 14 needlelike crystallitities of the very fine as-prepared particulates are evident. When the powders are calcined at 600 C^0 for 1 and 24 hrs. without tumbling and for 24 hrs. with

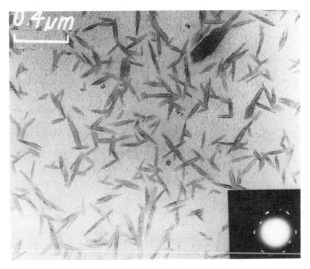

Figure 14. Electron Micrograph of As-Prepared Mullite Powder.

tumbling in Fig. 15, crystallite growth occurs (agglomeration of small particles followed by rapid growth) leading to a larger but well-defined acicular or prismatic particulates. Electron diffraction patterns are also included in Fig. 14 and 15. The as-prepared powders are extremely active, with a surface area of 550 $M^2/g \pm 10\%$; however, the surface area is reduced to 280 Fig. 15, $M^2/g \pm 10\%$ when the powders are calcined at 600 C^0 for 1 hr. This reduction in

surface area is attributed to nucleation and growth of the finer
particles to larger networks of needlelike crystallites.

In the alkoxy-derived mullite at near-room temperature conditions and calcined to higher temperatures, the needlelike morphology
of the crystallites is retained in the absence of a liquid phase,
contrary to the finding of Lohre, et al [49]. The intimate mixing
of highly active fine particulates in the alkoxide decomposition

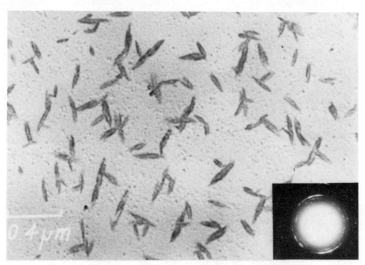

Figure 15.   Electron Micrograph of Mullite Powder Calcined at
600 C^0 for 1 hr.

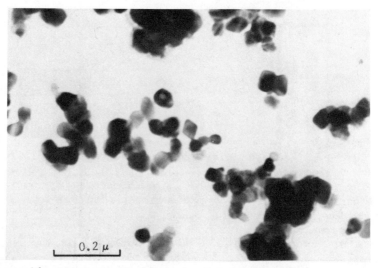

Figure 16. Electron Micrograph of 1 M% $Y_2O_3 \cdot HfO_2$ Powders.

process described is apparently responsible for formation of the acicular morphology which is characteristic of crystalline mullite.

Mazdiyasni and Brown [50] examined the particle size of $HfO_2 \cdot Y_2O_3$, powders by electron microscopy of a powder ultrasonically dispersed in a solution of absolute ethyl alcohol. After the powder was dispersed, a small amount of the dispersion was placed in a Freon nebulizer and sprayed onto carbon substrated on a Cu mesh screen. The as-prepared powders were extremely fine (particle size 4nm) Fig. 16. Because of this small particle size, a meaningful particle size distribution curve could not be determined accurately for any of the as-prepared powders. However, after they were calcined at 550-600 C^0 for 1 hr. nucleation and crystallite growth occurs leading to larger crystallite. In Fig. 17 several of the particles contain dark parallel bands. This phenomenon was observed to a greater or lesser degree in all powders examined. Rau [51], who observed this effect in electron micrographs of BeO, attributed it to stacking faults within the crystals. The dark bands are produced by the interaction of the electron beam with the fault zones in properly oriented crystallites.

A typical particle size distribution plot for the powder of Fig. 17, as measured by electron microscopy and calculated by

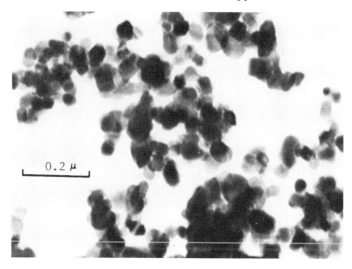

Figure 17.   Electron Micrograph of Fully Stabilized $HfO_2 \cdot Y_2O_3$ Showing Stacking Faults.

computer, is shown in Fig. 18. This powder had an arithmetic mean particle size of 230nm; 90% of the particles were 35nm in diameter. Rhodes and Haag [52] independently studied particle size distribution of alkoxy-derived Y_2O_3 stabilized ZrO_2. A typical cumulative particle size distribution curve for two different batches of powders as shown in Fig. 19 which is a common characteristic of most ceramic powders prepared by this technique.

Using the continuous flow technique of Nelsen and Eggersten [53] and the single-point BET method with an accuracy of 10 to 15%, BET specific surface areas of the $HfO_2 \cdot Y_2O_3$ powders were measured by Mazdiyasni and Brown [50]. Several runs were made for each sample at different temperatures. Fig. 20 illustrates the effect of calcination temperature on the specific surface area of these powders. The as-prepared powders are extremely active, with surface areas 210 M^2/g. Up to 400 C^0, surface area is greatly reduced, to a large extent because of the nucleation and growth of the finer particles to larger crystallites. From 400 to 750 C^0, the surface area decreases only slightly to 120 M^2/g. A similar phenomenon

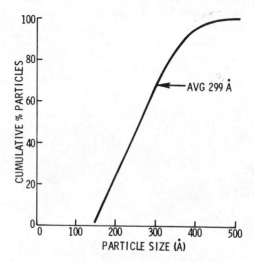

Figure 18.   Particle Size Distribution Plot of Y_2O_3 Stabilized Hafnia (Calcined 750 C^0 for 24 hrs.).

was observed by Rau [51], who attributed it to a temperature range in which crystallite growth was suppressed while the misoriented portions of the faulted lattice began to align themselves. Above 800 C^0, the surface area decreases rapidly with increasing temperature to 10 M^2/g at 1200 C^0. This curve suggests that calcination temperature of 750 C^0 are desirable to retain a relatively high surface active powder for best sintering results.

Mazdiyasni and Brown [44] studied the particle sizes of four typical rare earth oxide powders by electron microscopy. The powders were dispersed ultrasonically in absolute ethyl alcohol; the solution was then placed in a Freon nebulizer and sprayed onto a C substrate on a Cu mesh screen. The crystallite size and morphology of representative as-prepared alkoxy-derived lanthanide oxides

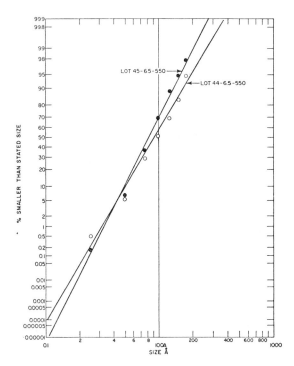

Figure 19. Particle Sized Distribution of 6.5 M% Y_2O_3. ZrO_2
Courtesy of W. H. Rhodes.

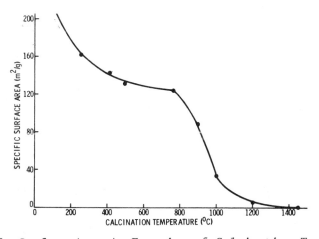

Figure 20. Surface Area As Function of Calcination Temperature
for $Y_2O_3 \cdot HfO_2$ Powder.

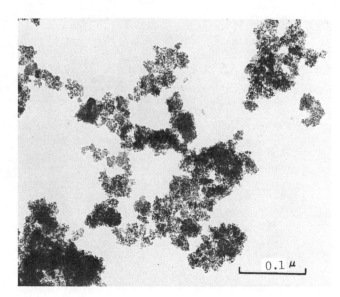

Figure 21. Typical Electron Micrograph of As-Prepared Lanthanide
Oxide.

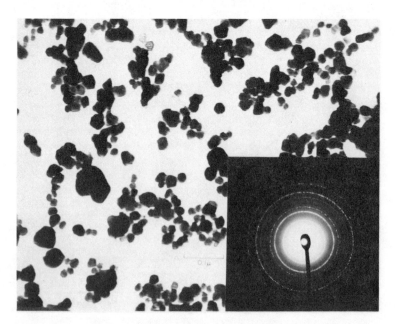

Figure 21A. Electron Micrograph of Ln_2O_3 Calcined With
Tumbling at 800 C^0 for 24 hrs.

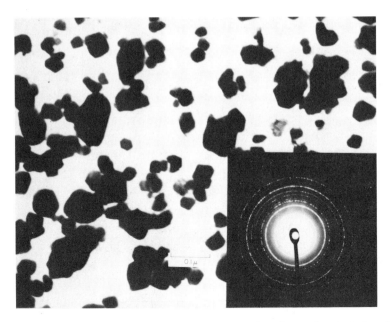

Figure 22.   Typical Electron Micrograph of Ln_2O_3 Calcined
Statically at 800 C^0 for 24 hrs.

Fig. 21 and calcined at 800 C^0 for 24 hrs. with and without tumbling
were determined by electron microscopy (Figs. 21A and 22).   The
particles are cubic and hexagonal.   The electron diffraction patterns for these powders compared favorably with the X-ray data.   It
is interesting to note that after the powders were calcined at
800 C^0 with tumbling for 24 hrs. the particles grew to 18 to 28nm.
Particle size distribution curves for the powders calcined with and
without tumbling at 800 C^0 for 24 hrs. for particles >1nm in diameter as measured by electron microscopy and calculated by computer
are given in Figs. 23 and 24, respectively.   The largest particle
diameter range observed in the powder calcined without tumbling at
the same temperature for the same length of time, the particle size
increased to 60 to 100nm in diameter.   Similarly, the arithmetic
mean particle sizes of these powders differed considerably, giving
further evidence that tumbling is an effective means of controlling
growth of primary crystallites into larger, more uniform particles.
The surface areas measured for the alkoxy-derived powders as-prepared, calcined, and calcined with tumbling are listed in Table VI.
These data illustrate that the heat treatment of the

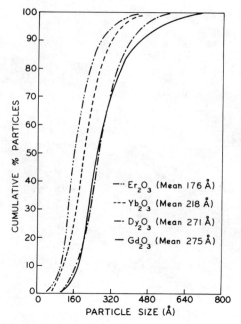

Figure 23. Particle Size Distributions of Some Lanthanide Oxides
Calcined, with Tumbling, at 800 C^0 for 24 hrs.

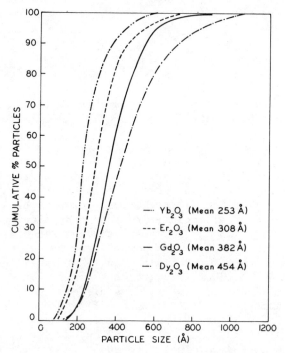

Figure 24. Particle Size Distributions of Some Lanthanide Oxides
Calcined without Tumbling at 800 C^0 for 24 hrs.

Table VI   Surface Area Measurements of Some Lanthanide Oxides

Compound	Specific Surface Area (m²/g)		
	As Prepared	Calcined*	Calcined* - Dynamic
Gd_2O_3	72, 98	10. 26	27.83
Dy_2O_3	110. 33	7. 11	21. 23
Er_2O_3	51. 06	14. 79	42. 50
Yb_2O_3	172. 14	24. 29	31. 02

*800°C for 24 h.

powders (for BET studies) in a tumbling furnace resulted in much smaller particles and correspondingly larger surface areas compared with those of the powder calcined statically under the same conditions. The great reduction in surface area with calcination is, to a large extent, a result of crystallite growth, the particles having grown from 1 to 3nm to 60 to 100nm in size.

Mazdiyasni, et al [15] determined the particle size of the as-prepared $BaTiO_3$ powder by electron microscopy of a dispersion of the powder in a solution of 0.75% Parlodion in amyl acetate sprayed onto a carbon substrate on a copper mesh screen. The results are shown in Fig. 25. Rectilinear symmetry of several of the very fine as-prepared particulates is evident. Fig. 26 is an electron micrograph of the same powder after calcination at 700 C° for 60 min. Here the rectilinear symmetry of the particulates is more apparent. The electron diffraction patterns, showing the diffraction lines for the as-prepared and calcined $BaTiO_3$ powders, are also included in Figs. 25 and 26.

The particle size distribution curves for particle 1nm in diameter as measured with the electron microscope are given in Fig. 27. The largest particle observed in the as-prepared powder was 15nm in diameter. The mean particle size was approximately 5nm, and 90% of the particles were less than 10nm in diameter. The mean particle size for the calcined powder (Fig. 27), however, increased to ∿30nm, and 80% of the particles were less than 35nm in diameter. The as-prepared $BaTiO_3$ powder has a specific surface area greater than 150 M^2/g, and the calcined powder has a specific surface area of approximately 50 M^2/g.

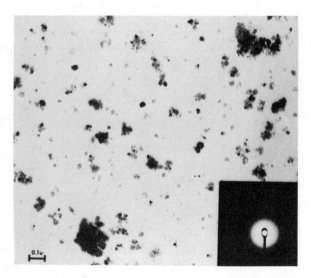

Figure 25. Electron Micrograph of As-Prepared BaTiO$_3$.

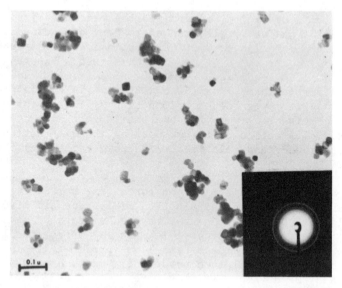

Figure 26. Electron Micrograph of BaTiO$_3$ Calcined at
700 C^0 for 1 hr.

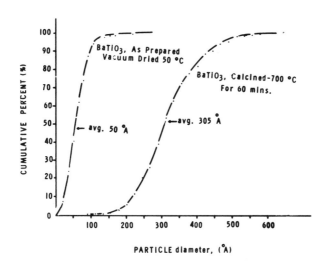

Figure 27. Particle Size Distribution of High-Purity BaTiO$_3$.

AGGLOMERATION

Rhodes and Haag [52] have shown that agglomerate size distribution is more dependent on the method of preparation of alkoxy-
derived powders. Fig. 28 shows the distribution curves for two
batches of "zyttrite," 6.5m% Y$_2$O$_3$ stabilized ZrO$_2$ which indicate
that the alcohol washed powder yields aggregates which are only
about one-half as large as those prepared with a water wash.

Hoch and Nair [54] and Haberko [55] have demonstrated that
calcination of gels washed with alcohol resulted consistantly in a
powder composed of agglomerates which were much more porous with low
powder density and mechanically weaker than agglomerates obtained
from gels washed with water. Apparently these weak agglomerates
could be easily crushed under compaction pressure, in contrast to
the behavior of the strong agglomerates obtained from gels washed
with water. Haberko [55] suggests that, in compacts of the strong
agglomerates two populations of pores very different in size are
present; this characteristic of the strong agglomerates is perhaps
related to their resistance of crushing. The presence of the two
different inter- and intra-agglomerates pore size lead to severe
fluctuations of the sintering driving force across the body and the

resultant densification of the body is poor. However, the collapse
of weak agglomerates observed on compaction results in a very fine
uniform microstructure of the compact.

    Rhodes and Haag [52] have also shown the effect of temperature
calcination upon the size of the crystallite size. Fig. 29 shows
the mean crystallite size as determined by electron microscopy and
by X-ray diffraction of several Zyttrite powder samples. The
crystallite size increases nearly exponentially with

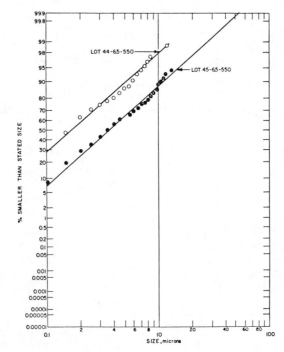

Figure 28.   Electron Microscopy Agglomerate Size Distribution for
             Zyttrite Powder (Courtesy of W. H. Rhodes).

calcination temperature in the temperature range indicated. De-
agglomeration can be accomplished by grinding in a mortar and pestle
or in a Waring blender or by dispersion by ultrasonic vibration in
an aqueous suspension. Ball milling is relatively ineffective in
altering the agglomerate size distribution in these powders.

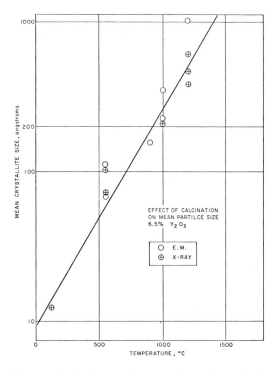

Figure 29. Effect of Calcination on Mean Particle Size (Courtesy of W. H. Rhodes).

AGGLOMERATE AND PARTICLE SIZE EFFECTS ON SINTERING

A calculation of particle size effects on initial stage sintering of two stabilized zirconia powders based on volume diffusion control came close to predicting the observed differences in sintering temperature for equivalent sintering rates, Rhodes [56]. However, Rhodes [56] claims that "the final stage sintering experiments on a de-agglomerated powder, as well as the slope of the initial stage kinetics experiment, suggested that: (1) agglomerates interfere with initial stage sintering, and (2) grain boundary or surface diffusion may dominate initial stage sintering. Therefore, it is clear that the Herring scaling law is appropriate for predicting the potential benefit in sintering from particle size reduction. The benefit may be larger or smaller than predicted if the dominant sintering mechanism changes concurrently. This clearly is a limitation in the application of the Herring scaling law rather than the law itself."

The full potential of fine powders in terms of increased reactivity and reduced sintering cycles for final state sintering is not realized unless agglomerates are eliminated and high green

Figure 30.   -0.2μm Grain Size Achieved by Sintering De-agglomerated
             Powder at 1100 C^0 for 1 hr. (Courtesy of W. H. Rhodes).

densities achieved. It is probably not possible to achieve high
green density without elimination of agglomerates.

     The initial, intermediate, and final stage sintering of fine
crystallite yttria stabilized zirconia was studied by Rhodes [56].
Experiments were conducted on powder lots of differing agglomerate
size, as well as one specially prepared agglomerate-free powder.
Initial stage sintering kinetics were compared with a sintering
study on larger crystallite size calcia stabilized zirconia to access
the Herring scaling law. It was found that agglomerates limit
attainable green density, interfere with the development of microstructure, impede initial stage sintering kinetics, and limit the
potential benefit of fine crystallites on final stage sintering.
An agglomerate-free powder centrifuge cast to 74% green density by
Rhodes [56] was sintered to 99.5% of theoretical density in a 1
hr-1100 C^0 cycle, Fig. 30, which is ∿300 C^0 lower than necessary
for an agglomerated but equal crystallite size powder.

LOW TEMPERATURE SINTERING OF ALKOXY-BASED POWDERS

Yttria Stabilized Zirconia and Hafnia

     Lynch, et al [29] have shown in Table VII that as low as a 2 M/O
addition of yttria to zirconia produces substantially a tetragonal
solid solution from these powders. The calcination temperature for

Table VII Phase Stabilization of Spectrographic Grade ZrO_2

Additive	Mole %	Estimated stabilization (X-ray data)		Heat treatment
Dy₂O₃	1	96% M	4% T	2000°C, 30 h
	3	62% M	38% T	2000°C, 22 h
	3	10% M	90% T	2000°C, 30 h
	3	5% M	95% T+C	2000°C, 44 h
	5	50% M	50% T+C	1700°C, 15 h
	5	10% M	90% T+C	1700°C, 16 h and 1900°C, 4 h
Yb₂O₃	1	70% M	30% T	2000°C, 28 h
	3	3% M	97% T	2000°C, 28 h
	4	1% M	99% T	2000°C, 28 h
	5	20% M	80% T+C	1700°C, 16 h and 1900°C, 4 h
Y₂O₃	4	50% M	50% T	2000°C, 72 h
	6	25% M	75% T	2000°C, 24 h
	8	100% T+C		2000°C, 24 h
	10	100% T+C		1800°C, 24 h
	12	100% C		1800°C, 24 h
(Alkoxy) Y₂O₃	2	20% M	80% T+C	800°C, 30 min
	3	10% M	90% T+C	800°C, 30 min
	4	100% T+C		800°C, 30 min
	5	100% T+C		800°C, 30 min
	6	100% C		800°C, 30 min

M = Monoclinic phase   T = Tetragonal phase   C = Cubic phase

this powder was only 800 C^0 for 30 min. From 3 to 5 M/O additive the tetragonal phase is replaced by the cubic phase and the tetragonal is barely detectable. At 6 M/O only cubic phase is present. Because of similar covalent radii and electronegativities, Y, Dy, and Yb form similar compounds and, with ZrO_2 or HfO_2, their stabilization behavior is virtually the same within experimental error. It is interesting to compare, therefore, some data on the stabilization of commercial spectrograde ZrO_2 powders with DY_2O_3 and Yb_2O_3. The difference in heat treatments for commercial and alkoxy-based powders to achieve apparent equilibrium conditions in phase stabilization is quite evident. An X-ray diffraction pattern of the 6 M/O Y_2O_3 stabilized ZrO_2 is seen in Fig. 31. It shows that the alkoxy-based powder calcined at 500 C^0 for only 30 minutes (dotted line curve) is already in the cubic phase with no evidence of the monoclinic phase present.

Fig. 31 also show results (solid line curve) after a body either cold pressed at 20-25 ksi in a steel die or cold isotatic pressed at 10 ksi from this powder was sintered at 1450 C^0 for 3-4 hrs. The diffraction peaks are sharper indicating a higher degree of crystallinity (than after calcining) and full stabilization of the cubic phase with no evidence of any residual monoclinic phase.

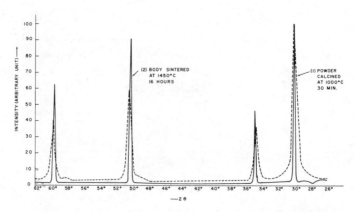

Figure 31. X-Ray Diffraction of "Zyttrite" Powder and Sintered
Body.

In Figs. 32 and 33 the microstructures of a 6 M/O Y_2O_3
stabilized ZrO_2 and a 7 M/O Y_2O_3 stabilized HfO_2 body respectively
prepared by this technique. The electron micrographs show that
there are negligible internal porosity and insignificant porosity
at grain boundaries of the materials. Both materials are fine
grained having the microstructure in which the grains are from 2-5µm
across in their longest dimensions. Such a grain structure
ordinarily cannot be obtained for an oxide as refractory as ZrO_2 or
HfO_2 without heat treatment at temperatures ranging from 1800-
2200 C^0 and firing times of 24 to 48 hrs. or longer. This makes
achievement of low porosity concomitant with maintaining a fine,
uniform grain size extremely difficult. It is apparent that the
mixed oxide approaches in behavior, in the solid state, reaction
conditions indicative of high mobility, resulting in extremely rapid
reactions. Aggregates which are equivalent to several molecules of
the particular oxide are possibly the particles entering into
reaction.

PRESSURE, GREEN DENSITY RELATIONSHIPS:

Rhodes and Haag [52] Mazdiyasni, et al [19] and Hoch and Nair
[54] have observed that in practice no advantage is gained in final
density of a green compact made of ultrafine ceramic powders by
exceeding 45,000 psi. Sintered parts pressed at 10-45 ksi all reach
∿95% of theoretical density in a matter of a few minutes at several
hundred degrees lower than necessary for the conventional ceramics.
However, duplex structure and porosity next may arise from an
incorrect rate of decomposition of the mixed alkoxides to the mixed
oxides or the conditions of calcination of the as-prepared powder.
In either case, particle size uniformity is not achieved before
processing, and grain size in the sintered body varies widely.

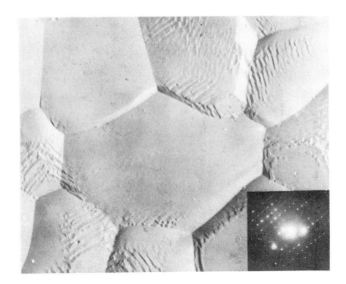

Figure 32. Electron Micrograph of Cold Pressed and Sintered 6 M% $Y_2O_3 \cdot ZrO_2$.

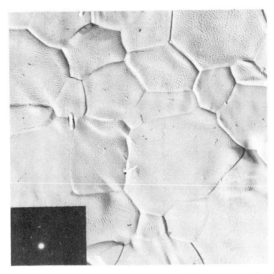

Figure 33. Electron Micrograph of Cold Pressed and Sintered 7 M% $Y_2O_3 \cdot HfO_2$.

In general the high surface activity associated with these
powders make possible relatively low temperature sintering, reactive
hot pressing and vacuum hot pressing of the powder compact to near
theoretical density and uniform fine grain size bodies. The
influence of residual concentrations of alloying elements as major
impurities and their effect on microstructure of both high temperature refractory and electronic oxide ceramics are studied in great
detail by Mazdiyasni and co-workers. Optical, transmission electron
and scanning electron microscopy are used to show the unique microstructural feature observed in these materials Figs. 34-37.

ALKOXY-DERIVED SOLS AND GELS

Yoldas [57] has shown that the process of making sols and gels
from alkoxide involve three basic steps, 1) Hydrolysis of metal
alkoxide, 2) Peptization of the resultant metal hydroxide to a clear
sol, and 3) Gel formation. In the first step, the most important
consideration is the prevention of stable hydroxide formation to
assure the success of subsequent steps. For example, when the
aluminum alkoxides are hydrolyzed with hot water, a stable crystalline
monohydroxide forms. However, when hydrolysis is made with cold
water the monohydroxide formed is largely amorphous and easily
connected to tri-hydroxide, bayerite on aging. Only the monohydroxide forms of aluminum can be peptized to a clear sol. The
requirement for the formation of a gel is sufficient concentration
of the sol. As the sol is concentrated the particle movement passes
from a transitory to an oscillatory one and later ceases to be
evident - as a state where the sol has set to a gel. According to
Yolda's [55] observation in some cases, electrolyte addition, hydra-
tion, and aging also result in gel formation.

Thomas, et al [58], Yoldas [59-60], have developed a process
based on cohydrolysis of silicon alkoxide and metal alkoxides to
produce oxide glass precursors which sinter in the temperature
range of 300-600 C^0 and melt to uniform glass at temperatures below
860 C^0. Fig. 38 is a monolitic alkoxy derived sol-gel specimen of
porous alumina sintered at 1000 C^0 and Fig. 39 is an alkoxy-derived
25% titania-silica glass made by low temperature chemical polymerization which shows no crystallinity until above 1200 C^0. The "low
temperature" glass synthesis offers unlimited potential for the
production of high purity specialized glass films and no doubt have
a considerable impact on glass technology in an energy scarce
future.

CONCLUSION AND SUMMARY

The alkoxide decomposition process for the preparation of
single phase and homogeneous mixed oxides of high purity and fine
grain size has been demonstrated. The extremely high purity in
excess of 99.95% can be obtained for a wide variety of oxides.

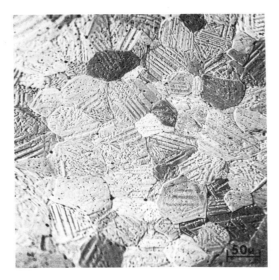

Figure 34. Scanning Electron Micrograph of BaTiO$_3$ Showing 90^0
and 180^0 Domain Pattern.

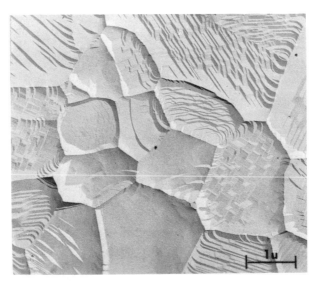

Figure 35. Electron Micrograph of La$_2$O$_3$ Doped BaTiO$_3$ Cold Pressed
and Sintered at 1300^{0}C^0 for 4 hrs.

Figure 36. High Resolution Electron Micrograph of the La_2O_3 Doped $BaTiO_3$ Grain Boundary Showing No Evidence of Porosity or Impurities.

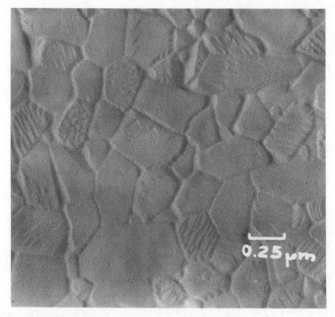

Figure 37. Electron Micrograph of Reactively Hot Pressed Mullite at 3.5 MPa and 1500 C^0 for 1/2 hr.

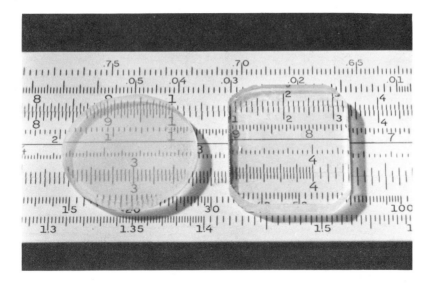

Figure 38.   Transparent Porous Alumina Prepared by Chemical
             Polymerization of Alkoxide, Courtesy of Yoldas.

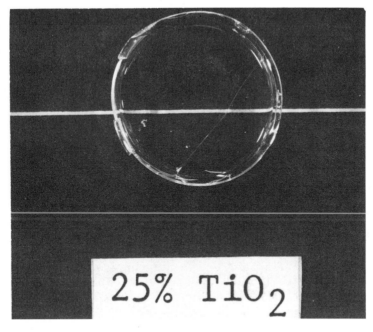

Figure 39.   75% SiO$_2$, 25% TiO$_2$ Glass Produced by Chemical
             Polymerization of Alkoxide, Courtesy of Yoldas.

Particle size range below 50nm are easily achieved while maintaining
this exceptionally high level of purity for refractory and electronic
oxide ceramics.   The oxide powders have extremely high activity during
subsequent solid state reactions making feasible low-temperature
firing processess to prepare high density oxides ceramics.   The
alkoxy technique is applicable to all mixed oxides where the metal
alkoxides are readily prepared or commercially available.   Advantages
of the process, lie in the intimate mixing of the highly active
powders as prepared facilitating further processing steps while
maintaining stoichiometry and purity throughout the forming process.

Thermal decomposition of metal alkoxide provides a particularly
unique approach to the production of thin films on various substrates.
And hydrolytic decomposition of metal alkoxides provide an inexpensive, economical route toward the production of high purity fine
particulate powders on a large scale.

ACKNOWLEDGEMENTS

The author would like to thank all of his co-workers for their
contribution and special thanks to Miss Sylvia R. Hatch for her work
in preparation of this paper.

REFERENCES

1.   Bradley, D. C., Metal-Organic Compounds, Advances in Chemistry
          Series 23, pp. 10-36 American Chemical Soc. (1959)
2.   Hunter, E. A., U. S. Patent 2, 570, 058 (1951)
3.   Hunter, E. A. and Kimberlin, C. N., U. S. Patent 2,656,321 (1953)
4.   Kimberlin, C. N., U. S. Patent 2,953,314 (1952)
5.   Bradley, D. C., Progr. Inorg. Chem. 2, 303 (1960)
6.   Bradley, D. C., in Jolly, Preparative Inorganic Reaction, Vol
          II. p. 169, Wiley Intersci. Publ. New York (1962)
7.   Mehratra, R. C., Inorg. Chem. Acta, Rev. 1, 99 (1967)
8.   Brown, L. M., Mazdiyasni, K. S., Anal. Chem. 41, (10) 1243
          (1969)
9.   Mazdiyasni, K. S., Lynch, C. T., Smith, J. S., J. Amer. Ceram.
          Soc. 50, 532 (1967)
10.   Mazdiyasni, K. S., Proceeding of the Vl International Symposium
          on Reactivity of Solids, Eds. Mitchell, R. W., Devries, R. C.,
          Roberts, R. W., Cannon, P. J., Wiley, Intersci. Publ. pp.
          115-125, Aug. 25-30, (1968)
11.   Mazdiyasni, K. S., Brown, L. M., J. Amer. Ceram. Soc. 53 (1) 43
          (1970) Mazdiyasni, K. S., Lynch, C. T., in P. Popper, Special
          Ceramics p. 115 Academic Press, New York (1965)
12.   Mazdiyasni, K. S., Lynch, C. T., Smith, J. S., J. Amer. Ceram.
          Soc. 48 (7) 372 (1965)
13.   Aboaf, J. A., J. Electrochem. Soc. Solid State Science 114 (9)
          948-52 (1967)

14. Harwood, J. H., Chem. Process Eng. 48, 100 (1967)
15. Mazdiyasni, K. S., Dolloff, R. T., Smith, J. S., J. Amer. Ceram. Soc. 52 523 (1969)
16. Mazdiyasni, K. S., Lynch, C. T., Smith, J. S., J. Amer. Ceram. Soc. 49 (5) 286 (1966)
17. Mazdiyasni, K. S., Lynch, C. T., Smith, J. S., Inorg. Chem. 5, 342 (1966)
18. Bradley, D. C., Mehratra, R. C., Gaur, D. P., Metal Alkoxides 411 pp. Academic Press, New York (1978)
19. Mazdiyasni, K. S., Lynch, C. T., Smith, J. S., J. Amer. Ceram. Soc. 50 (10) 532 (1967)
20. Bradley, D. C., Mehratra, R. C., Wardlaw, J., Chem. Soc. (London) 1963 (1953)
21. Herman, D. F., U. S. Patent 2, 654, 770 (1953); 2, 655, 523 (1953)
22. Anand, S. K., Singh, J. J., Multani, B. D. Bain, Israel J. Chem. 7, 171 (1969)
23. Mehratra, R. C., J. Amer. Chem. Soc. 76, 2266 (1954)
24. Pauling, L., The Nature of Chemical Bond, 3rd Ed. pp. 65-105, Cornell University Press, Ithaca, New York, (1960)
25. Hammy, M. B., Smith, C. P., J. Amer. Chem. Soc. 68, 171 (1964)
26. Brown, L. M., Mazdiyasni, K. S., Inorg. Chem. 9 2783 (1970)
27. Adkins, H., Cox, J., J. Amer. Chem. Soc., 60, 1151 (1938)
28. Mazdiyasni, K. S., Lynch, C. T., ASD-TDR 63-322 AFML, WPAFB, OH (1963)
29. Lynch, C. T., Mazdiyasni, K. S., Smith, J. S., Proceeding of the 3rd International Symposia on High Temperature Technology, Eds, Thompson, H. W., Weedon, B.C.L., Cullis, C. F., Gujral, P.D., Butterworth 393 (1967)
30. McDevitt, N. T., Baun, W. K., J. Amer. Ceram. Soc. 47 (12), 622 (1964)
31. Phillipi, C. M., Mazdiyasni, K. S., J. Amer. Ceram. 54 (5), 254 (1971)
32. Hair, M. L., Infrared Spectroscopy in Surface Chemistry pp. 200-201, Marcel Dekker, Inc., N. Y. (1967)
33. Barker, A. S., Phys. Rev. 132 (4) 1474 (1963)
34. Ruh, R., Wysong, E. D., Tallan, N. M., Ceramic Age Now 44-48 (1967) Senha, R. N. P., Sci. Cult. (Calcutta) 25 (10) 594 (1960)
35. Garvi, R. C., J. Phys. Chem. 69 1238 (1965)
36. William, H. J. and Shackly, W., Phys. Rev., 57, 178 (1949)
37. Deckker, A. J., Solid State Phys. 540 pp. 475-476 (1957)
38. Burke, J. E., Turnbull, D., Progress in Metals Physics 3, 220 (1952)
39. Coble, R., J. Applied Phys. 32, 787 (1961)
40. Anderson, P. J., Morgan, P. L., Trans Farad., Soc. 60, 930 (1964)
41. Mazdiyasni, K. S., Brown, L. M., J. Amer. Ceram. Soc. 55 (1), 548 (1972)
42. Smith, J. S., Dolloff, R. T., Mazdiyasni, K. S., J. Amer. Ceram. Soc. 53, 91 (1970)

43. Bradley, D. C., Warlaw, W., Nature 165, 75 (1950) Bradley, D. C., Wardlaw, W., J. Chem. Soc. (London) 208 (1951)

44. Mazdiyasni, K. S., Brown, L. M., J. Amer. Soc. 54 (11) 539-543 (1971)

45. Brown, L. M., Mazdiyasni, K. S., J. Amer. Ceram. Soc. 55 (1) 541 (1972)

46. Haertling, G. H., Land, C. E., J. Amer. Ceram. Soc. 54, (1) 1 (1971)

47. Haertling, G. H., J. Amer. Ceram. Soc. 54, (6) 303 (1971)

48. Mazdiyasni, K. S., Proceeding of the Second International Conference Fine Particles, Ed. Kuhn, W. E., Electrochem. Soc. Princeton, N. J. pp. 3-27 (1974)

49. Lohre, W. and Urban, J., Ber Dent. Keram. Ges., 6, 249-51 (1960)

50. Mazdiyasni, K. S., Brown, L. M., J. Amer. Ceram. Soc. 53 (11), 590 (1970)

51. Rau, R. C., J. Amer. Ceram. Soc. 47 (4), 179 (1964)

52. Rhodes, W. H., Haag, R. M., Tech Report AFML-TR-70-209 Sep. 1970

53. Nelsen, F. M., Eggertsen, F. T., Anal. Chem. 30, 1387 (1958)

54. Hoch, M., Nair, K. M., Ceramurgia, International 2, n.2 88 (1976)

55. Haberko, K., Ceramurgia, International 5, n.4 148 (1979)

56. Rhodes, W. H., J. Amer. Ceram. Soc. (1981)

57. Yoldas, B., Búll. Amer. Ceram. Soc. 54 (3) (1975)

58. Thomas, I. M., U. S. Patent 3,799,909 (March 26, 1974)

59. Yoldas, B. E., J. Materials Science 12, 1203 (1977)

60. Yoldas, B. E., J. Non-Crystalline Solids 38 and 39, 81 (1980)

"SHS" SELF-SINTERING OF MATERIALS IN THE

TITANIUM-BORON-CARBON SYSTEM

N. D. Corbin, T. M. Resetar, and J. W. McCauley

Materials Characterization Division
Army Materials and Mechanics Research Center
Watertown, Massachusetts 02172

ABSTRACT

Refractory ceramics (e.g., borides, carbides, and nitrides) are
families of materials having desirable properties for a variety of
DoD applications. Their key properties are high hardness and strength,
along with resistance to heat, corrosion, and wear. Conventional
methods for producing these materials as monolithic ceramics can be
very expensive. An alternative to these conventional methods consists of using the heat generated in exothermic reactions to simultaneously form the desired phases and densify the product. This
processing technique is termed "Self-Propagating High-Temperature
Synthesis (SHS)." This technique results from the large exothermic
reaction between the elemental constituents of a refractory compound.
Several key advantages of SHS over conventional processing methods
are: 1) a high rate of synthesis (seconds versus hours or days);
2) attainment of equilibrium; 3) improved purity due to loss of volatiles; and 4) no need for expensive energy-consuming furnaces.

The Ti-B-C system contains at least four high-temperature refractory phases (TiB_2, TiB, TiC, and B_4C) which can be produced by
SHS. In particular, compositions along the Ti-B_4C composition join
have been critically investigated. Through extensive characterization of precursor powders our studies show that small variations in
the titanium particle size distributions have a large effect on the
processing aspects of SHS (reaction initiation, rate, weight losses)
and influence the final product characteristics (microstructure,
density, phase composition).

INTRODUCTION

Conventional fabrication of refractory materials generally involves two processing steps. First, powders having the desired chemistry, phase, and physical characteristics must be produced. Second, the powder must be formed into fully dense, highly bonded material. This second step is generally carried out by using high temperature heat treatment (1500°C to 2200°C) in controlled atmospheres and pressures for extended periods of time. These conventional methods for producing monolithic ceramics are very expensive, especially when carried out in batch modes.

An alternative to these conventional methods consists of using the heat generated in exothermic reactions to simultaneously form the desired phases and densify the product. This processing technique termed "Self-Propagating High-Temperature Synthesis (SHS)" has been investigated in the Soviet Union since 1967; over 200 phases have been produced by this technique.[1] These exothermic, gasless reactions have been utilized for many years in the United States for a variety of heat requirements (e.g., thermal batteries) and pyrophoric applications.[2] However, it is only recently that they have been explored as a method for processing dense material.[3] Earlier works by Walton and Poulos[4] utilized primarily thermite (oxidation-reduction) type reactions for similar applications.

SHS processing results from the large exothermic reaction between elemental constituents of a refractory compound. An example of this is the production of titanium diboride (TiB_2) from titanium and boron powders. The enthalpy of reaction (ΔH^o_{298}) for this case is 66.8 Kcal/mole or 962 calories per gram of TiB_2 produced. Table 1 lists similar thermodynamic data[5] of other refractory materials. The adiabatic temperatures are calculated by assuming that all heat generated by the reaction is used to increase the temperature of the reaction products. Hence, the heat generated by reacting titanium and boron to form TiB_2 is enough to heat the TiB_2 product to its melting point (3190°K).

The method utilized involves pressing compacts of the mixed precursor particulate components, and then igniting a local area on the sample by heating very rapidly (electrical spark, hot wire, ion beam, etc.). After local initiation of the reaction the liberated heat sustains a reaction front or zone which propagates through the mixture (Figure 1). Not all exothermic reactions are self-sustaining. Experimental data has shown that self-sustaining reactions occur only when the product phase is liquid at the reaction temperature and the precursor powders are fine enough to limit heat transfer into unreacted material.[6] The heat generated is also available for sintering the newly formed phase or phases.

Table 1. THERMODYNAMIC INFORMATION ON PHASES PRODUCED BY "SHS"

PHASE	Tmp (OK)	Tad (°K)	ENTHALPY ΔH_{298}^{0} (Kcal/mole)	HEAT OF REACTION Q (cal/g)
TiB_2	3190	3190*	66.8	962
TiC	3210	3210*	44.13	736
SiC	3100	1800	16.5	412
B_4C	2620	1000	13.8	250
WC	3070	1000	8.4	43
AlN	2500	2900*	64.0	1560
Si_3N_4	2170	4300*	176.0	1260
TiN	3220	4900*	80.75	1300

*Tad ≥ Tmp

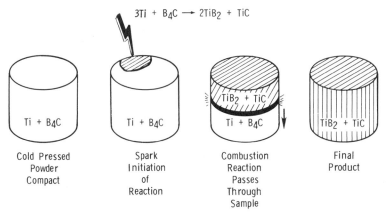

Figure 1. SHS reaction scheme.

Several key advantages of SHS processing over conventional methods are as follows: a high rate of synthesis (seconds versus hours or days), attainment of equilibrium, improved purity due to loss of volatiles, and no need for expensive energy-consuming furnaces.

Currently, there are two general areas of research in SHS processing. One is to develop an understanding of the reaction process, including theoretical and experimental evaluations of reaction initiation, rate, and mechanisms. The second is to study the effects of processing variables (powder size, composition, etc.) on the product characteristics (density, microstructure, phases).

The primary interest of the authors concerns a critical evaluation of the influence of precurser powders on the product characteristics of materials in the titanium-boron-carbon system.

Experimental Method

Four different compositions along the Ti-B$_4$C composition join
were investigated (Figure 2). All mixtures used the same boron
carbide powder which had a median particle diameter of 17.4 μm.
Three different titanium powders were used, each having a different
particle size distribution (Figure 3). Titanium 12-80 was not used
and is only shown for comparison purposes. All three "as-received"
powders were purchased from the same company, thus, this figure
shows the kinds of variation expected from different powder lots.
Scanning electron micrographs (SEM) of two different lots of titanium
are illustrated in Figure 4. Note that these particles have dramatically different morphological features (rough-spherical versus
smooth-irregular). Chemical analysis indicates that lot 6-81 contains a higher percent of impurities.

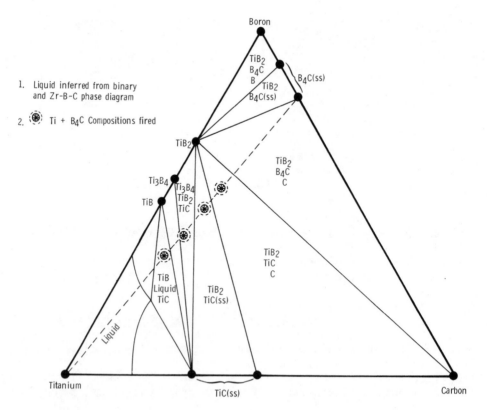

Figure 2.   Inferred Ti-B-C phase diagram at 2000°C.

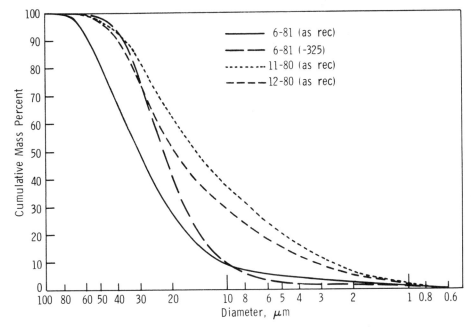

Figure 3.   Titanium powder size distributions.

The titanium-boron carbide mixtures were dry mixed with an alumina mortar and pestle. Disc-shaped pellets approximately 0.75 in. in diameter and weighing approximately 2.0 g were uniaxially compacted at 10,000 lb. All reactions were carried out in air and initiated with an electrical spark. Reactions were filmed using high speed photography to allow measurement of reaction rates. The final products were characterized by weight loss measurement, X-ray diffraction, and scanning electron microscopy.

RESULTS/DISCUSSION

Table 2 lists results along the Ti-B$_4$C composition join. The weight-percent gain increases as the titanium content is increased. This is consistent with the partial reaction between titanium and air. The initiation of reaction is easiest when the finest titanium powder is used. Large percentages of coarse titanium powder would not react. This is consistent with the delicate balance of heat generation by reaction and its dissipation by conduction. A powder compact containing coarser particles tends to be a better conductor of heat.[6] Our results show that the reaction propagation rate (speed of reaction wave front) is dependent on the size of the original titanium particles. Coarse particles produce faster propagation due to the better thermal conductivity of the pellets. The coarse titanium particles make these mixtures difficult to initiate reaction,

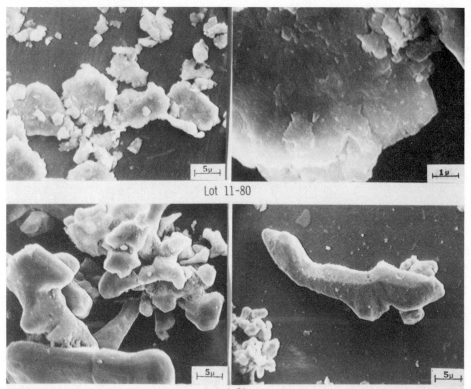

Lot 11-80

Lot 6-81

Figure 4.   Scanning electron micrographs of various titanium powders.

but once started, propagation is more rapid. Mercury porosymmetry
data show that a sample using finer titanium has more open porosity.
Different compositions along the Ti-B₄C join produce a variety of
phases as also shown in Table 2. The results agree fairly well with
the isopleth of the Ti-B₄C composition join (Figure 5) with the
exception of the formation of Ti_3B_4. This diagram was drawn using
the information of Rudy. [7]

The resulting macrostructures of the reacted samples are illustrated in Figure 6. Note that the 74 w/o Ti composition has large
pores on the surface which are not seen in the 85 w/o Ti sample.
The microstructure of two compositions are shown in Figure 7; both
are relatively porous. It is interesting to note that the 81% Ti
sample shows unreacted B₄C particles while the 74% Ti sample does
not. This is contrary to what might be expected since the 81% Ti
sample contains less B₄C (19% versus 26%). These macrostructures
and microstructures suggest that the 74 w/o mixture had probably

Table 2.   Ti + B$_4$C RESULTS

WT % Ti	SAMPLE	Ti[†]	% WT GAIN	REACTION INITIATION*	COMBUSTION PROPAGATION RATE (g/sec)	MERCURY POROSYMMETRY (cc/g)	MAJOR PHASES
68	16	6-81 (as)	-	DIFFICULT	-	-	TiB$_2$
68	24	6-81 (-325)	+2.48	DIFFICULT	0.34	-	TiC
74	3	11-80 (as)	-	EASY	0.14	0.072	
74	21	6-81 (as)	+1.08	DIFFICULT	1.14	-	TiB$_2$
74	25	6-81 (-325)	+2.59	DIFFICULT	0.42	-	TiC
81	4	11-80 (as)	-	EASY	0.11	-	TiB$_2$
81	22	6-81 (as)	+6.24	DIFFICULT	0.25	-	Ti$_3$B$_4$
81	32	6-81 (-325)	+6.34	DIFFICULT	0.19	-	TiC
85	19	11-80 (as)	+0.14	EASY	0.21	0.069	TiB$_2$
85	23	6-81 (as)	-	NONE	-	-	TiB
85	31	6-81 (-325)	+6.67	DIFFICULT	0.19	0.039	TiC

*INITIATION DIFFICULTY:

EASY	→	SPARK ONLY
DIFFICULT	→	STARTER POWDER USED
NONE	→	NO IGNITION

[†]MEDIAN PARTICLE SIZE

11-80 as	14.8 μm
6-81 as	30.0 μm
6-81, -325	22.0 μm

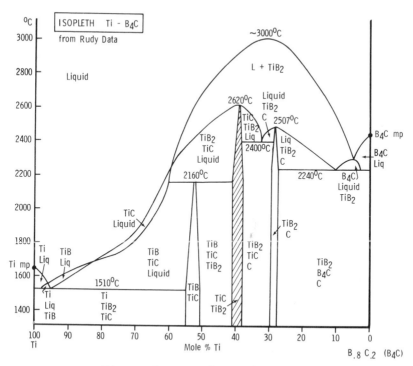

Figure 5.   Isopleth Ti-B$_4$C.

COMP.		TITANIUM LOT		
w/o Ti	w/o B$_4$C	11-80	6-81	6-81 (-325)
74	26	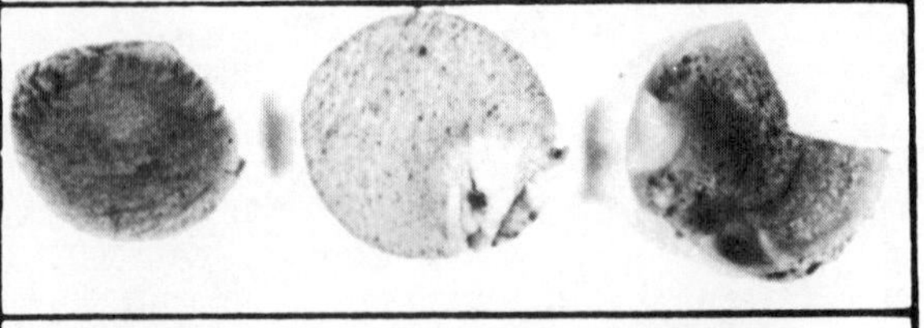		
85	15	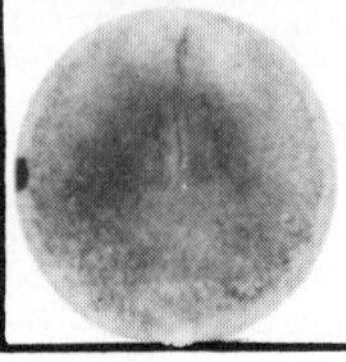	No Reaction	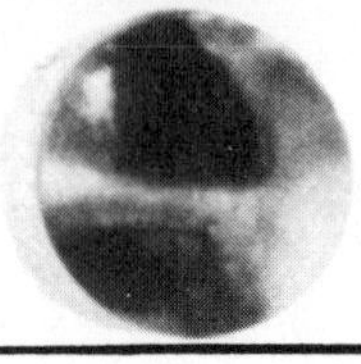

Figure 6.   Titanium influences on macrostructure.

reached a higher temperature during reaction since more vaporization had occurred (large voids) and the sample was fully reacted. Some fully dense regions did occur in these samples as shown in Figure 8.

SUMMARY

    The production of materials in the titanium-boron-carbon system by SHS is influenced by overall sample composition and titanium particle size. The starting composition dictates the final product phase assemblage, the overall macrostructure, and the weight changes during reaction. Mixtures consisting of the finest titanium powder are easiest to initiate, have slower reaction propagation rates, and are more porous than using coarser powders. Further study on the relationship of precurser particle sizes to reaction kinetics and final product characteristics should result in the capability to produce dense refractory materials by SHS without the use of pressure.

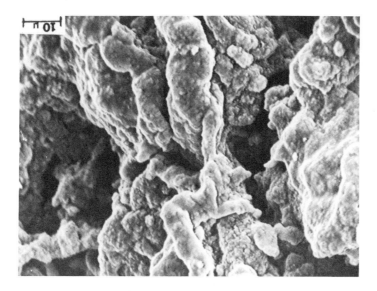

74% Ti (11-80)
26% B$_4$C

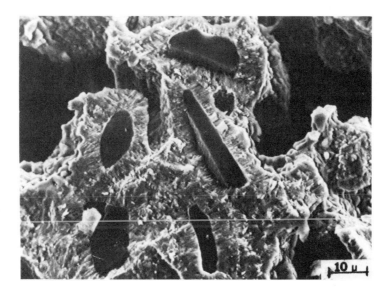

81% Ti (11-80)
19% B$_4$C

Figure 7.   Composition effects on microstructure.

Figure 8.   SEM of 74 w/o Ti (6-81) sample.

REFERENCES

1. J. F. Crider, Self-Propagating High-Temperature Synthesis, in
    "Ceramic Engineering and Science Proceedings," 3, (1982),
    p. 519-528.
2. J. W. McCauley, N. D. Corbin, N. E. Rochester, J. J. DeMarco,
    L. Schioler, and P. Wong, Key Physical Characteristics for
    Predicting Zr Powder Burn Times, in "Proceedings of the 29th
    Power Sources Symposium," (1981), p. 19-23.
3. J. W. McCauley, N. D. Corbin, T. Resetar, and P. Wong, Simulta-
    neous Preparation and Self-Sintering of Materials in the
    System Ti-B-C, in "Ceramic Engineering and Science Proceed-
    ings," 3, (1982), p. 538-554.
4. J. D. Walton, Jr., and N. E. Poulos, Cermets from Thermite
    Reactions, J. Am. Ceram. Soc., 42, (1959), p. 40-49.
5. N. P. Novikov, I. P. Borovinskaya, and A. G. Merzhanov,
    "Combustion Processes in Chemical Technology and Metallurgy,"
    Chernogolovka, (1975), p. 174-188.
6. A. P. Hardt, and R. W. Holsinger, "Propagation of Gasless
    Reactions in Solids - II. Experimental Study of Exothermic
    Intermetallic Reaction Rates, Combustion and Flame," 21,
    (1973), p. 91-97.
7. E. Rudy, "Ternary Phase Equilibria in Transition Metal-Boron-
    Carbon-Silicon Systems, Part V," AFML-TR-65-2, (1969),
    p. 599-618.

VIBRATIONAL FUSING OF PLASTIC PARTICLES

Robert P. Fried

Rte. 9G
Staatsburg, New York   12580

BACKGROUND OF THE INVENTION

VIM molding relates to a process and apparatus for molding plastic articles from particulate plastic material.

There are various techniques for molding plastic articles known in the art, each of which has its own particular advantages and disadvantages. The most common of these techniques is injection molding, wherein liquid plastic is injected into a mold and subsequently cooled. While injection molding is the least expensive process available for forming plastic articles in large quantity, the high cost of producing a mold makes its use with low volume work prohibitively expensive. In addition, the injection mold must be quickly cooled to permit its repetitive use at high volume, and such cooling creates stresses within the plastic article that may result in cracks.

Another technique for producing molded plastic articles is called thermoforming. In this process, a plastic sheet is heated and stretched onto a mold. While thermoforming is exceptionally useful for shallow articles, plastic articles of any substantial depth, exhibit internal stresses and weaknesses which may result in cracks. In addition, the sheet plastic used as the raw material is somewhat more expensive than the pellets or powder used in injection molding.

A still further technique for producing molded plastic articles is known as rotational molding. In this process, a fixed charge of plastic is placed within a female mold, and the mold is rotated. During rotation, heat is applied to the outside of the mold to

347

cause the plastic to melt. With this process, energy use is relatively high and the mold is expensive to produce. In addition,
the plastic article, although smooth on the outside which contacts
the mold, may have a relatively rough surface on the inside. Such
a rough surface is a disadvantage when the article is to be used as
a container.

It is an objective of VIM molding to provide a process and
apparatus for forming plastic molded articles for which the tooling
cost for molds is lower than for other plastic molding processes
known in the art.

It is a further objective to provide a process and apparatus
for forming molded plastic articles:

- which requires a minimum of capital expenditures for a
  complete molding machine.

- which permit the wall thickness of each articles to be
  controlled within fine tolerances.

- such that the articles exhibit very little internal stress.

These objectives are achieved by means of the VIM process as follows:

A hollow-heat-conductive mold is surrounded with particulate
plastic material. Heat is applied to the interior of the
mold until the temperature of the outer surface is above
the fusion temperature and below the melting temperature of
the particulate plastic material. This temperature is maintained until an article of desired thickness is formed.
Thereafter, loose particulate material is removed from the
outer surface of the molded article and heat is again
applied to the interior of the mold so as to cause the outer
surface of the molded article to become smoother. Finally,
the article and mold are cooled and the article is removed
from the mold.

This process lends itself to an extremely low tooling cost for
each different article to be formed. The heat conductive male mold
may be formed of sheet material which is welded together and sanded
smooth. Aircraft quality aluminum is the preferred material,
although other heat-conductive materials may also be used for the
mold.

Because no stresses appear in the material as the article is
forming, internal stresses do not develop unless the article is
cooled quickly. The wall thickness of the molded article formed by
this process may be precisely controlled by controlling the time
during which heat is applied to fuse the particulate plastic
material together.

Finally, the cost of the apparatus for carrying out the process is extremely modest in comparison with the capital investment required to purchase apparatus to carry out other, prior art, plastic molding processes.

VIBRATIONAL FUSING OF PLASTIC PARTICLES

VIM (Vibrational Microlamination) is a new technology in the physical sciences. In VIM development, the physical, the chemical, and engineering technologies have been combined to contribute effective new knowledge.

In preparing plastics for use (plasturgy), our biggest single problem is the interface between limiting process parameters and use requirements. Traditional thermoplastic processes involve temperature, stress (shrinkage and flow), ejection (draft) requirements, flow and orifice limitations, and mold requirements, as examples of limiting parameters. The intellectual direction for arriving at the VIM method recognized these limits and, deriving from vapor deposit of metal on plastics, determined that plastics could not be easily deposited on metals if the plastic were in the vapor or in the molten state. Molding in the plastic state was essential.

Since "plastics" involve flow (creep), or change of shape, the studies converged on the kinematic variables and thus became dynamic research involving quantification of the kinetic parameters of plastic state molding. If F still equals M_xA, the "A" became the rate of fusion of the materials involved, and research centered around how to get rapid fusion, if indeed fusion could be achieved at all.

One solution to the problem came from the vibrational sciences. Various means of achieving vibration were studied, as well as potential distribution of vibrational forces. Thus, began a continuing parade of investigation. The evolution of "vibrational microlamination" thus began, leading to a process where plastic particles were rapidly and sequentially fused to the mass, the mass being deposited on a conductive metal substrate. The process created strong, stress free parts from plastic materials, and was made effective by applying mold release to the metal substrate (making the reusable "mold") while at the same time vibrational microlamination was made more generic by utilizing the name "VIM" molding, or vibrational molding.

VIM today involves the intercoupling effects of multiple variables.

Resultant "breakthru" technology has given advancement in strength (no-stress), shapes (no draft needed), materials (molded PVDF, etc.) to the plastics industry.

I appears that the VIM process provides design and material freedoms. The VIM process of fusing plastic particles with heat and vibration is not a panacea, nor does it obsolete other processes. But VIM does add new dimensions of strength, freedom from stress, freedom of design, and capability for broadened use of plastics.

VIM DESIGNS

        Weight: Up to 200 pounds or more
        Size: Up to 10' x 6' x 5'
        Low cost tools: Male, female or combined molds
                         Mold the size of a desk is approximately
                         $4,000
        Uniform wall thickness
        Normal tapers, or no draft parts, can be molded by the VIM
        process
        Wall thicknesses from .060 to over ½"
        Surfaces can be smooth or textured
        Composite and laminate embedment freedom

VIM MATERIALS - Broad freedom of choice

        Freedom of material choice:

                Polyethylene
                Polypropylene
                Polycarbonate
                Polysulfone
                Polyphenylene Oxide
                Polyethersulfone
                Polyvinylchloride

        Freedom of composite choice:

                Glass
                Kevlar
                Graphite, etc.

        Freedom of laminate choice:

                Plastics
                Metal
                Cloths
                Fibers, etc.

FEATURES AND BENEFITS OF VIM MOLDING

VIM Molded Plastic	VIM products are heavy duty industrial type units which are weather proof, chemical proof, have high strength and are highly resistant to impact and cracking. Units in the field under constant abuse have now been tested for over five years without failure.
High Strength	The inherent strength of VIM products is due to NOW's unique molding process. This unique molding process starts with powder (not molten plastic) and fuses the powder under heat and vibration in such a way that inherent stress is avoided. The resultant containers thus do not crack in normal use. In addition, VIM molding can make thick or heavy duty sections without setting up distortions in the product.
Crack Resistant	Thermoplastic materials (polyethylene, polypropylene, etc.) melt when heated. They consist of long chain molecules which inherently have good strength and flexibility. They are excellent for container use. The VIM process molds these materials with no degradation of the basic polymer strength.
Lowest Tooling	VIM products are molded from a single surface aluminum mold which is the most cost effective. Normally VIM costs are lower than your engineering costs.
Short Lead Time	Average tooling time ranges between 4 to 8 weeks. Mold revisions are also timely.
Stress Free Products	VIM molding provides uniform wall thickness for resistance to stress cracking and impact shock loads.
Unique Design Features	VIM molding permits special tote features not possible by other methods; i.e., square corners, straight sides, 0^0 taper, molded in holes, etc.
Ideal for Automatic Systems	Surface finish and design can vary to accommodate most systems applications; i.e., rough bottom for conveyor use, rigid lip for monorail use, minimum lip for shelving applications, etc.

#1 in   NOW has combined a quality product with low cost
Specials   tooling, unique design features and wide material
               options to maximize your cost effectiveness.

Why Other   Thermoplastic materials have normally been molded
Methods   molded by injection, extrusion or thermoforming.
Fail   In injection or extrusion molding, the plastic is
               melted, shot through a nozzle and is cooled either
               in space or in an enclosed mold. The cooling
               process does not occur evenly throughout the prod-
               uct so that the resultant product has inherent
               stress in its finished form. Thermoforming heats
               sheet plastic so it will flow and then stretches
               it either in a die or by vacuum. This stretching
               process also results in inherent stress in the
               product. Containers made by these methods have a
               tendency to crack or break in use along stress
               lines (e.g. garbage containers in cold weather).

Long   Heavy duty products made by NOW take advantage
Life   of the VIM process to produce containers with
               inherent crack resistance and strength which will
               outlast others available for customer use.

CONCLUSION: VIM molding adds new dimensions to plastic molding
               capabilities. It is not a panacea that solves all
               problems. Nor does it obsolete the other molding
               methods. But, VIM is a means for fusing plastic
               particles with lower cost tooling, higher quality,
               less stress and greater design freedom than any other
               process.

Table I

HOW THE PROCESSES COMPARE

	VIM	Injection Solid	Injection Str foam	Thermo-forming	Blow molding	Rotational molding
Production quantity						
Low (100–1,000)	Yes	No	No	Yes	No	Yes
Med (1,000–20,000)	Yes	Sometimes	Yes	Yes	Sometimes	Sometimes
High (2,000 +)	Sometimes	Sometimes	Yes	Yes	Sometimes	No
Tooling cost	Low	High	High	Low	High	Moderate
Materials						
Polyolefins	Yes	Yes	Yes	Limited	Limited	Limited
Engineering thermoplastics	Yes	Yes	Yes	Limited	Limited	Limited
Part weight (lb)	To 180	Under 10	To 100	To 40	To 20	To 200
Part shape	Open one side	Broad range	Broad range	Open one side	Closed hollow	Open or closed hollow
Design						
Draft required	No	Yes	Yes	Yes	No	Sometimes
Wall thickness (in.)	0.005 to 1.0	To 3/8	½ to 3/8	Varies	Varies	To 3/8
Corner angles	Can be square	Can be square	Can be square	Needs radius	Needs radius	Needs radius
Ribs possible	Yes	Yes	Yes	Yes	No	Yes
Inserts possible	Yes	Yes	Yes	No	No	Some
Stress	Low	High	Low	High	High	Low
Tolerances (in.)	± 0.005 mold side or closer	± 0.005 or closer	± 0.010 or closer	± 0.015	± 0.005	± 0.032

SYNTHESIS OF CERAMIC POWDERS AND SURFACE FILMS FROM

LASER HEATED GASES

John S. Haggerty

Senior Research Scientist
Massachusetts Institute of Technology
Cambridge, MA 02139

ABSTRACT

Two new processes have been developed that are based on laser
heated gases. Both permit unusually precise levels of process
control and thereby materials having superior properties.

The powder process yields Si, Si_3N_4 and SiC powders that are
uniform in size, non-agglomerated, small diameter, spherically
shaped and high purity. Manufacturing cost analyses show that submicron powders can be made with an energy cost of approximately 2
kWhr/kg and a dollar cost of 2-3.30 \$/kg exclusive of the costs of
feed materials. This type of process should be capable of producing
technically superior, lower cost submicron powders than existing
processes.

The laser induced chemical vapor deposition process (LICVD)
causes reactant gases to be heated by absorbing IR light from a
laser beam that passes parallel to the substrate surface. Laser
heating permits independent control of gas and substrate temperatures
while operating in a conventional thermally activated CVD mode. For
hydrogenated amorphous silicon films, this is particularly important
because deposition rates are determined by the high gas temperatures
and film properties by the low substrate temperatures. Spin density,
hydrogen content, electrical conductivity and mobility gap properties
show the LICVD process capable of producing very high quality films.

INTRODUCTION

Two processes based on laser induced gas phase reactions are
discussed in this paper. Powders are made in one[1,2] and monolithic

films[3] in the other. Although neither process is as technically
mature as most of the others discussed at this conference, they are
presented as case studies of processes successfully developed to
yield materials having highly specific properties. The materials
are needed to address complex issues including improved system performance and reliability, use of strategic materials and energy
conservation at an acceptable cost.

The ceramic powders are designed to conform to the requirements
of a new particulate ceramics process under development at MIT.[4]
This particulate ceramic process is based on creating ordered dispersions, possible only with particles having equal diameters.
Virtually defect free green bodies with uniform, high coordination
numbers can be produced with these powders permitting improved
reliability, lower densification temperatures and times and the
achievement of highly specific materials properties.[5] The primary
application we are addressing is ceramic components in heat engines;
although, there are many other important military and civilian applications for the high purity, flaw-free ceramic parts that result
from using these types of powders.

The hydrogenated amorphous Si films will be used as photovoltaic
devices. The independent control of laser heated gas and substrate
temperatures possible with this CVD process permits the most favorable
combination of high temperature and low temperature characteristics.
High growth rates can be combined with the superior microstructural
quality of CVD films[6] and the excellent electrical characteristics
of sputtered[7] or glow discharge[8] a-Si:H films deposited on low
temperature substrates. These developments illustrate that
"intrinsic" materials problems can in fact be resolved with funda-
mentally new processes.

We are modeling both processes because resulting material
characteristics are totally dependent on the nucleation and growth
processes. Modeling has proceeded from calculating time-temperature
histories of the gases subjected to laser radiation.[1] While
substantial progress has been made, there is little basis for calculating nucleation and growth rates from existing physical models or
emperical results. With the powder process, we have achieved more
important results by measuring nucleation and growth rates;[9,10]
these results are good basis for real time process control and for
definition of kinetic models. Film growth rates have been measured[3]
as a function of calculated temperatures[11] to provide a basis for
a physical model of the process.

When electromagnetic radiation is directed into an absorbing
medium, it causes heating as is the case with microwave coupling to
water. Our synthesis research has been based primarily on SiH_4
chemistries and IR radiation with wavelengths near 10.6 µm.
Adsorption of the light energy is inherently complex, depending on

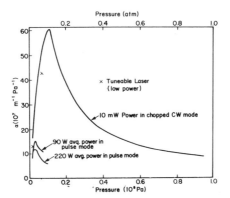

Figure 1.

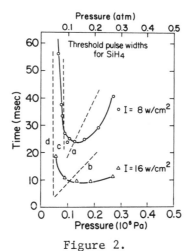

Figure 2.

the gas molecule, mixture, and pressure and the laser wavelength,
intensity and bandwidth among other factors. The absorptivity we
have measured[1] for SiH$_4$ as a function of pressure for the CO$_2$ P(20)
line (10.591 μm) is shown in Figure 1. At the high pressures used
for powder synthesis, (P > 500 torr), α is relatively constant. At
the low pressures used for the CVD process, (P ≈ 10 torr), α is
extremely sensitive to many variables and is approximately linearly
dependent on the SiH$_4$ pressure for pure SiH$_4$ gas. We have shown
that the P(20) line is most strongly absorbed by SiH$_4$ at moderate
and high pressures.[12]

    One of the consequences of the pressure dependence of α is
shown in Figure 2. Threshold laser pulse lengths that just caused
a reaction in SiH$_4$ gas were determined as a function of pressure for
two different power levels.[1] Two limiting cases are evident. At
high pressures, the heat capacity of the gas and the laser energy
by the gas both increase proportionally with pressure. The time
needed to reach the reaction temperature on this side of the curves
(lines a and c) is governed by sensible heat. We operate the powder
process on the transient side of the curves. At low pressures, the
thermal conductivity of the gas is approximately constant and the
energy absorbed decreases with decreasing pressure. A lower pressure
limit exists, corresponding to an infinite pulse length (lines b
and d), where the energy absorbed by the gas just equals the heat
lost to the cold walls with the gas at the reaction temperature.
We operate the CVD process in this steady-state condition.

    The powder and CVD laser driven reactions permit important
process attributes to be achieved. For both, these include:

    1. highly controlled atmospheric conditions
    2. cold wall reaction vessels
    3. unusually rapid heating and cooling rates
    4. precise control of process variables
    5. accessibility of the reaction to process diagnostics.

For the laser CVD process, the ability to establish the substrate
temperature at a level that is different from the gas reaction
temperature is an important feature.

THE PROCESSES AND THEIR PRODUCTS

A. Powders from Laser Heated Gases

    The powders required for the ceramic process based on ordered
dispersions must exhibit several important characteristics that are
defined by the synthesis process. The physical requirements include
small size, narrow size distribution, absence of agglomerates,
equiaxed shapes and purity. Small size and narrow size distribution
require high initial nucleation rates that terminate before the

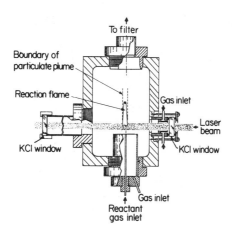

Figure 3.

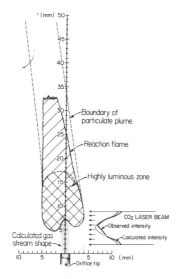

Figure 4.

growth process has proceeded to an appreciable extent. Absence of
agglomerates requires a minimum exposure to elevated temperatures
where strong necks can form between colliding particles. Equiaxed
shapes result from high growth rates. High purities are achievable
with cold wall reaction vessels. The laser heated gas phase powder
synthesis process achieves all of these process attributes; consequently, the powders can have the desired characteristics.

    1. Synthesis Apparatus. Powders have been synthesized with
crossflow and counterflow gas stream - laser beam configurations
and with both static and flowing gases. Most of the process research
has been conducted with the reaction cell shown schematically in
Figure 3.

    In the crossflow configuration, the laser beam having a
Gaussian shaped intensity profile orthogonally intersects the reactant gas stream possessing a parabolic velocity profile. The
laser beam enters and exits the cell through KCl windows. The
premixed reactant gases, under some conditions diluted with an inert
gas, enter through a 0.75-1.5 mm ID stainless steel nozzle located
2-3 mm below the laser beam. A coaxial stream of argon is used to
entrain the particles in the gas stream. Argon is also passed
across the inlet KCl window to prevent powder build-up and possible
window breakage. Cell pressures are maintained between 0.08 to
2.0 atm with a mechanical pump and throttling valve. At present,
the powder is captured in a microfiber filter located between the
reaction cell and vacuum pump. In the future, the particles will
be collected by electrostatic precipitation or by separation with
the fluid used for ultimate dispersion.

    The counterflow geometry, with laser beam and reactant gas
streams impinging on each other from opposite directions, has the
important advantages of exposing all gas molecules to identical
time-temperature histories and absorbing all of the laser energy.
However, the achievement of a stable reaction requires that the
reaction and gas stream velocities must be equal and opposite to one
another. While this is really accomplished once process conditions
are defined, it is an impractical geaometry for experimental work in
which process conditions are varied over wide ranges.

    2. Chemistries. Principally, Si, Si_3N_4 and SiC powders have
been made from appropriate combinations of SiH_4, NH_3, and C_2H_4
gases. CH_4 was investigated but we found that it did not react to
form SiC with the laser heated process conditions. Alternative
reactant gases such as $CH_3 \cdot SiH_3$ and Cl_2SiH_2 have been investigated
for SiC. B_2H_6 was used to add boron to the Si and SiC powders as
a sintering aid.[13] While most of this process research has focused
on a limited set of compounds and reactants, it is apparent that
laser induced reactions are applicable to a broad range of materials.

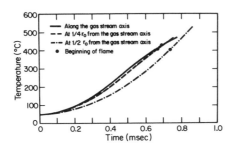

Figure 5.

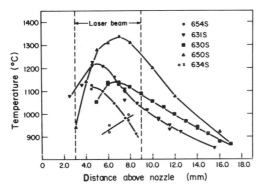

Figure 6.

3.   Process Modeling. A typical reaction zone, located where
the laser beam and gas stream intersect, is shown[1] schematically
in Figure 4. Average heating rates, nucleation rates, growth rates
and temperatures can be estimated based on measured zone and particle dimensions, calculated gas velocities and uncorrected pyrometrically determined temperatures. For typical reaction conditions
employing a total reactant gas flow rate of approximately 100
cm^3/min and a pressure of 0.2 atm, the velocity of the gas decreases
from approximately 500 cm/sec at the nozzle to 350 cm/sec at the
center of the laser beam. With a NH_3/SiH_4 flame, the reaction
commences approximately 3 to 5 mm into the laser beam. Thus, the
exposure time needed to initiate the reaction is nominally 10^{-3} sec.
For a reaction temperature of approximately 1000°C, this indicates a
gas heating rate of approximately 10^6°C/sec. With growth occurring
primarily in the hot zone, growth rates are approximately 10^6 Å/sec
and particles are exposed to the maximum reaction temperatures for
less than 10^{-3} seconds.

Using calculated gas stream velocities, measured optical
absorptivities, and measured laser intensities, the time-temperature
histories of the reactant gases can be calculated[2] by assuming all
absorbed energy is converted to sensible heat. A typical calculated
profile is shown in Figure 5 for gas molecules flowing along stream
lines located at the centerline, r/4 and r/2 of the reactant stream.
The reaction times were determined from the measured lower boundary
of the reaction zone and calculated gas velocities. Other calculations have studied the effects of gas composition, pressure and
velocity as well as laser intensity. Qualitatively the results
agree with expectations and observations. Quantitatively they
predict reaction temperatures that are as much as 400°C below actual
reaction temperatures; the error is probably attributable to the
assumed optical absorptivity. The principal conclusion is important
and it appears valid; all gas molecules can be made to experience
uniform time-temperature histories up to the point of the initial
reaction.

Diagnostics based on light scattering and transmittance
measurements permit nucleation and growth processes to be analyzed
in terms of temperature and reactant partial pressures with < 1 mm
spacial resolution throughout the reaction zone and particulate
plume. They also permit important observations about the crystalline
state of the reaction product and the formation of agglomerates.
Using an experimental technique developed by Flint[9] and Marra[10] and
modifications of computational techniques reported by Sarofim[14],
it is possible to calculate the local emissivity, particle diameter
and number density from the extinction and scatter-extinction ratio
results. Mass balance calculations, based on local number density
and particle diameter, permit the local reactant concentrations to
be calculated. These, combined with local temperatures, permit

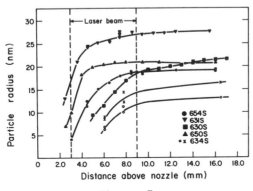

Figure 7.

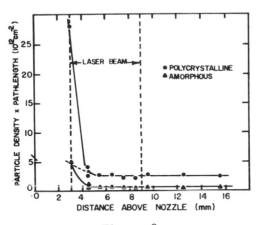

Figure 8.

reaction kinetics to be interpreted in terms of rate controlling models. They also provide a basis for real time process control. Typical results[9,10] are given in Figures 6,7 and 8 for Si powders made from SiH_4. The temperatures shown in Figure 6 are calculated from the pyrometrically determined brightness temperatures and the calculated emissivities.

These calculations employ the complex index of the particulate materials which are strongly dependent on crystallinity, temperature, stoichiometry and wavelength. This variance can be used to gain insights about the reaction process because the actual number densities and diameters are measured independently by characterizing the powders. For either amorphous or polycrystalline indices, Figure 8, the calculated number densities decrease rapidly over a distance of approximately 1 mm ($\sim 3 \times 10^{-4}$sec) and then remain constant. Supported by direct STEM observations and annealing studies[10], these results show that the Si particles form with an amorphous structure then crystallize when the gases become hotter as they penetrate into the laser beam.[9] The actual number density follows the dotted line from the amorphous curve ($\sim$ 3 mm) to the polycrystalline curve ($\sim$ 4.5 mm) remaining essentially constant ($N \approx 4 \times 10^{12}$cm^{-3}) throughout the growth process. The cessation of nucleation is an essential requirement for achieving the desired uniform particle size. Equally important, this result shows that agglomeration does not occur within the hot region where hard agglomerates could form in agreement with the constant particle diameter regions shown in Figure 7. This result shows that typically observed chainlike agglomerates probably result from powders being captured in the filter, their extraction and the preparation of TEM samples. Other results have shown that exposures to excessive temperatures can cause rigid interparticle necks and probably melting.

In close agreement with the simplified analyses, these direct observations show that the process is characterized by rapid heating rates ($\sim 10^6$°C/sec), short reaction times ($\sim 5 \times 10^{-4}$sec) and rapid cooling rates ($\sim 10^5$°C/sec). The high heating rates cause high nucleation rates and consequently small particle sizes. Growth occurs at a constant rate until reactants are consumed then ceases at which point the temperature begins to decrease. The short time at high temperatures minimizes problems with forming hard agglomerates when particles collide through Brownian motion. The cooling rates are fast enough to quench in non-equilibrium crystalline or amorphous structures. The maximum exposure temperature can also be precisely controlled largely independent of heating and cooling rates. This unusual feature permits the powder's crystalline state to be manipulated. In combination, the attributes of this process constitute a very precise means of producing ceramic powders.

These kinetic studies provide the basis for defining the physical nature of the nucleation and growth processes. Combined,

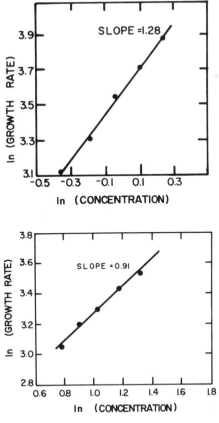

Figure 9.

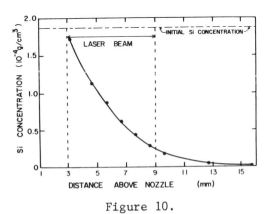

Figure 10.

the number density and particle diameter results permit the SiH$_4$ partial pressure to be calculated[10] throughout the reaction zone as shown in Figure 9. With measured temperatures, these results are used to define physical reaction models. The growth rates in Figure 10 indicate[10] growth occurs by molecular bombardment. This type of analysis is being extended to a more complete description of silicon nucleation rates and complete kinetic studies with Si$_3$N$_4$ and SiC.

4. Synthesis Conditions. The specific process conditions employed to make the powders have been manipulated over an extensive range. The range of investigated conditions are summarized in table 1. More recent synthesis runs have employed higher pressures, higher mass flow rates and near zero argon flow rates for the window stream. Fortunately, these latter conditions combine more desirable powder characteristics with economically more attractive process parameters.

5. Powder Characteristics. The Si, Si$_3$N$_4$ and SiC powders all exhibit the same general features and appear to match the idealized characteristics we sought. Table 2 summarizes the range of properties that have been achieved with these particulate materials.[15,16]

The particles are spherical and uniform in size. The Si$_3$N$_4$ and SiC powders are usually smaller and have a narrower size distribution than the Si powders. After being captured in a filter, the powders were observed in chainlike agglomerates. Neck formation has been observed by TEM between Si particles, but not with the Si$_3$N$_4$ and SiC powders. Dispersion results[17] have shown primary bonding does not always exist at necks between Si particles because light scattering and photon correlation spectrometer characterizations of the sols showed that the dispersed particle sizes can be equal to those of the individual particles.[15,17,18] It is presumed that the Si$_3$N$_4$ and SiC powders will be dispersable because they do not exhibit interparticle necks.

The BET equivalent spherical diameter, and the average TEM diameter corrected for the size distribution measured from micrographs, have always been nearly equal.[10,19] This indicates that the particles have smooth surfaces, no porosity accessible to the surface, nearly spherical shapes, and relatively narrow size distributions. Powder densities, measured by He pycnometry, indicated the particles had no internal porosity.[2]

Chemical analyses[15,16] indicate that the oxygen content is generally less than 1.0% by weight and some powders are as low as 0.05 wt%. The total cation impurities are typically less than 200 ppm. For Si$_3$N$_4$ and SiC powders, the stoichiometry can be made to vary substantially depending on the process conditions. For Si$_3$N$_4$ powder, the stoichiometry ranged from nearly pure Si$_3$N$_4$ (< 1.0 wt%

TABLE 1

PROCESS CONDITIONS USED FOR Si, Si_3N_4 AND SiC

Process Variable	Si	Si_3N_4	SiC
Cell Pressure (atm)	0.2-0.9	0.2-0.9	0.2-0.8
SiH_4 Flow Rate (cc/min)	5.4-110	5.4-40	11-45
NH_3 Flow Rate (cc/min)	0	44-110	0
C_2H_4 Flow Rate (cc/min)	0	0	9.0-45
Ar Flow Rate to Annulus plus window (cc/min)	560-1450	1000-1100	1000
Laser Intensity (W/cm^2)	$176-5.4 \times 10^3$	$530-1 \times 10^5$	$530-5.2 \times 10^3$
Reaction Zone Tem. (^0C)	750-1390	675-1390	865-1930

TABLE 2

SUMMARY OF POWDER CHARACTERISTICS

Powder Type	Si	Si_3N_4	SiC
Mean Diameters (A)	190-1700	75-500	200-500
Standard Deviation of Diameters	46	23	∿25
(% of Mean) Impurities (ppm by wt.) O_2	680-7000	3000	3300-13,000
Total Others	<200	<100	NA
Major Elements	Ca,Cu,Fe	Al,Ca	NA
Stoichiometry	---	0-60% (excess Si)	0-10% (excess C or Si)
Crystallinity	crystalline -amorphous	amorphous -crystalline	crystalline Si and SiC
Grain Size:Mean Diameter	1/5-1/3	∿1/2	1/2-1

excess Si) to Si_3N_4 + 60 wt % excess Si. Near stoichiometric values
are caused by increased laser intensity, increased pressure, and
lower gas stream velocities. C_2H_4/SiH_4 runs produced SiC powders
with up to 10 wt% excess C or Si. The more stoichiometric powders
were produced with increased laser intensity and close to stoichiometric reactant gas ratios. It appears the excess Si and C are
distributed uniformly throughout the individual particles in both
the Si_3N_4 SiC powders.

Except when synthesized under very low laser intensities, the
Si powders are crystalline. In all cases, the crystallite size was
a fraction (1/5 - 1/3) of the BET equivalent diameter, indicating the
individual particles are polycrystalline. Marra found[10] that the
particle size to grain size ratio reflected the nucleation and growth
of crystals in the amorphous particles. Virtually all process conditions produced Si_3N_4 powder which was amorphous. High laser
intensities and high reactant pressures resulted in stoichiometric,
crystalline Si_3N_4 powders. Nearly all SiC powders were crystalline.
Most process parameters produced polycrystalline SiC particles with
the BET equivalent spherical diameter approximately twice as large
as the crystallite size measured by X-ray line broadening.

6. Manufacturing Cost. The costs of manufacturing powders by
the laser heated processes have been estimated and compared to
conventional powders. The results are very encouraging.

Material, utility, energy and labor costs are treated conventionally. The major assumption of the Aries and Newton[20] simplified
estimation technique is making the plant cost a simple multiple
(4.74) of the capital equipment costs; the annualized plant cost is
0.202 x total plant cost. Labor costs are treated as 1.79 x direct
salaries and annual building costs as 0.052 x total building costs.

This analysis was based on a single gas tip since this will be
the scaling unit for a plant. The operating conditions and yields
summarized in table 3 are the same as the laboratory scale process.
The assumed 2.0 cm diameter gas tip is approximately the maximum
laser beam penetration depth that can produce uniform local intensities with multiple, radially opposed laser beams. Preheated gases
were presumed to minimize laser heating; but, no heat recovery was
assumed. The plant was scheduled for full time operation (8760
hrs/yr maximum) and assumed to actually produce 91% of available
time (8000 hrs/yr). Four full shifts having one person for three
tips was assumed. Current bulk costs for NH_3 (0.52 $/kg) and C_2H_4
(0.5 $/kg) were used for these calculations. Silane costs were
estimated informally from several sources.[21,22,23] The cost analysis
was done based on presumed silane costs of 2.0 and 20.0 $/kg. A
figure of 10-15 $/kg appears reasonable for silane with somewhat
higher impurity levels than semiconductor grades.

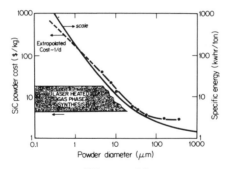

Figure 11.

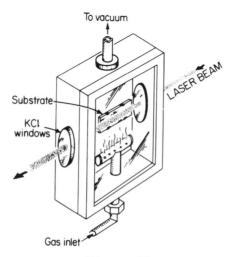

Figure 12.

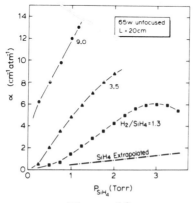

Figure 13.

The results of the manufacturing cost analysis are given in table 3. They show that the assumed process configuration is well matched to available CO_2 laser equipment. The presumed gas tip diameter and gas flow rate require lasers in the 1.5 kW range. Also, the output per tip is 40-57 tons per year; a level that is well into an acceptable manufacturing scale. Finally, calculated manufacturing costs range from 4.30 to 25.00 \$/kg depending mostly on the silane cost. Generally we feel that this projected manufacturing cost is conservative and may well be much lower.

Figure 11 summarizes present sales prices of Acheson SiC in 5000 kg lots as a function of particle size.[24] The price is essentially constant for powder sizes greater than 100 μm. Below 100 μm, the cost increases rapidly with decreasing particle size. A l/d extrapolation was used to project prices into the submicron range needed for sintering because it is the most conservative projection if prices are dominated by comminution energy per unit mass[25] as suggested by the results in Figure 11. These powders will have the typical broad particle size distribution and contaminations that result from comminution processes.

The range of our SiC manufacturing costs is shown in Figure 11 for comparison with the commercial powders. These costs are independent of particle diameter because control of the nucleation process has no obvious consequence to equipment costs or mass flow rates. Even with probably higher than actual cost figures, the laser heated gas phase synthesis process has a much lower manufacturing cost than the sales price of present submicron SiC powders. Also, these powders have a narrow size distribution and are free of unwanted contaminates.

B.   Films from Laser Heated Gases

Thin film semiconductor materials must also satisfy a rigorous set of criteria to be useful. Hydrogenated amorphous (a-Si:H) is a leading candidate for solar cell and thin film transistor applications;[26,27] the principal difference between the two are the permissible processing costs and processing energies. The common requirements include precisely defined electrical, optical and physical properties. This processing research has concentrated on a-Si:H from SiH_4 because it is so important in its own right and it serves as a model for other materials.

In this material, hydrogen saturates dangling bonds and relieves strains resulting in a reduced defect level and the ability to modulate the Fermi level by substitutional doping. a-Si:H is usually produced by the glow-discharge decomposition of silane[8] (SiH_4) or by reactive sputtering in argon-hydrogen mixtures.[7] Structurally superior amorphous films are produced by chemical vapor deposition (CVD) processes,[6] but they contain insufficient H (< 1 at%) to

TABLE 3

MANUFACTURING RATES AND COSTS

Assumptions
Gas tip diameter   = 2.0 cm
Gas velocity   = 500 cm/sec
Gas pressure   = 1 atm
Reaction efficiency   = 100 %
Capture efficiency   = 100 %
Operational efficiency = 91 % (8000 hr/yr)

Heating -

        0 - 700^{0}C by resistance
        700 - 1000^{0}C by laser
        no heat recovery

Results	Si	Si$_3$N$_4$	SiC
Production rate: kg/hr	7.2	5.1	6.8
ton/yr	57.6	40.8	54.4
Laser power requirement (kW)	1.5	1.3	1.5
Total process energy (kWhr/kg)	1.8	2.1	1.9

Costs ($/kg)

Labor	0.77	1.09	0.82
Plant	1.66	2.34	1.76
Building	<0.01	<0.01	<0.01
Utilities	0.19	0.25	0.21
Raw materials	2.28-22.80	1.62-13.95	1.77-16.17
Total	4.29-25.44	5.32-17.65	4.57-18.97

achieve good electronic properties because high substrate temperatures ($T_s \sim 600^0C$) are needed to obtain reasonable deposition rates.
In the laser induced CVD (LICVD) process, high deposition rates can
be achieved at low substrate temperatures leading to films containing
adequate H.

1. Thin Film Apparatus. The apparatus used for the laser
induced CVD process (LICVD) is shown[3] in Figure 12. Substrates are
mounted on a temperature controlled stage in a hermetic reaction
cell; substrate temperature is monitored by means of a thermocouple
in contact with the stage. Although substrate heating is not
necessary for deposition to occur, the substrate temperature has a
significant effect on the properties of the resultant films and is
an important variable in process optimization. Generally, substrate
temperatures have been in the 200 to 400^0C range for these experiments.[11] Untuned CO_2 and grating-tuned CO_2 lasers have been used
in the cw mode. The axis of the laser beam is made to pass through
the cell, parallel to the substrate plane at a distance of approximately one beam diameter. The unfocused beam diameters are 6 mm.
A lens with a focal length of 12.7 cm was used in the experiments
involving a focused beam. Incident laser intensities have been in
the 100 to 1000 watts/cm^2 range.

Pure silane (4 to 10 torr), has been used in most of the experiments. SiH_4/He and SiH_4/H_2 mixtures have been used in a few experiments. Films have been deposited with both vertical and horizontal
substrate orientations, and under both constant volume (static cell)
and constant pressure (flowing gas) conditions. In the flowing gas
experiments, silane is admitted through a nozzle below the substrate
stage. The gases pass through the laser beam and out the top of the
cell through a filter and throttling valve assembly that maintains
a constant pressure. A variety of substrate materials including
borosilicate glass, aluminum, aluminum oxide, single crystal silicon
and borosilicate glass coated with platinum, molybdenum or indium-
tin oxide has been used. Deposition rates were measured by an
interferrometric technique employing reflected He-Ne laser light.

The LICVD technique has similarities to the HOMOCVD technique
reported by Scott et al.[28] in that the reactant gas and substrate
temperatures can be different from one another and also it proceeds
as a homogeneous thermal reaction. However, LICVD is not subject
to parasitic reactions on the hot walls nor to contamination from
heated surfaces. Bilenchi et al.[29] have recently reported the
deposition of a-Si:H films by a technique very similar to the one
described here.

2. Process Modeling. The molecular mechanisms by which SiH_4
decomposes to a-Si:H films are not certain; but, they probably
involve a step which produces SiH_2(g) and H_2(g) intermediate
products.[30] For both conventional and HOMOCVD processes, the

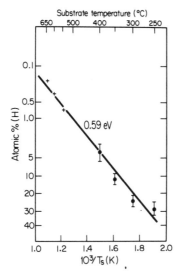

Figure 14.

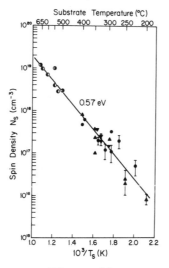

Figure 15.

overall deposition rate is linearly dependent on SiH_4 gas pressure
and exponentially dependent on gas temperature. Our modeling[31]
has proceeded on this basis, concentrating on the calculation of the
laser heated gas temperatures and measurement of growth rates.

Local gas temperatures are calculated by equating the local
power absorbed to the rate heat is lost to surrounding surfaces.
The heat losses are dominated by regions where surfaces are close
to the beam or where they are far from the beam. The close regions
include the entrance and exit windows and the temperature controlled
substrate. The far regions are those points along the beam path
that are distant from the window regions and the substrate. The
heat loss calculations, though tedious, are straightforward and are
subject primarily to the errors in thermal conductivity resulting
from uncertain gas compositions. Definition of the local laser
intensity and local absorptivity is much more uncertain. Figure 1
showed that absorptivity is sensitive to pressure and beam power.
Figure 13 shows that it is also extremely sensitive to gas composition in a counter intuitive direction. As the SiH_4 is consumed and
diluted with H_2, absorptivity actually increases due to pressure
broadening effects at low total gas pressures. Local laser intensities and average absorptivities are estimated by measuring input and
exiting CO_2 laser beam powers. Calculated gas temperatures within
the CO_2 laser beam path are 440-550°C under most growth conditions.[31]

An Arrhenius plot of growth rate is linear in the region over
which our estimate of the gas temperature is expected to be accurate,
yielding an activation energy of 48 kcal/mole. This value is close
to the 50-55 local/mole reported for conventional pyrolysis of
SiH_4.[32] Actual growth rates up to 160 Å/min. have been achieved.
Higher rates can be anticipated now that the transition to homogeneous nucleation of powders caused by excessive temperatures at
points far from heat sinks is understood better.

Growth rates are defined by the gas temperature and SiH_4 partial
pressure. The substrate temperature effects growth rate indirectly
through its influence on the laser heated gas temperature in the
proximity of the substrate. Growth rates increase with SiH_4 flow
rate by an as yet undetermined effect. With that exception, the
growth process and thresholds to powder synthesis are reasonably
defined.

3. Film Characteristics. Adherent films up to 0.5μm thick
have been deposited on all substrate materials. Adhesion and
stresses have been toublesome for thicker films ($\geqslant$ 2μm) on substrates with poorly matching thermal expansion coefficients. All
films were determined to be amorphous by TEM.

The hydrogen content in the $a-SiH_4$ films has a dominant effect
on spin density, optical gap, and electrical conductivity. The

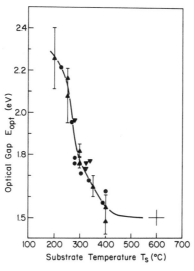

Figure 16.

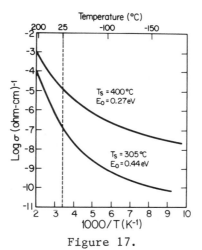

Figure 17.

hydrogen concentration was determined by the effusion technique.[33] These results are plotted as a function of the recriprocal of the absolute substrate temperature in Figure 14. The hydrogen content in the LICVD films depends primarily on the substrate temperature over a wide range of concentrations. The absolute concentrations are much higher than conventional CVD films and can be made higher than is achieved by other deposition processes.

Electron spin resonance (ESR) measurements indicated the existence of neutral dangling-bonds (g = 2.0055). The unpaired-spin concentration, N_s, is a sensitive function of only T_s. Figure 15 shows an Arrhenius plot of N_s versus T_s for both LICVD and conventional CVD a-Si:H films.[33] All the data are fit by the relationship, N_s= (1.5 x 10^{22}cm^{-3} exp(-0.57 eV/kt$_s$) emphasizing the basic similarities of the two processes. However, the LICVD process permits films to be deposited at T_s values that are inaccessable to conventional CVD. The N_s values, $\geqslant$ 8 x 10^{15}cm^{-3}, equal other high-quality films and importantly do not exhibit the minimum in N_s observed near T_s = 300^{0}C for glow-discharge[34] or HOMOCVD[35] deposited a-Si:H. These LICVD results show that an equilibrium number of dangling bonds form during growth at T_s with a formation energy of 0.57 eV.

The variation of the optical absorption coefficient (α) as a function of photon energy (hν) provides information about the "optical" gap (E_{opt}) defined by the expression, $(\alpha h\nu)^{1/2}$ = (hν – E_{opt}).[36,37] Figure 16 shows E_{opt} as a function of T_s for films produced at laser powers of 40 and 60 watts using a 6mm diameter beam. Values of E_{opt} between 1.5 and 2.0 eV are in agreement with those of typical glow-discharge films.[8] The decrease of E_{opt} with increasing T_s is most likely due to the decreasing H concentration in the films. We expect that the admixture of the stronger Si-H bonds ($\approx$ 3.4 eV) to the Si matrix (the Si-Si bond energy is $\approx$ 2.4 eV) will tend to increase E_{opt}.

Figure 17 represents an Arrhenius plot of the electrical conductivity (σ) for two films deposited at T_s = 400^{0}C (curve A) and 305^{0}C (curve B) made by using a planar sample configuration with a distance of 5 x 10^{-2} cm between two aluminum contacts. The activation energies of E_0 $\sim$ 0.27 eV (T_s = 400^{0}C) and E_0 $\sim$ 0.44 eV (T_s = 305^{0}C) deduced from the high-temperature regions (T $\geqslant$ 50^{0}C) most likely represent the energy difference between the Fermi level and the nearest mobility edge, presumably that of the conduction band. The increase of E_0 with decreasing T_s is probably due to the effects of hydrogen which both increases the gap and tends to lower the Fermi level. The curvature of the Arrhenius plot in the lower temperature region (T $\leqslant$ 25^{0}C) can be explained by the onset of electronic transport by hopping between localized states near the Fermi level, both between the nearest neighbors and more distant sites (variable-range hopping).[38] At low temperatures, log σ is proportional to $T^{-1/4}$.

CONCLUSIONS

Two new laser induced gas phase synthesis processes have been developed successfully. Powders are made by one and thin films by the other. Each was developed to produce materials having highly specific properties that are not achievable with other available processes. For both, process variables can be established at precisely defined levels and the processes can be monitored, serving as a basis for exercising real-time process control.

Both processes permit stated objectives to be met. For the powders these include small and uniform diameters, freedom of agglomerates, spherical shapes, high purity and acceptable projected manufacturing cost. This success provides the basis for proceeding to a pilot scale process. The CVD process is less developed; but, we have shown that high deposition rates and superior microstructural quality, characteristic of high temperature CVD processes, can be combined with superior electrical characteristics normally achievable only with glow discharge or sputtering processes. Both permit product characteristics to be manipulated over wide ranges through varied process conditions and both are applicable to many materials. We anticipate both will see commercial utilization.

ACKNOWLEDGEMENTS

Many students and staff have made important individual contributions to the research discussed in this paper; hopefully, all are referenced properly. Also research support has been contributed by several sponsors. These are the U.S. Department of Defense through DARPA, ONR and AROD, NASA-Lewis and the 3M Company. Both contributions are gratefully acknowledged.

REFERENCES

1. W. R. Cannon, S. C. Danforth, J. H. Flint, J. S. Haggerty, R. A. Marra, "Sinterable Ceramic Powders from Laser Driven Reactions; Part I: Process Description and Modeling", J. Am. Ceram. Soc., 65, 7:324-30 (1982).
2. W. R. Cannon, S. C. Danforth, J. S. Haggerty, and R. A. Marra, "Sinterable Ceramic Powders from Laser Driven Reactions; Part II: Powder Characteristics and Process Variables", J. Am. Ceram. Soc., 65, 7:330-5 (1982).
3. T. R. Gattuso, M. Meunier, D. Adler, and J. S. Haggerty, "IR Laser-Induced Deposition of Thin Films", in Laser Diagnostics and Photochemical Processing for Semiconductor Devices, R. M. Osgood, S. R.J. Bruek, H. R. Schlossberg, Eds., North-Holland, 215-221, (1983).
4. E. Barringer, N. Jubb, B. Fegley, R. L. Pober and H. K. Bowen,

"Processing Monosized Powders", Int. Conf. on Ultra-structure Processing of Ceramics, Glasses and Composites, Eds. L. L. Hench and D. R. Ulrich, Wiely & Sons, New York (1983).

5. E. A. Barringer and H. K. Bowen, "Formation, Packing, and Sintering of Monodisperse TiO_2 Powders", J. Am. Ceram. Soc., 65, 12:C199-C210.

6. M. Hirose, "Physical Properties of Amorphous CVD Silicon", J. de Physique, C4:705-14 (1981).

7. W. Paul and D. Anderson, "Properties of Amorphous Hydrogenated Silicon with Special Emphasis on Preparation by Sputtering", Sol. En. Mat., 5:229-316 (1981).

8. H. Fritzsche, "Characterization of Glow-Discharge Deposited a-Si:H", Sol. En. Mat., 3:447-501 (1980).

9. J. H. Flint, "Powder Temperature in Laser Driven Reactions", M. S. Thesis, M.I.T., February 1982.

10. R. A. Marra, "Homogeneous Nucleation and Growth of Silicon Powder from Laser Heated Gas Phase Reactants", Ph.D. Thesis, M.I.T., February 1983.

11. M. Meunier, T. R. Gattuso, D. Adler, and J. S. Haggerty, "Hydrogenated Amorphous Silicon Produced by Laser Induced Chemical Vapor Deposition of Silane", Appl. Phys. Lett., 43:273 (1983).

12. J. S. Haggerty and W. R. Cannon, "Sinterable Powders from Laser Driven Reactions", MIT-EL 79-047, ARPA Order No. 3449, Dept. of Navy, Office of Naval Research, Arlington, VA., Contract No. N00014-77-C-0581, July 1979.

13. G. Greskovich, and J. H. Rosolowski, "Sintering of Covalent Solids", J. Am. Ceram. Soc., 59:285-8 (1976).

14. A. D'Slessio, A. Dilorenzo, A. F. Sarofim, F. Beretta, S. Masi, and C. Venitozzi, "Soot Formation in Methane-Oxygen Flames", Fifteenth Symposium (International) on Combustion 1427, (1975), The Combustion Institute, Pittsburgh, PA.

15. J. S. Haggerty, "Sinterable Powders from Laser Reactions", MIT-EL 82-002, Final Report N00014-77-C-0581, September 1981.

16. Y. Suyama, R. A. Marra, J. S. Haggerty, and H. K. Bowen, "Synthesis of Ultrafine SiC Powders by Laser Driven Gas Phase Reactions". Submitted for publication to the J. Am. Ceram. Soc., October 1982.

17. S. Mizuta, W. R. Cannon, A. Bleier, and J. S. Haggerty, "Wetting and Dispersion of Silicon Powder Without Deflocculants", Am. Ceram. Soc. Bull., 61:872-5 (1982).

18. S. C. Danforth, (Rutgers University, NJ), and M. Dahlen, (Volvo, Sweden), unpublished results.

19. G. Garvey, M.I.T., unpublished results.

20. R. S. Aries, and R. D. Newton, Chemical Engineering Cost Estimation, Chemonomics, Inc., N.Y., April (1951).

21. R. H. Baney, Dow Corning, private communication.

22. F. Chambers, Standard Oil Co. (Indiana), private communication.

23. P. Orinsnshky, Union Carbide Co., private communication.

24. R. Cannon, M.I.T., (Berkeley University, CA), private communication.

25. "Comminution and Energy Consumption", National Materials Advisory Board (NAS-NAE), PB81-225708, May 1981.
26. Y. Hamakawa, "Recent Advances in Amorphous Silicon Solar Cells", Solar Energy Materials, 8:101 (1982).
27. P. G. LeComber, A. J. Snell, K. D. Mackenzie and W. E. Spear, "Applications of a-Si Field Effect Transistors in Liquid Crystal Displays and in Intrigated Logic Circuits", J. de Physique, C4:423-32 (1981).
28. B. A. Scott, R. M. Placenik, and E. E. Simonyi, "Low Defect Density Amorphous Hydrogenated Silicon Prepared by Homogeneous Chemical Vapor Deposition", Appl. Phys. Lett., 39:73 (1981).
29. R. Bilenchi, I. Gianinoni, M. Musci and R. Murri, "Laser Induced Chemical Vapor Deposition of Hydrogenated Amorphous Silicon",Laser Diagnostics and Photochemical Processing for Semiconductor Devices, R. M. Osgood, S. R. J. Brueck, H. R. Schlossberg, Eds., North-Holland, 199-205 (1983).
30. M. Meunier, J. H. Flint, D. Adler and J. S. Haggerty, "Laser Induced Chemical Vapor Deposition of Hydrogenated Amorphous Silicon", Elect. Mat. Conf., Burlington, Vermont (1983).
31. M. Meunier, J. H. Flint, D. Adler and J. S. Haggerty, "Hydrogenated Amorphous Silicon Produced by Laser Induced Chemical Vapor Deposition of Silane" Proceedings of the 10th International Conference on Amorphous and Liquid Semiconductors, Tokyo, Japan, August 22-26, 1983.
32. C. G. Newman, H. E. O'Neal, M. A. Ring, F. Leska and N. Shipley, "Kinetics and Mechanism of the Silane Decomposition", Int. J. Chem. Kinetics, XI 1167 (1979).
33. P. Hey and B. O. Seraphin, "The Role of Hydrogen in Amorphous Silicon Films Deposited by the Pyrolytic Decomposition of Silane", Sol. En. Mat., 8:215-30 (1982).
34. H. Fritzsche, "Characterization of Glow-Discharge Deposited a-Si:H", Sol. En. Mat., 3 447-501, (1980).
35. B. A. Scott, J. A. Reimer, R. M. Placenik, E. E. Simonyi and W. Reuter, "Low Defect Density Amorphous Hydrogenated Silicon Prepared by Homogeneous Chemical Vapor Deposition", Appl. Phys. Lett., 40:973 (1982).
36. J. Tauc., "Optical Properties of Non-Crystalline Solids" in Optical Properties of Solids, Ed. F. Abeles, North-Holland, Amsterdam, p. 227-313, (1970).
37. C. D. Cody, C. R. Wronski, B. Abeles, R. B. Stephens and B. Brooks, "Optical Characterization of Amorphous Silicon Hydride Films" in Solar Cells, 2:227-243 (1980).
38. N. F. Mott, and E. A. Davis, Electronic Processes in Non-Crystalline Materials, 2nd Ed., Clarendon Press, Oxford, (1979).

DYNAMIC COMPACTION OF METAL AND CERAMIC POWDERS

Vonne D. Linse

Battelle-Columbus Laboratories

Columbus, Ohio 43201

INTRODUCTION

The advent of powders having unique properties derived from special processing such as rapid quenching offers the potential for materials with significantly improved performance capabilities. The fabrication of these powders into useful structures or components by conventional powder fabrication techniques without unfavorably altering or even destroying their unique properties is oftentimes impossible. As a result, new approaches and techniques for fabricating these and other powders with unique properties are being sought and investigated.

One of the powder fabrication techniques currently being investigated for its potential to fabricate rapidly quenched and ceramic powders is dynamic compaction. Dynamic compaction utilizes a combination of high pressures and velocities to rapidly densify powders. It is applicable to both metals and ceramics or their combination. Although the dynamic compaction process has not been used extensively, particularly in the U.S., it offers the potential to compact these unique powders close to theoretical density without compromising the unique microstructures and/or properties of the powder. While this potential was recognized in a recent National Materials Advisory Board Committee assessment[1], it was also recognized that there was a serious lack of understanding of the process and, as a result, a lack of principal applications. Since that time, however, there has been a significant increase in interest and research efforts to better understand and apply the dynamic powder compaction process.

While at first the process appears to be simple, the phenomena occurring in high-velocity, short-time duration compaction waves are numerous and complex. In addition, a wide number of controlling process variables and parameters are important in the achievement of a structurally sound, high-density powder compact. A broad spectrum of technical disciplines is required to thoroughly understand and utilize the dynamic compaction process.

This review will cover a number of areas that are important to providing a better overall understanding of our current knowledge of the dynamic powder compaction process and its potential for future application in the powder fabrication field. The initial areas covered will include a discussion of the energy source for dynamic compaction, the compaction wave, and the controlling process parameters and factors which will first of all give a better understanding of how the process mechanistically works and is controlled. The remaining areas will then provide an understanding of the potential for future application through the discussion of materials and structures that have been compacted as well as practical and potential applications of the process.

THE ENERGY SOURCE FOR COMPACTION

The high-pressure, high-velocity wave required to compact a powder can be generated by various methods. The most widely used method is by the detonation of an explosive in direct contact with the powder system. The pressure generated in the powder is a function of the explosive detonation pressure (characteristic of the specific type of explosive being used) and the material in which the powder is contained that is in direct contact with the explosive. The duration of the pressure pulse is also a function of the characteristics of the particular explosive being used and its areal mass.

The direct explosive contact method is generally used to make powder compacts with symmetrical geometries such as cylinders (solid or hollow), or in some instances it has also been used successfully to compact flat plates. The explosive is almost always detonated parallel to the surface of the powder compact.

The other method of energy delivery to the powder compact is the impact technique in which a plate or projectile is driven onto the powder system. In this instance, the pressure is determined by the velocity of the projectile at the time it impacts the powder container surface and the high-pressure characteristics of the projectile and container materials. The pressure duration is primarily a function of the thickness of the projectile.

Projectiles are usually accelerated or launched by gas guns which can produce a range of projectile velocities that result in low to moderate impact pressures. The powder compact geometry is limited to that of a thin disk with diameters presently in the range of 5.0 to 7.5 cm due to the limiting practical size of existing gas guns. In spite of these size and geometric limitations, a significant amount of dynamic compaction work has been conducted using gas guns.[2,3,4]

Impacting plates are usually driven by an explosive such that they impact a flat powder compact in a planar fashion. Although rarely used for powder compaction due to the unfavorable shape of shock wave that it generates in the powder, this technique is capable of producing extremely high pressures in a powder compact.

THE COMPACTION WAVE

In order to understand the nature of the compaction wave in the dynamic powder compaction process, it is first necessary to understand the basic nature of the powder that is being compacted as well as the requirements for taking it from the powder state to a fully dense body. Materials can be divided into three categories according to their type and nature: solid, porous or distended solid, and powder. Solid materials are continuous in nature while porous or distended solids are composed of a continuous solid with either continuous or discontinuous pores or voids. Powders on the other hand are distinctly different in that they are a combination of continuous void with discontinuous solid.

During the dynamic compaction process, the powder is trans-formed within the compaction wave from its initial powder state through an intermediate porous solid to a final solid state if full densification is to be accomplished. In order for this to be accomplished, the powder particles must undergo a significant amount of deformation at very high rates within extremely short time periods. It is the precise understanding of the nature of this transformation zone in which the powder consolidation takes place and its relationship to the input pressure and time function that is needed to ultimately understand how to effectively control and utilize the process.

The general nature of the phenomena occurring in the powder compact during the dynamic compaction process has been known for some time[5]. A schematic and high-speed radiograph for conformation in Figure 1 show the general nature of the process with respect to the powder (before and after compaction), the powder container and its movement during compaction, the explosive and its detonation front, and the location of the compaction front. Obviously missing is any definition or resolution of the true particulate nature of the mechanisms of powder densification and interparticle bonding that are occurring in the compaction zone behind the compaction front. Until recently, little attempt has been made to improve this understanding.

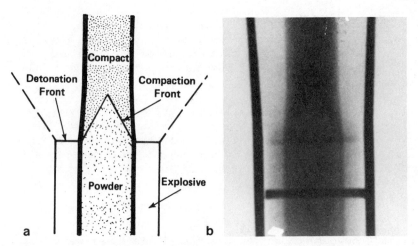

Figure 1. (a) Schematic and (b) High Speed X-ray of the Events that Occur During the Explosive Compaction of a Cylindrical Powder Sample.

If one looks at the classical shock wave in a solid, continuous material as shown schematically in Figure 2, the observer at the shock front would see the material jumping in an instantaneous and discontinuous fashion from its initial state in front of the shock front to its final state immediately behind the front. These conditions can be mathematically defined by the jump equations defining the states of the material ahead of and behind the shock front[1]. There is no allowance for a transformation zone of finite width as would be logically required in the compaction of a powder. This concept is valid only for strong shock waves where the material can be treated like a compressible fluid. At lower shock pressures, most materials behave as an elastic-plastic solid with a yield point that results in the formation of a two-phase shock wave consisting of an elastic precursor followed closely by a slower shock wave at the peak pressure. Even in this latter case, the shock wave is discontinuous and its width and time are usually small.

If one considers a distended solid that contains continuous or discontinuous pores or voids in which the solid material is still continuous, the velocity of the shock wave, U, decreases approximately as a linear function of increasing porosity[6]. Due to the continuous nature of the material, however, the shock wave still

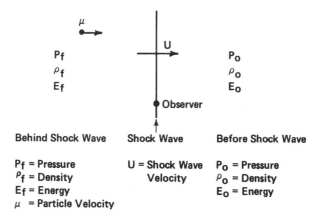

Figure 2.   Classical Shock Wave in a Solid Material.

maintains its narrow, discontinuous nature. Furthermore, to completely close and eliminate the pores or voids will require pressures in excess of the Hugoniot elastic limit of the material. As a result, significant energy must be expended to totally deform the continuous solid matrix and eliminate the voids to accomplish complete consolidation. An indication of the energy absorbed during the complete densification of a porous or distended solid structure by a shock wave in comparison to the energy dissipated by the passage of a shock wave through a solid material is schematically shown in Figure 3[7]. The difference in the areas bounded by the Rayleigh line and the release isentrope of the Hugoniot for each form of the material represents the relative difference in energies dissipated.

Due to its entirely different nature of being a discontinuous solid, a powder causes a shock wave to become extremely complex in nature. This is due to the fact that the shock wave cannot be transmitted through a continuous solid, but must be transferred from particle to particle either through point/small area contacts or by particle-particle collisions. When low to medium pressure shock waves typically used during dynamic compaction are introduced into the powder, the powder is only capable of transmitting a very weak or low pressure wave through these interparticle contacts or collisions. The distinct, discontinuous nature of the shock wave described above for a solid or distended solid material no longer occurs. Instead, the shock wave becomes a slower moving pressure zone having a rather broad width in which the pressure and density increase more gradually over the width of the zone as shown in

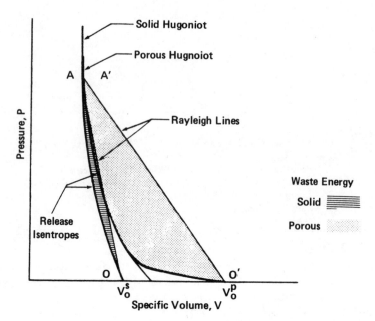

Figure 3. Comparison of Energies Expended During the Passage of a Shock Wave Through Solid and Porous Media. (Jones)

Figure 4. The time period covered by the passage of the zone now becomes much more significant. This is the consolidation zone in dynamic compaction. The events that occur on a particulate or microscale within this zone are critical to understanding the dynamic compaction process.

In conventional powder densification, conducted basically under static pressure conditions on metal powders, Donachie and Burr[8] noted three distinct stages of powder densification shown by the three regions in Figure 5. The first of these stages is particle rearrangement in which significant relative particle movement occurs at low pressing pressures and when the void content is greater than approximately 30-40 percent. As pressure increases, the densification passes into the second stage - particle yield. In this stage the particles must yield and begin to exhibit significant deformation with increasing consolidation. The third and final stage, particle fracture, is then reached. In this stage very high pressures are required with little increase in void elimination realized. Continued densification in this region results in particle fracture and will eventually lead to bulk fracture of the material.

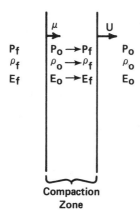

Figure 4.   The Nature of a Weak Shock/Compaction Wave in a
Powder Assemblage.

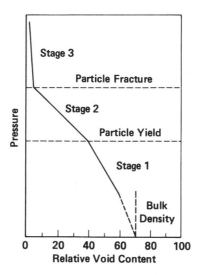

Figure 5.   Regions of Densification During Static Pressing of
a Powder.

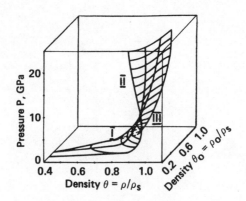

Figure 6.   Three Regions of Densification During Explosive
            Compaction of Powders.

        A similar three-stage mechanism of densification has also been
noted in explosive compaction[9] and is shown graphically in Figure 6.
The first region or stage (I) is likewise noted to be one of particle
rearrangement in which significant densification is achieved at
relatively low pressure levels with little interparticle bonding.
The second region (III) is the region in which the majority of the
particle deformation and interparticle bonding occurs. The third
and final region (II) is indicated to be one in which consolidation
may actually be reversed due to the presence of excessively strong
shock waves. This would cause severe cracking and breakage of the
interparticle bonds which would account for the claimed loss of
density.

        Recent studies at Battelle have confirmed the occurrence of
three mechanistic regions of densification in the compaction zone
that agree in general with both the static and dynamic mechanisms
described in the paragraphs above. These are shown schematically
on a particulate scale in Figure 7. The first region of particle
rearrangement occurs rapidly in the beginning of the compaction
zone and is quite narrow, having a width of only two to three
particle diameters. Relative particle movement is initiated but
takes place with little deformation as the particles rearrange at
low pressures in this region. Almost no interparticle bonding is
observed.

        In the second region, the initial point contacts between
particles change into areas of contact which grow rapidly as densification increases. The frictional heating caused by the relative
movement or sliding of the contact areas creates localized thermal
softening of the particle surfaces. As a result most of the particle deformation and void elimination during consolidation takes

place in these thermally softened outer surface layers and not in the bulk volume of the particle. It is desirable that this second region be as wide and long in time span as possible to allow for maximum depth of thermal softening without particle melting. It is also desirable that consolidation be completed in this region before the third and final region is reached and entered.

If the thermally softened layers are not sufficient in depth to accommodate the total flow necessary to complete densification, before the third region is reached, fragmentation of particles and the compact results. The third region is one in which the particles lock up and relative particle movement stops. The compact now has become a true distended solid structure as described earlier. Further consolidation can now only be achieved through bulk deformation at very high pressures with the attendant strong, damaging shock waves. Bulk consolidation in this region to achieve a high density material with structural integrity is impossible.

Based on the above description, the pressure-specific volume path followed and its approach to the solid Hugoniot during dynamic compaction will be similar to that shown in Figure 8. Due to the localized thermal softening and surface deformation without bulk deformation of the particles or total structure, the solid Hugoniot is approached at relatively low pressure levels with low levels of energy dissipation in the compaction wave. These arguments demonstrate the reason why the solid Hugoniot and full densification should be reached in the second region. The dynamic compaction energies must be distributed primarily into consolidation and not fracture with the solid Hugoniot being approached in the second region and not in the third.

THE CONTROLLING PROCESS PARAMETERS AND FACTORS

The dynamic compaction process is controlled or affected by a wide number of parameters and factors. The degree of influence of each of these parameters and factors varies considerably with some of them having a major influence while others are negligible and can be ignored in most instances. The general categories and some of the parameters and factors that control the process are:

- Bulk properties and characteristics of the material (mechanical and physical properties)

- Geometry and size of powder sample

- Powder characteristics (particle size, size distribution, shape, etc.)

- Powder container (container material, mechanical properties, dimensions, mass, etc.)

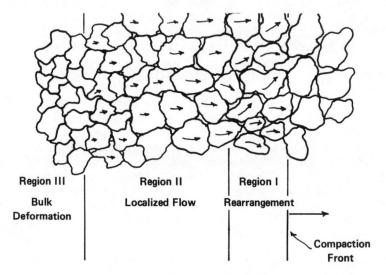

Figure 7.   Mechanisms of Densification in the Dynamic
             Compaction Process.

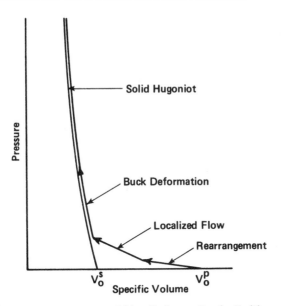

Figure 8. Pressure-specific Volume Path Followed During
           Dynamic Compaction.

- Nature of the powder in the container (green density,
  orientation factor, evacuated or unevacuated, etc.)

- Explosive system (explosive type, density, detonation velocity, detonation pressure, loading, etc.)

- Nature of the compaction wave (shape, pressure, time duration, etc.)

- Initial powder temperature

- Post compaction preservation

    A few investigators have in the past attempted to correlate
the results in dynamic compaction to the input parameters and
material factors. Leonard, et al,[5] were the first to investigate
the effect of input parameters on the dynamic compaction process
itself. They recognized that the ratio of the mass of the explosive
to the mass of the powder (E/M) was a controlling factor in compaction and that it appeared to correlate to the static compressive
yield strength of the material. They further noted that extreme
pressures were not required to yield good compaction. Pressures
slightly in excess of the static yield strength were sufficient.

Later studies by Prummer[10] confirmed the importance of the mass
ratio (E/M) as an important parameter in the dynamic compaction
process and further recognized it as a measure of pressure duration.
He also showed a relationship between the mass ratio (E/M), the
detonation pressure in the explosive, and the final density achieved
in the powder compact. A final important finding shown by his data
revealed a correlation between the initial particle size of the
powder and the final achievable density in the compact with larger
particle sizes yielding higher final densities. Similar findings
on the effect of particle size were also indicated in studies by
Roman, et al[11]. Almost all other investigators have attempted to
relate compaction results strictly to the shock pressure in the
powder without consideration of any other parameters such as
particulate characteristics.

In the previously mentioned studies recently conducted at
Battelle, a number of the more important explosive compaction
parameters were studied to more precisely establish their effect on
the consolidation of powders. Those parameters which were studied
and found to have the most significant impact on the explosive
compaction process were:

- Pressure

- Explosive/material mass ratio (E/M)

- Powder particle size

- Starting powder temperature

The following summarizes the findings and results of explosive compaction experiments conducted on cylindrical or rod-shaped powder
samples with a number of metals and ceramics.

Pressure

Accurate constitutive equations and the availability of reliable data to describe low to moderate pressure shock waves in
powders have been limited. The densification path followed as a
function of increasing pressure is important to controlling the
dynamic compaction process from a parametric viewpoint. Beyond the
basic particle rearrangement caused by the initial weak shock wave
at the front of the compaction zone, almost all of the compaction
is brought about by the powder container wall pushing on and transmitting pressure to the powder over an extended time period during
which the reaction zone in the explosive is acting on the container.
The pressure developed in the container wall by the explosive is,
therefore, the critical pressure parameter.

As indicated from earlier studies[5], the pressure necessary to compact a powder is not very high and needs to be only high enough to initiate and sustain localized flow and thermal softening during the duration of the compaction process. Although difficult to accurately quantify, the actual pressure in the powder during dynamic compaction is probably near or slightly higher than the static compressive yield strength of the material being compacted. It may, in certain cases, be even lower than the compressive yield strength.

Mass Ratio (E/M)

In determining the E/M for a given explosive compaction operation, the mass, M, must include both the powder and the powder container acted on by the explosive. The E/M is a direct measure of the time that a given type of explosive applies pressure to the container wall and hence the powder. The time relates to the reaction zone length in the explosive. Once the threshold pressure to produce the desired flow and thermal softening is reached, the E/M must be sufficient to provide the pressure dwell time to complete consolidation without proceeding into the third region of bulk deformation.

The combination of the pressure and the E/M, therefore, determines the total energy input to the powder compact and its resultant density. When the percent theoretical density achieved in a powder compact is plotted as a function of the pressure in the container and the E/M applied, a three-dimensional surface is generated as shown in Figure 9.

Particle Size Effect

Extensive explosive compaction studies have recently been conducted at Battelle with alpha alumina (Al_2O_3) to more specifically determine the effect of starting powder particle size on the final density achievable in the compact. Four different lots of alumina powder having mean equivalent spherical diameters of 0.57, 4.0, 59, and 300 microns were explosively compacted at the appropriate parameters to identify the maximum practical densities that could be achieved (second region of densification). All of the samples had a cylindrical (rod) geometry and were compacted at room temperature. The maximum practical densities achieved as a function of the mean particle size are shown in Figure 10 along with the equivalent pertinent data generated by Prummer[10] also for alumina and Bhalla and Williams[12] work on selected metals. Although slightly offset from each other with respect to maximum density achieved, the curves resulting from the alumina data generated both at Battelle and by Prummer show a distinct break in density achievable at a particle size of about five microns. Below this break density increases rapidly with increasing particle size, while above

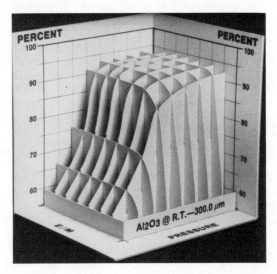

Figure 9. Plot Showing Final Achievable Density (% of T.D.)
as a Function of the Pressure in the Container and the Explosive/
Material Mass Ratio (E/M) Applied to 300 Micron Particle Size
Alumina Powder During Explosive Compaction at Room Temerature.

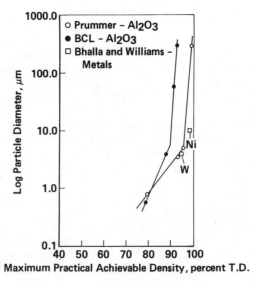

Figure 10. Plot of Maximum Practical Achievable Density in
Powders as a Function of the Mean Particle Size.

the break the increase in densification achieved is less dramatic with further increase in particle size. From this and similar data observed from the explosive compaction of other ceramics and metals in the Battelle studies, the minimum particle size should be five microns or preferably larger particularly if higher-strength materials such as ceramics are being compacted. Most metal powders appear to compact to somewhat higher densities than ceramic powders having an equivalent particle size.

The reason for the slight offset in maximum density achieved in the Battelle study and reported by Prummer for the compacted alumina particularly at the larger particle sizes has not been resolved. The specific nature of the powder (apparent density, size distributions, etc.) and the method(s) used to measure final compact densities used by Prummer are not known.

The effect of particle size on the pressure-specific volume path followed during densification and the resultant density achievable can be seen in Figure 11. In the first region of particle rearrangement, the particle size has almost no effect on the densification. In the second region of localized thermal softening and particle deformation, the effect of particle size is most dramatic. Smaller particles experience less relative movement, thermal soft-

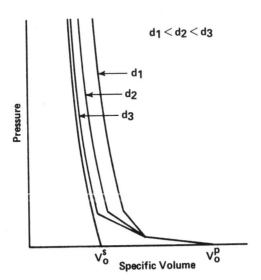

Figure 11. Effect of Powder Particle Size on the Pressure-Specific Volume Path Followed and Density Achievable During Dynamic Compaction.

ening, and deformation. As a result, they lock up and enter the third region of bulk deformation at a much lower density than larger particles.

Temperature Effect

One of the major problems found in the dynamic compaction of high strength, low-ductility materials is the difficulty in achieving high densities (near theoretical) with good interparticle bonding and without severe cracking in the compact. One of the solutions applied to this problem has been elevated temperature compaction in which the powder is preheated prior to compaction[13,14]. The powder needs to be preheated to a temperature which is sufficiently high that the strength level of the powder material decreases and the ductility increases. Preheating allows greater thermal softening to occur in the powder during compaction resulting in greater densification at lower pressures with less tendency for cracking. The resulting higher temperature generated at the particle boundaries by the preheating with the added friction induced thermal excursion also enhances the interparticle bonding.

An example of the improvement in maximum achievable density with increasing compaction temperature for aluminum nitride is shown in Figure 12. While the aluminum nitride could only be compacted to a practical density of slightly less than 90 percent of T.D. at room temperature, a density of almost 98 percent of T.D. could be achieved at 1100 C. In a similar but more dramatic comparison, titanium diboride, which could not be explosively compacted with satisfactory interparticle bonding and structural integrity at room temperature, was compacted into a crack-free, high-integrity compact having a density of 98.4 percent T.D. when compacted at a temperature of 1100 C. This compacted sample had a fracture toughness (K_{Ic}) value of 7.73 MPam$^{1/2}$ using the chevron notch beam test.

The effect of temperature on the pressure-specific volume path followed during dynamic compaction and the density achievable is shown in Figure 13. The effect of elevating the powder temperature during compaction can be observed even in the first region of particle rearrangement particularly at the higher temperatures. This results from the particles being softened sufficiently to allow some of the high points on the particles to be slightly deformed even at the low pressure levels experienced in this region. In the second region, particle deformation and flow is much more extensive even at considerably lower pressure levels than are required in room temperature compaction. In most materials, it is easy to approach theoretical density and the solid Hugoniot curve at reduced compaction pressures when the temperature is raised sufficiently.

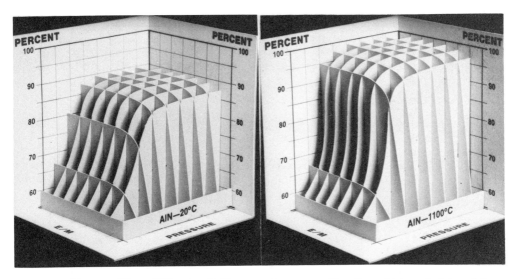

Figure 12. Effect of Temperature on the Densities Achievable
in Aluminum Nitride at (a) Room Temperature and (b) 1100 C.

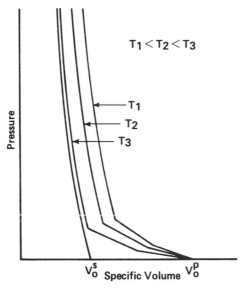

Figure 13. Effect of Temperature on the Pressure-Specific
Volume Path Followed During Dynamic Compaction.

MATERIALS COMPACTED

The materials that have been compacted by explosive compaction are extensive and diverse[1]. The types of materials compacted can be divided into the following catagories:

- metals

- metal-metal composites

- ceramics

- ceramic-ceramic composites

- ceramic-metal composites

- other materials (organics, amorphous materials, etc.)

While the list of materials compacted is extensive, almost all of them have been compacted on an investigative or demonstration basis only. They have not been or are not being compacted on a production or commercialized basis. This situation is primarily due to the incomplete understanding of the basic process phenomena and the resultant inability to solve the problems (cracking, density and property gradients, etc.) associated with the dynamic compaction process in the past. Current research and development efforts, however, are rapidly correcting this situation which will allow more extensive practical use of the process in the near future.

MICROSTRUCTURES OF COMPACTED MATERIALS

The microstructures that can be observed in dynamically compacted materials can be widely varied depending on the specific nature of both the materials being compacted and the compaction conditions being applied to the materials. In most instances, the general nature of the original particles or grains for the most part will be retained in the microstructure as shown in Figure 14. This is particularly true for metals where the initial temperature is sufficiently low such that bulk recrystallization and grain growth does not occur. While the general nature of the particles is retained, their character may be changed significantly in that they may be distorted and exhibit both strain and thermal gradient effects within the individual grains. The latter aspect of thermal gradient effect will be particularly evident near the particle boundaries where the frictional heat and thermal softening is caused by the relative movement of the particles during compaction.

Although much more difficult to metallographically discern ceramic microstructures will also tend to retain evidence of their original particle structure but with less deformation and thermal

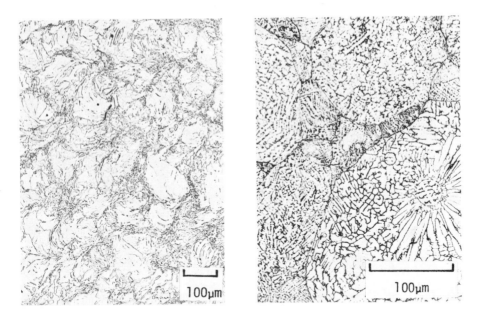

Figure 14. Microstructures Showing Original Particle Structure Retained During Explosive Compaction (a) Tungsten and (b) Nickel-base Superalloy.

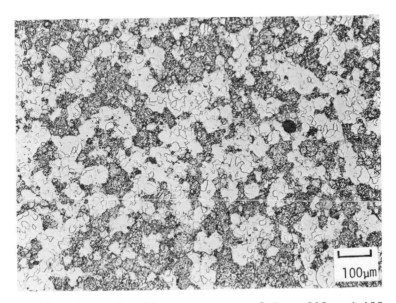

Figure 15. Composite Microstructure of Type 310 and 405 Stainless Steel with the Identity of each Component Being Retained.

gradient effects and greater particle fracturing.

The most important feature of this type of microstructure that is associated with the dynamic compaction process is the fact that the thermal excursion has been sufficiently restrained in intensity and time so that little or no diffusion has occurred between particles. It is for this reason that the dynamic compaction process has excellent potential for the fabrication of composite structures of two or more primary materials in which the identity and nature of each component is retained in the compacted microstructure as shown in Figure 15.

In some instances, it may be desirable to apply the dynamic compaction process to eliminate all evidence of the original particle structure or alter the original microstructure within the particles. Such was the case in the titanium alloy microstructure in Figure 16 where the starting temperature of the powder combined with the thermal excursion imparted during compaction caused the formation of a totally recrystallized microstructure. All evidence of the original particle structure has been eliminated. The input energy in combination with the starting temperature must be controlled closely in this type of compaction to assure that recrystallization occurs yet the localized temperature excursion must not be sufficiently high to cause melting at the particle boundaries. If desirable, this approach can also be used to cause diffusion or a reaction between the components during the compaction of a composite structure.

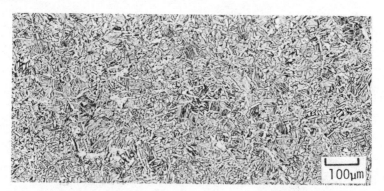

Figure 16. Microstructure in Explosively Compacted Ti-6Al-4V Where the Original Particle Structure has been Completely Eliminated.

STRUCTURES COMPACTED

To date most structures compacted by dynamic compaction have in general been simple configurations. Examples of typical structures compacted by the explosive compaction process are shown in Figures 17 and 18. Rod or cylindrical (solid or hollow) compacts similar to those shown in Figure 17 have traditionally been used in developmental and exploratory studies while plate-type structures as shown in Figure 18 have only been investigated on an occasional basis. The explosive compaction of complex structures having non-symmetric configurations such as the hemisphere in Figure 17 or near-net shapes have received little attention in the past primarily due to an insufficient understanding of the basic nature of the compaction wave and how it can be successfully applied to these difficult to compact, asymmetrical shapes. The overall size of powder compacts that can be fabricated by explosive compaction are at present primarily limited by knowledge and experience with the process. Once these limitations are removed, the process capability will then be more limited by the quantity of explosive that can be safely employed for compaction.

As indicated earlier, compacts fabricated by gas guns will be much more limited in geometry and size than explosive compacts. The geometric capability of a gas gun will be almost totally restricted to thin disk shapes (low l/d ratios) or near disk shapes with maximum possible diameters in the 10-15 cm range.

Figure 17. Structures Compacted by Explosives (Clockwise from the Top: Titanium Alloy Rod, Stainless Steel Hemisphere, Beryllium Rod, and Tungsten Nozzle).

Figure 18.   Nickel Alloy Flat Plate Compacted by Explosives (Scale in Inches).

PRACTICAL AND POTENTIAL APPLICATIONS

Applications of the dynamic compaction process to date have been almost nonexistent. The principal reason for this has been: (1) the lack of a sufficiently good understanding of the process to achieve technically useful fabricated materials and (2) the inability of the process to compete economically with the more conventional powder fabrication processes. As the demand for the more difficult to fabricate, high-value materials increases and our understanding and capability to apply the process in a controlled fashion continues to improve, dynamic compaction will then begin to find the applications for which it will be best suited in the future. The two major areas of highest potential application will be in the powder consolidation and the synthesis and transformation areas.

Although the powder consolidation area has drawn the most interest, it has not had any truly practical applications to date. A number of potential future applications have been identified[1] as having significant potential once the process mechanisms are better understood. One of the potential materials applications for dynamic compaction currently being considered is the compaction of rapidly quenched and solidified materials, both amorphous and microcrystalline. The major driving force for this consideration is the potential for achieving a dense body having structural integrity without subjecting it to a temperature sufficiently high to degrade the unique microstructures and properties generated in the powder by the rapid solidification process. The success of this potential application could in turn lead to a significant increase in the application of the rapidly solidified powders which in some cases currently cannot be fabricated into useful form with the desired properties being retained.

The area of synthesis and transformation, although considered a hybrid process of dynamic compaction, is being used for practical and commercial applications including the synthesis of diamonds and the transformation of graphitic boron nitride to the wurtzite phase[15]. The ultimate usage of both of these hard materials is for grinding, polishing, and cutting tool applications. This area holds considerable promise for a number of other materials and applications in the future.

ACKNOWLEDGEMENTS

The work discussed above to define the controlling parameters and their effect on the explosive compaction process was sponsored by the Defense Advanced Research Projects Agency. I wish to thank Drs. James H. Adair and Hayne Palmour III for their invaluable discussions on the nature of the compaction wave.

REFERENCES

1. Committee on Dynamic Compaction of Metal and Ceramic Powders, "Dynamic Compaction of Metal and Ceramic Powders," National Materials Advisory Board Report Number NMAB-394, Washington, D. C. (1983).
2. D. Raybould, The production of strong parts and non-equilibrium alloys by dynamic compaction, in: "Shock Waves and High-Strain-Rate Phenomena in Metals," M. A. Meyers and L. E. Murr, editors, Plenum Press, NY (1981).
3. D. Raybould, The dynamic compaction of aluminum and iron powder, in: "Proceedings of the 15th International Machine Tool Design and Research Conference," S. A. Tobias and F. Koenigsberger, editors, MacMillan Press Ltd., London (1975).
4. D. Raybould, Wear-resistant aluminum-steel mixtures produced by PM technique which avoids sintering, Powder Metallurgy 23, 1, p. 37 (1980).
5. R. W. Leonard, et al., Advances in explosive powder compaction, Second International Conference of the Center for High Energy Forming, Estes Park, CO (1969).
6. W. H. Gust and E. B. Royce, Dynamic yield strengths of B_4C, BeO, and Al_2O_3 ceramics, J. of Appl. Phys., 42(1), pp. 276-295 (1971).
7. O. E. Jones, Metal response under explosive loading, Engineering Design, pp. 125-148 (1972).
8. M. J. Donachie, Jr. and M. F. Burr, "Effects of Pressing on Metal Powders," J. of Metals, 15(11), pp. 849-854 (1963).

9. O. V. Roman and V. G. Gorobtsov, Fundamentals of explosive
    compaction of powders, in: "Shock Waves and High-Strain-
    Rate Phenomena in Metals," M. A. Meyers and L. E. Murr,
    editors, Plenum Press, NY (1980).

10. R. A. Prummer, Latest results in the explosive compaction of
    metal and ceramic powders, Fourth International Conference
    of the Center for High Energy Forming, Vail, CO (1973).

11. O. V. Roman, et al., Influence of the powder particle size on
    the explosive pressing process, in: "Fizika Goreniya i
    Vzryva," Vol. 15, No. 5, pp. 102-107, U. S. S. R.,
    September-October, 1979.

12. A. K. Bhalla and J. D. Williams, The role of the container in
    the consolidation of powders by direct explosive compaction,
    Fifth International Conference on High Energy Rate
    Fabrication, Denver, CO (1975).

13. V. D. Linse, Dynamic compaction of ceramics and metals. Pre-
    sented at the H. J. Kraner Award Symposium: Innovative
    Forming Methods in Ceramic and Metal Systems, Lehigh Valley
    Section of the American Ceramic Society, Bethlehem, PA
    (1980).

14. V. G. Gorobtsov and O. V. Roman, Hot explosive pressing of
    powders, International J. of Powder Metallurgy and Powder
    Technology, 11(1), pp. 55-60, (1975).

15. L. Davison and R. A. Graham, Shock compression of solids,
    Physics Reports (Review Section of Physics Letters), 55,
    No. 4, pp. 255-379 (1979).

HIGH METAL REMOVAL RATE (HMRR) MACHINING

Adam M. Janotik and Jonathan S. Johnson

Research Staff
Ford Motor Company
Redford, Michigan 48239

THE CONCEPT

The need for improving the productivity of machining operations has encouraged the development of new concepts and techniques, such as thermal machining, as well as high speed and
ultra-high speed machining. High metal removal rate machining
shares the goal of improving productivity with the others, but its
approach is more evolutionary in nature; it attempts to take advantage of step advances in machine and cutting tool technologies,
as they occur, and to incorporate them into machining operations
with a minimum of delay.

To accomplish this, the concept uses a structured optimization
procedure designed to find the most productive cutting conditions
for any combination of machine, workpiece and cutting tool. Productivity gains are achieved by bringing together a number of
elements, each of which contributes its share to the total improvement. Some of the more important elements are:

o   Higher cutting speeds and feeds (subject to constraints dis-
    cussed later).

o   Use of stiffer, dynamically stable machines, allowing deeper
    cuts and larger feeds.

o   More power per spindle.

o   Better utilization of available power in multi-cut operations,
    by varying feed from cut to cut within predetermined accep-
    table limits.

405

o   Constant spindle speeds, where possible, to minimize accelera-
    tion/deceleration losses.

o   Extensive use of optimization software backed by machine and
    tool data bases.

The concept was developed, mainly, for rough turning of high
production volume parts. This explains its emphasis on the optimization of cycle timing -- down to a fraction of a second. The
same techniques, however, could be applied to other types of
operations as well.

Figure 1 shows the HMRR cutting domain for turning and compares it to that associated with "conventional" cutting. It can
be seen to occupy the region immediately above that for "current
production" cutting, but below those for high and ultra-high speed
machining.

Although, theoretically, optimum cutting speeds can be as high
as 30 m/sec, they are generally lower in practice. On typical
automotive cast iron parts, speeds range from 10 to 20 m/sec.
Steels require somewhat lower speeds, aluminum higher speeds.

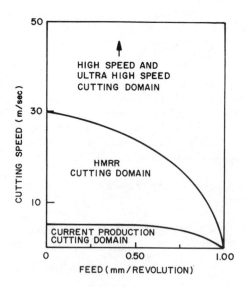

TYPICAL CUTTING SPEEDS AND FEEDS FOR
VARIOUS MACHINGING STRATEGIES

Figure 1

CONSTRAINTS

Ideally, productivity of HMRR would increase indefinitely with cutting speed. In reality, however, cutting conditions are bounded by three constraints imposed by machine, tool life and inertia.

Machine imposed constraints can manifest themselves in one of several ways:

o   Spindle speed capacity of the machine may be too low.

o   Machine may be underpowered, restricting feeds to relatively low levels.

o   Dynamic instability may be present, preventing operation outside of a limited speed/feed envelope.

Tool life imposed constraints include:

o   Excessive tool wear at higher speeds, which causes tool cost to rise to a point where further increases in speed/feed are no longer cost effective.

o   Tool change frequency increases, which reduces time available for cutting, and eventually decreases output.

The inertial constraint affects operations involving rotating workpieces and stationary tools, such as turning, facing, grooving, single point boring, etc. Higher speeds bring with them longer acceleration/deceleration periods in each cycle, until productivity gains due to speed are outweighed by the losses caused by longer acceleration/deceleration times.

OPTIMIZATION

To determine the most productive and/or cost effective production conditions under existing constraints, a systematic optimization procedure is used which calculates cutting conditions based on machine, tool and workpiece characteristics. The necessary inputs may be fed in directly from a data base, or entered individually line by line in response to prompting by the optimization program. Defaults are built into the system for data which may not be readily available to the user.

Three sets of optimum conditions are calculated in terms of speed for a range of feed values:

.   Minimum Cost
.   Maximum Output
.   Optimum Inertial

Cost Optimization

The total cost of an operation can be expressed as:

$$Cost = \frac{(M+U+O)}{P} + E + T$$

where:

M = Amortized machine cost per hour of operation
U = Machine upkeep cost per hour of operation
O = Operator cost per hour of operation
E = Energy cost per part
T = Tool cost per part
P = Output (parts per hour)

Of these, upkeep, energy and tool costs are functions of cutting speed. Unfortunately, upkeep costs vs. speed are not always well documented. As a result, they are generally considered to be speed independent. Under this assumption, M, U and O can be combined into a single value F (fixed cost per hour).

Energy costs are calculated on the basis of type and amount of material removed.

Tool costs are estimated using a modified Taylor relation for tool life, and tool cost per cutting corner:

$$T = \sum_{i=1}^{j} Q_i L_i$$

where:

Q_i = cost of i_{th} tool per corner
L_i = fraction of life of a corner of the i_{th} tool used up in operation

The cost C can, therefore, be expressed as:

$$C = \frac{F}{P} + E + \sum_{i=1}^{j} Q_i L_i$$

This relation is optimized numerically to obtain minimum cost for a range of feeds.

Output Optimization

Optimum output conditions are calculated using the relation

$$P = \frac{60}{t + \sum\limits_{i=1}^{j} L_i Z_i}$$

where:

P = output (parts per hour)
t = Cycle time for operation (minutes), including load/unload, tool approach retract and transfer times
L_i = Fraction of life of a corner of the i_{th} tool used up in operation
Z_i = Tool change time for i_{th} tool (minutes)

Numerically optimized, this relation yields the maximum output constraint.

Inertial Optimization

System rotating inertia constrains only those operations having rotating parts and non-rotating tools. Assuming a simple turning operation, the cutting time (in minutes) can be expressed[1,2] as:

$$t_1 = \frac{Y}{fn}$$

where:

Y = linear length of cut (mm)
f = feed (mm/rev.)
n = spindle speed (rpm)

In this ideal mode, output -- which is a reciprocal of t -- is proportional to speed. Productivity increases indefinitely.

Adding nonproductive cycle time, such as loading, unloading tool transfers etc., cycle time becomes:

$$t_2 = U + \frac{Y}{fn}$$

where:

U - nonproductive cycle time per operation (min.).

Productivity increases with speed at a declining rate, approaching a maximum value which depends on U. This maximum value is the ultimate optimum speed for processes utilizing rotating tools.

For rotating parts, an inertial term must be added to account for spindle acceleration and deceleration to and from the operating speed. Assuming a constant-torque prime mover and an optimum transmission ratio at each speed, total cycle time (rotating parts) becomes:

$$t_3 = U + \frac{Y}{fn} + \frac{Kn^2 J}{H}$$

where:

K = conversion constant (KW - min/N - m - rev^2)
J = total rotating system inertia (kg - m^2)
H = available spindle power (KW)

This relation can be shown to have a minimum value at:

$$n_o = \sqrt[3]{\frac{CHY}{fJ}}$$

where C is another conversion constant. n_0, the optimum inertial speed, represents the ultimate constraint on a machining operation which involves rotating parts. For a given part and preselected feed, the inertial constraint is a function of

$$\left(\frac{H}{J} \right)$$

, the existing power to inertia ration.

Interpretation of Optimization Results

Graphics are used extensively to interpret optimization results. One of the most useful graphic aids is the speed/feed diagram in Log-Log coordinates. An example is shown in Figure 2:

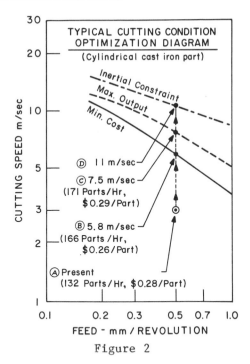

Figure 2

Under current practice a hypothetical turning operation on a cylindrical part might be carried out at 3 m/sec, using a feed of 0.5 mm/revolution. This is represented by Condition A. Total cost of this operation can be estimated to be $0.28 per part, at an output of 132 parts per hour.

By going to HMRR cutting, using a 37.5 kW machine and maintaining the same feed, the minimum cost condition of $0.26 per part is reached at about 5.8 m/sec. The output, now, is 166 parts per hour (Condition B).

The maximum output point of 171 parts per hour is reached at 7.5 m/sec. Cost increases to $0.29 per part (Condition C).

The inertial limit of 11 m/sec (Condition D) represents the ultimate limit beyond which no further productivity improvements are possible with that particular system configuration.

HMRR EXPERIENCE AT FORD

Since the inception of HMRR investigations, considerable knowledge has accumulated in several areas:

o   A high performance chucking machine was built and commissioned
    to provide the capability for experimentation.

o   A large scale test was conducted to evaluate machine as well
    as concept under quasi-production conditions. A gray iron
    automobile transmission part was used in the test.

o   Tool life tests were run at HMRR conditions -- using TiC,
    coated WC and silicon nitride tools -- on gray cast iron, low
    carbon steel and pearlitic nodular iron.

These developments will be discussed in more detail next.

High Performance Chucking Machine

    A high performance laboratory chucker was built. Its major
design objectives were: high dynamic stiffness, 4000 rpm operational capability for the spindle and spindle power in the 37.5 to
75 kW range.

    The machine features a heavy duty cast iron frame, a custom-
built heavy duty 4000 rpm spindle equipped with low compliance
roller bearings, and a 37.5 kW variable frequency controlled AC
motor. A distributed processing microprocessor based CNC controls
the machine.

Large Scale Testing

    To obtain hands-on experience with HMRR cutting, a large scale
test was conducted on the transmission part shown in Figure 3.
Over fifteen hundred of the gray iron parts were machined at
speeds ranging up to 30 m/sec and feeds of up to 1 mm/rev to
verify machine capabilities and to determine tool lives of various
cutting tools at HMRR conditions. The largest fraction of the
tests was run at the inertial speed limit for the part/machine
configuration of about 15 m/sec -- above theoretical optimum cost
and optimum output conditions -- to collect as much experience as
possible at higher cutting speeds.

    Results obtained in the test series are compared to actual
plant results in Table 1.

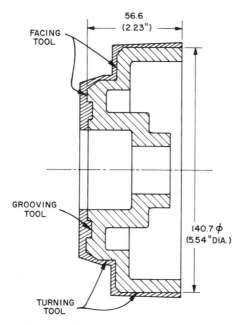

Figure 3:   Transmission Part

Table 1

Comparison of Results
(Large-Scale Turning Test)

	Plant	Laboratory
Power (KW)	15	37.5
Tooling	Coated WC	Coated WC, Silicon Nitride
Feeds (mm/rev):		
Turning	0.3 - 0.5	0.75
Facing	0.3 - 0.4	0.75
Grooving	0.13	0.07
Spindle Speed (rpm)	400 - 600	2140
Cutting Speed (m/sec)	2 - 3.2	5 - 16
Cycle Time (sec)	58	18

Production rate increased from 56 pieces/hr at the plant to 135 pieces/hr in the laboratory, at HMRR conditions, for a productivity gain of 140 percent.

Tool Life Results

Tool life results on gray cast iron are given in Table 2:

Table 2

Tool Life Results – Gray Cast Iron
(Pieces per Corner)

	Plant (Coated WC)	Laboratory (Coated WC)	(Si_3N_4)
Turning	300	75	225*
Facing	90	40	117
Grooving	200	150	200*

*Test suspended

As expected, coated WC tools had shorter lives at HMRR conditions in the laboratory. The higher productivity in the laboratory, however, more than compensated for higher tool usage.

Silicon nitride tools were not tested at plant feeds and speeds. At HMRR conditions, their lives actually exceeded those of coated WC tools in the plant.

SUMMARY

High metal removal rate machining relies on higher than customary speeds, feeds, more powerful machines, cycle time saving devices and accessories as well as extensive optimization techniques to improve the productivity of machining operations, particularly of those involving turning of ferrous parts at high production rates.

The highest productivity improvements, ranging up to 250 percent, were achieved on gray cast iron. Smaller gains (up to 50%) were realized on low carbon steels.

High hardness pearlitic nodular as well as some malleable irons were found to be unsuitable for HMRR machining because of excessive tool wear.

REFERENCES

1. J. E. Mayer, Jr. and D. G. Lee, "Future Machine Tool Requirements for Achieving Increased Productivity," The Carbide and Tool Journal, Society of Carbide and Tool Engineers, Vol 13, No. 3, May–June 1981, p 4-8.
2. A. M. Janotik and J. E. Mayer, Jr., "Optimum Spindle Speeds for High Production Turning," Proceedings 10th North American Manufacturing Research Conference, May 1982, p 342.

HIGH-SPEED MACHINING

D. G. Flom

General Electric Corporate Research & Development
Schenectady, New York

INTRODUCTION

The term "high-speed machining" (HSM) is a relative one from a materials viewpoint because of the vastly different speeds at which different materials can be machined with acceptable tool life. For example, it is easier to machine aluminum at 6000 surface feet per minute (sfm) than titanium at 600 sfm. Because of this difference, and the fact that speed determines to a significant degree whether a material will form continuous chips or segmented, shear-localized chips, one way of defining high-speed machining is to relate it to the chip formation process. Localized shear occurs when the negative effect on strength of increasing temperature due to intense plastic deformation is equal to or greater than the positive effect of strain hardening. In this context, high-speed machining for a given material can be defined as that speed above which shear-localization develops completely in the primary shear zone.

While attractive from a concise technical standpoint, the foregoing is not very useful as a practical definition. For this reason, it is generally preferable to define machining speeds quantitatively in terms of specific ranges. A suggestion by von Turkovich is that 2000-6000 sfm should be termed high-speed machining, 6000-60,000 sfm very-high-speed machining, and greater than 60,000 sfm ultrahigh-speed machining.[1] In the case of very difficult-to-machine alloys, it is preferable to utilize the term "high-throughput machining" (HTM) rather than "high-speed machining" (HSM) in order to keep a proper focus on realistic goals in machining.

417

During the past few years General Electric Corporate Research and Development (GE-CRD) and a number of subcontractors have been engaged in the study of advanced machining for the Department of Defense (DOD) including, as one aspect, high-speed machinining.[2-8] The rationale for pursuing this work is evident when one considers the funds expended in metal removal annually in the U.S. (Table 1)[9]. Of the more than $100 billion spent for this purpose, approximately 75% can be attributed to the four conventional processes of turning, milling, drilling, and grinding. The materials of primary interest to DOD are aluminum alloys, nickel-base superalloys, titanium alloys, and special steels. While our program has been aimed specifically at these four critical classes of metals and alloys, the experimental matrices have included, for completeness, more than a dozen workpiece materials and over twenty grades of cutting tools. The intent of this paper is not to review the earlier work in high-speed machining but rather to summarize some of the findings in the more recent studies. A review of past work has been presented previously.[10]

Table 1. The Machining Problem

o Over $100 billion are spent annually in the U. S. on metal removal.

o Roughly 75% of this amount is spent on turning, milling, grinding, and drilling.

o The critically-needed high-temperature alloys and titanium alloys are among the most difficult to machine.

o Significant reductions in machining costs can only be achieved by including ways to increase metal removal rates.

HIGH-SPEED MACHINING FACILITIES

The high-speed machining facilities used in this program are shown in Figure 1. The GE-CRD HSM spindle shown in the upper left hand corner is a 50,000 RPM, 35 hp spindle mounted on a Springfield grinder modified for milling studies. This is a specially designed facility both because of its simultaneous high speed and high power capability and also because of the method of force measurement used. The shaft of the spindle is hollow and contains strain gages for measuring the dynamic components of force and torque. A frequency modulated (FM) telemetry system is used for transmitting the transducer signal from the rotating high speed spindle.

The facility shown in the upper middle photograph of Figure 1 is a GE-CRD gas gun which is used to propel a workpiece bullet, 1/4 inch in diameter and 1 inch long, past a cutting tool placed at the end of the gun tube. Light detectors at the end of the

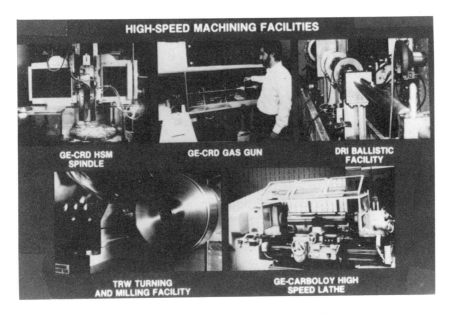

Figure 1. High-Speed Machining Facilities.

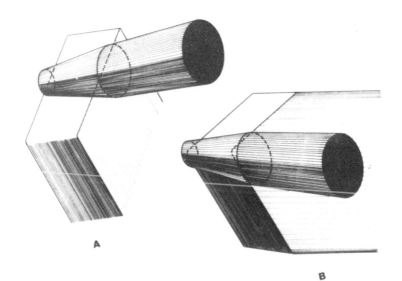

Figure 2. Sketches of the Two Gas Gun Machining Geometries.
(A) orthogonal (B) oblique.

barrel are used to measure the speed of the bullet. A high speed
framing camera can be used for the same purpose and also for
observing the bullet and chip movement. The cutting speeds attained
with this facility have been 20,000 to 60,000 sfm, i.e., within the
very-high-speed machining range. Two types of cutting geometries
used with the gas gun are shown in Figure 2, one corresponding to
orthogonal machining and the other to oblique machining.

A larger ballistic machining facility is that used at Denver
Research Institute (DRI) and shown in the upper right hand corner
of Figure 1. Details of this facility have been described by
Kottenstette and Recht.[11] Long cylindrical workpieces (typically
12 inches long by 0.7 inch diameter), with two premachined flats
on opposite sides, are accelerated in a gun tube, transferred to a
guide tube, and machined by two opposing tools as the workpieces
emerge from the end of the guide tube. The speeds attained with
the DRI facility in this program have been 20,000 to 80,000 sfm,
i.e., in the very-high-speed to ultra-high-speed range. Measurements

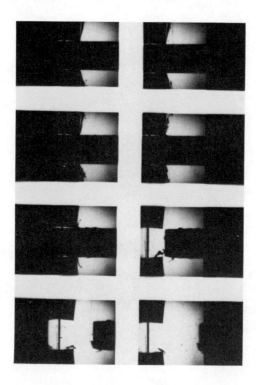

Figure 3. Machining of Mild Steel Workpiece with Carbide Inserts
           at 12,800 smm (42,000 sfm). Frames 1 through 8 at 40
           μsec Intervals.

Figure 4. Machining of Mild Steel Workpiece with Carbide Inserts
at 12,800 smm (42,000 sfm). Frame 6.

and observations have included chip type, cutting force, tool-
workpiece interface temperature, and tool wear. A typical photographic record of a machining event is shown in Figure 3.[2] This
was obtained with a Dynafax camera operating at 25,000 frames/
second; a 1 μs exposure was obtained every 40 μs. In this particular case the workpiece was mild steel being machined at 42,000 sfm.
Reading from bottom up, right to left, the first three frames in
Figure 3 show the workpiece approaching the space between two
carbide tools; the next five frames show the machining in progress.
A blowup of one of the frames is shown in Figure 4. The chips
forming at the two cutting edges are clearly evident.

Neither of the two facilities just described, i.e., the GE-
CRD gas gun and the DRI ballistic facility, have been designed for
practical machining on the manufacturing floor. They have been
used strictly for studying the effects of very high and ultra high
speeds on parameters associated with the cutting process. A more
conventional machine tool is the TRW facility shown in the lower

left hand corner of Figure 1. This facility incorporates two
interchangeable machining spindles to provide cutting speeds in
the range of 500 to 30,000 sfm.[3] The spindles are mounted on plates
permitting adaptation of the system to either turning or milling
modes. The particular photograph in Figure 1 shows turning of
6061-T6 aluminum alloy at 6000 sfm.

Another conventional machine tool used in this program is that
shown in the lower right hand corner of Figure 1. This is a high
speed, 20-inch, 150 hp LeBlond engine lathe at the GE Carboloy
Systems Department (GE-CSD). This lathe has a variable speed
capability of 250 to 5000 RPM which, with the appropriate size
workpiece, has allowed cutting speeds up to 25,000 sfm.

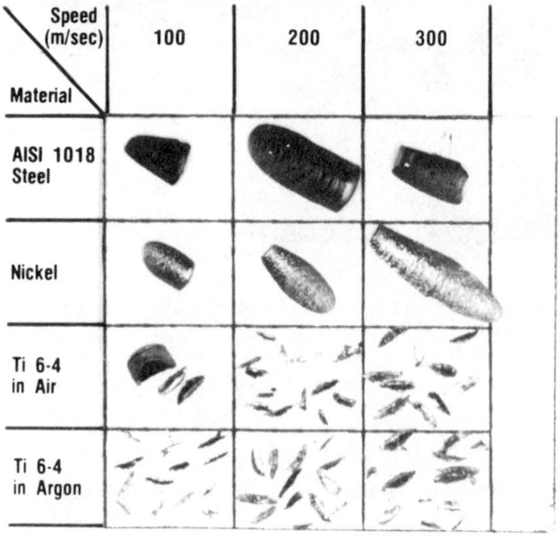

Figure 5. The Chips of Various Workpieces at Three Cutting Speeds.

CHIP FORMATION

The work in this program has resulted in experimental documen-
tation of chip morphology over a broad range of cutting speeds
- 0.004 to 80,000 sfm - and it has been possible to correlate
continuous-to-segmental chip transitions with the thermal diffusi-
vities, structures, and hardnesses of various metals and
alloys.[12-18] Representative chips of AISI 1018 steel, nickel and
titanium 6Al-4V alloy formed in the GE-CRD gas gun at speeds of
approximately 100, 200 and 300 meters/second ($\sim$ 20,000, 40,000 and
60,000 sfm) are shown in Figure 5.[3] At the three speeds noted, the
steel and pure nickel basically form the same type of chips as in
the ordinary machining operations whereas the chips from titanium

alloy form completely segmented and detached chips regardless of
their heat treatment conditions.

With several metals and alloys the degree of segmentation
depends directly upon cutting speed. An example is AISI 4340 steel
for which continuous chips are formed at 400 sfm (Figure 6) an
completely segmented and detached chips at 3200 sfm (Figure 7).
Similarly, Inconel 718 - a nickel-base superalloy, forms relatively
continuous chips below 200 sfm but within the range of 200 to 400
sfm segmentation begins and at higher speeds severe detachment
occurs (Figure 8). Titanium alloys such as titanium 6A1-4V are
unique; we have found them to form segmented chips over a range of
speeds from 0.004 sfm, in machining experiments inside a scanning
electron microscope, to 60,000 sfm, in gas gun experiments.

Some of the aspects of chip morphology observed are summarized in Figure 9. Basically, continuous chips form for many
materials within a particular range of speeds. At higher speeds a
segmental chip appears, composed of distinct trapezoidal segments
separated by thin long layers of intensely deformed material.
Because of this intense deformation - or shear localization - the
chips have come to be referred to as "shear-localized" chips. The
size and spacing of the segments depend on feed and rake angle but
appear to be independent of cutting speed. In general, continuous
chip formation is favored by high thermal diffusivity and (for a
given metal/alloy) low hardness. Chip segmentation is favored by
low thermal diffusivity, hexagonal close packed structure, and high
hardness. Many of the details of chip formation involving intense
deformation have been described in an early paper by Recht[20] and
more recently by Komanduri and Brown[20] and by Komanduri et al.[21]

An analytical model has been developed by von Turkovich relating
the onset of shear localization to cutting speed through the effect
of the latter on strain rate.[5] This analysis has led to the
formulation of a stability principle based on the properties of the
stress-strain-strain rate surface under adiabatic deformation
conditions. As depicted in Figure 10, such surfaces exhibit a
locus of instability which can be expressed by $\frac{\partial \tau}{\partial \gamma} = 0$ (where τ is
stress and γ is strain) for various strain rates. Since the strain
and velocity conditions in metal cutting are a coupled set of
variables, a speed can be identified at which the chip formation
mechanism becomes unstable and chip segmentation begins. The
processes in the secondary shear zone, i.e., the chip-tool interface,
are coupled with those in the primary zone. During the trapezoidal
segment formation, there is no bulk sliding; the sliding takes place
only during the shear band formation.

Some success has been achieved recently by Sedgwick, Read and
Wilkins in using the explicit version of a two-dimensional finite
difference computer program for modeling high-speed machining.[8] As

Figure 6.   AISI 4340 Steel Chips Formed at 400 sfm.

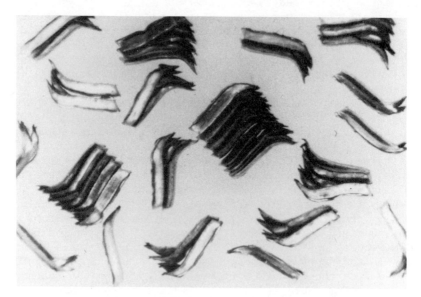

Figure 7.   AISI 4340 Steel Chips Formed at 3200 sfm.

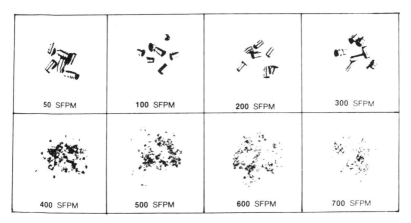

Figure 8.   Inconel 718 Chips Formed at Different Cutting Speeds.

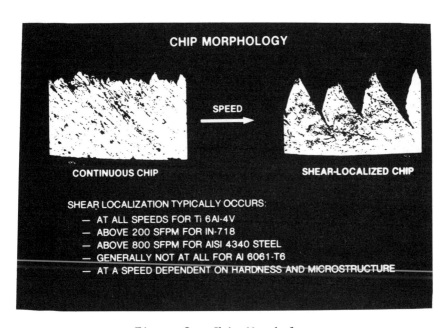

Figure 9.   Chip Morphology.

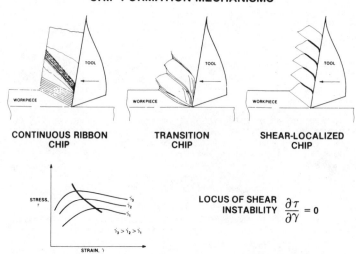

Figure 10.   Chip Formation Mechanisms.

input for such a model, D. Lee has used simple constitutive equations to describe the plastic flow behavior of materials over wide ranges of strain, strain rate and temperature.[8]

CUTTING FORCES

One of our objectives has been to resolve technical controversies arising out of prior work in high-speed machining.[10] For example, it has been confirmed that cutting force decreases with increasing speed until a minimum is reached at a speed characteristic of the given workpiece material. Be ond that characteristic speed, the force tends to slowly increase. For example, the decrease in force for AISI 4340 steel continues with increasing speed until about 5000 sfm at which point the force begins to increase with speed (Figure 11). Most of these data were generated on two lathes – the high speed, 20-inch, 150 hp, LeBlond engine lathe referred to earlier and a 32-inch, heavy duty, 125 hp LeBlond engine lathe. Both of these lathes are at GE-CSD and the work has been described by Schroeder and Hazra.[22] Additional confidence in these results is provided by force measurements for 4340 ballistic–

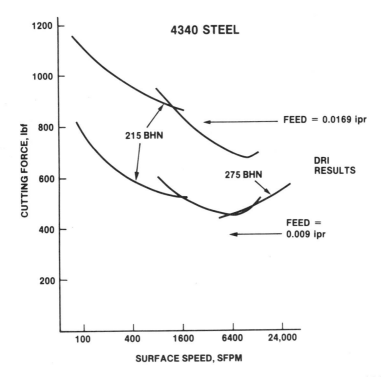

Figure 11. Cutting Force as a Function of Speed for AISI 4340 Steel.

ally-machined at DRI.[5] As shown in Figure 11, the two DRI points lie on an extrapolation of the curve generated at GE-CSD. Similar to 4340, aluminum 6061-T6 exhibits an initial decrease in force with increasing speed up to about 10,000 sfm beyond which the force increases slightly (Figure 12). This trend has been observed at both GE-CSD and by Stelson at TRW.[5] In contrast to the results for 4340 steel and aluminum, the cutting force for titanium is relatively unaffected by speed.

Some preliminary results indicate that transient dynamic components of force can be much greater than steady state components. The tests were conducted on 6061-T6 aluminum at 10,000 and 15,000 RPM using 1/2-inch solid carbide end mills. The dynamic components of force, measured with a strain gage dynamometer in the GE-CRD HSM spindle with FM telemetry, as described earlier, were up to 20 times greater than the steady state components, measured with a strain gage dynamometer under the workpiece and using a slip ring system.

The decrease in force with speed observed with several materials does not mean a lowering of horsepower requirements; higher material removal rates still require both higher horsepower and adequate rigidity. When these conditions exist, significantly increased metal removal rates can be achieved readily.

The benefits of power and rigidity have been demonstrated in this program by turning tests on Inconel 718 and titanium 6Al-4V on an extremely rigid, high power (500 hp), high-precision, heavy-duty, roll-turning lathe at Binns Machinery, Inc.[7] Material removal rates of over 50 cu.in./min. were achieved on both materials by increasing the depth of cut. The workpiece finish in general was excellent (30 to 100 microinches) indicating that for many applications it would be possible with such a machine tool to complete finishing cuts on the same tool as the roughing cuts.

With aluminum it has been shown that removal rates of 200 cu.in./min. are attainable. The practical top cutting speed for aluminum does not appear to be limited by lack of acceptable cutting tools; spindle speed and power are the controlling factors. It should be obvious, however, that high spindle speeds alone do not insure high metal removal rates. High feed rates and adequate depths of cut are also needed. Depending upon the kinds of cut being made (straight or contoured), the speed of response of the machine tool and its control may be critical.

TOOL WEAR AND INNOVATIVE CONCEPTS

Contrary to predictions based on the early work of others, there does not appear to be an upper speed range where cutting temperatures decrease with increasing speed. The general observation is that as speed is increased, the temperature approaches the melting point of the workpiece material as a maximum. Since tool wear in most cases is strongly dependent on temperature, this means that tool wear is the major factor limiting cutting speed – the main exception to this being aluminum machining. At ultra-high-speeds tool wear appears to be independent of speed; however, the wear is very high and well beyond that which could be tolerated in a practical machining environment.

Rapid tool wear is still a problem in the machining of titanium and other difficult-to-machine alloys even though cutting speeds for titanium have been recently increased 3 to 5-fold through judicious choice of cutter grades and geometries, fluids, and machining parameters. Partial solutions to the tool life problem lie in new tool geometries and use of rotating cutters. An example of a promising new geometry is the ledge tool which has a restricted contact flank face; cutting speeds for titanium alloys can be increased five times over conventional speeds, with long tool life.[5] Sparking can be eliminated by using an atmosphere of

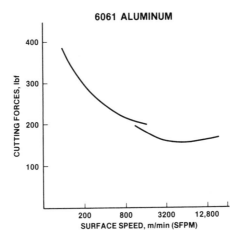

Figure 12.   Cutting Force as a Function of Speed for 6061-T6
             Aluminum.

inert gas (nitrogen), submerging the workpiece in a cutting fluid,
or combining a flood lubricant with an inert gas.

    The concept underlying the ledge tool is basically very simple:
the tool contains a ledge which is allowed to wear away at a
controlled rate with only minimal increase in force. Thus, tool
wear has not been reduced but tool life has been greatly increased.
A picture of a ledge tool is shown in Figure 13. It contains a
projecting ledge with a width equal to the depth of cut to be taken
and a thickness equal to the ultimate flank wear size to be tolerated. In turning, the square tool with the ledge side is brought
against the workpiece so that the edge of the ledge is almost
parallel (side cutting edge angle $\leq 1^\circ$) to the finished surface of
the workpiece (Figure 14). It is very important that the small but
finite clearance angle be maintained and the depth of cut be close
to the width of the ledge but not exceed it. With this arrangement
only the ledge portion of the tool does the cutting and wearing.
At increased speeds for a given material such as titanium 6Al-4V,

Figure 13.   Photograph of Ledge Tool.

the cutting edge of a ledge tool wears rapidly as might be expected.
However, the worn edge simply moves along the length of the ledge
and the tool continues to cut efficiently until most of the ledge
is worn away. Since the tip of the remaining ledge at any stage is
responsible for the surface finish, the finish is excellent no
matter how far the edge has worn back. Because of the present ledge
geometry, applications in turning are restricted to straight cuts.
Both a new and a worn ledge tool are shown in Figure 15.

The ledge tool can be used also in face and peripheral milling.
Typical results obtained by Gorsler of time and cost savings are
presented in Table 2.[5] The potential savings of 78% in face milling
titanium along with productivity increases of 3 to 4 times are
impressive. Of even more general applicability is the 6 times
increase in tool life in turning cast iron, as determined by GE-CSD.

As in turning, there are limitations in milling. For example,
in face milling, the axial depth of cut can be no more than the
projecting width of the ledge and in peripheral milling, the same
width limits the allowable wearback.

Another promising technique for extending tool life is the use
of rotary tool machining.[2] In this method a tool with a circular
cutting edge is allowed to rotate about its own axis, either self-
propelled by the cutting process as depicted in Figure 16 or driven
externally at the desired speed. As cutting proceeds, new portions
of the cutting edge are continuously brought into contact with the

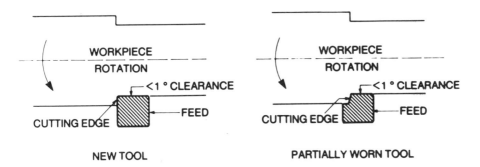

Figure 14.   Schematic Description of a Ledge Tool Application.

Figure 15.   New and Worn Ledge Tools.

workpiece at the cutting zone. Thus, increased tool life can be
anticipated because of lower temperatures and reduced chemical
reactions at the moving chip-tool interface, and reduced cutting
forces because of increased options for modifying the chip formation. A typical toolholder and insert for rotary tool turning is
shown in Figure 17 and a milling cutter with four rotary inserts
is shown in Figure 18.

Table 2.   Status of Ledge Tool Machining

o   Concept developed at CRD; further studied at AEBG and CSD.

o   Time and cost savings in milling titanium can be:

        - 78% in face milling
        - 26% in peripheral milling.

o   Productivity in milling titanium can be increased by a
    factor of 3 to 4.

o   High potential in airframe production indicated by Vought
    and Douglas Aircraft studies.

o   Carboloy tests on cast iron show 3 to 4x increase in cutting
    speed with 6x increase in tool life.

o   Thin wafer can be clamped to substrate to give ledge geometry
    with equal performance.

Rotary tool machining is not new, the technique having been
introduced in 1868. Also, Shaw et al. explored the field in some
depth in 1952.[23] However, the main problem until the present work
has been lack of adequate stiffness. Typically, with large, round
inserts ($\sim$ 1 inch diameter) and large depths of cut ($\sim$ 0.100 inch)
the radial forces have been very high. This has resulted in
unstable cutting leading to chatter, uneven chipping of the cutter
and poor surface finish on the work material. Komanduri and Reed
have shown recently that by increasing the dynamic stiffness of the
system, marked improvements in machining performance can be
achieved.[8] For comparable flank wear, the cutting time in Inconel
718 can be increased by a factor of over 20 by using a rotary insert
in place of a stationary tool; in the case of titanium 6A1-4V the
performance increase is about 7 times. Earlier results obtained
by Gorsler in milling show conservatively that cost savings of 47%
and 25% are achievable for Inconel 718 and titanium alloy, respectively (Table 3).[6]

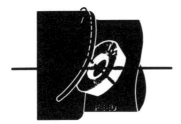

Figure 16. Rotary Tool Turning – Normal Rotation/Normal Feed.

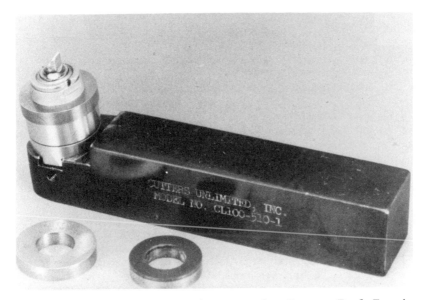

Figure 17. Toolholder and Insert for Rotary Tool Turning.

Figure 18. Milling Cutter with Four Rotary Inserts.

Table 3. Status of Rotary Tool Machining

o Concept not new but studied more in Russia than in U.S. to
   date.

o Increased rigidity key to improved performance.

o Cost savings are demonstrated to be:

     - 47% in milling Inconel 718
     - 25% in milling titanium alloy.

o Cutting speed can be increased (over conventional):

     - 4 to 6 times on Inconel 718
     - 3 times on titanium alloy.

o Rotary tool cutters now available commercially.

o Development of smaller rotary inserts on rugged bearing
   systems could broaden applications.

ECONOMIC ANALYSIS

   Throughout all of this work, economic models have proven useful in assessing the economic impact of a given process, in providing guidelines for further work, and in defining "opportunity

windows".[24-28] Such analyses indicate substantial payoffs for high-speed machining in airframe applications.

The introduction of any new technology can be considered to involve five stages of activity demonstrating: conceptual feasibility, technological feasibility, economic feasibility, production implementation, and technology transfer. A schematic of this sequence as presented by Tipnis, Watwe and Mantel is shown in Figure 19.[6] A preliminary application of economic feasibility analysis to high-speed machining of aluminum leads to a family of curves as shown in Figure 20.[5] The example chosen is the machining of airframe pockets utilizing realistic assumptions of number of airframes, milling cutters, types and depths of cut, feed rates, and speeds attainable. The lower curve in Figure 20 shows that if cutting speed can be increased 10 times over conventional, i.e., the minimum speed is in the range of 2600 to 13,500 sfm, then for an initial investment of $50,000, the payback period will be 10 years. For a 3-year recovery period, the maximum investment would be $27,500. This example is relatively conservative. Work in progress on other airframe applications indicate that initial investments much greater than this can be acceptable.

FUTURE NEEDS

High-speed machining can be cost-effective only if other aspects of machining, including reduction of non-cutting times and labor costs, can be improved also. For this reason automated machining is now receiving much attention. As a part of automated machining, dynamic in-process inspection (i.e., inspection during machining) is an integral part. Sensors and diagnostics have been identified as critical to effective in-process inspection. Recently, proximity sensors for measuring tool wear, vibration sensors for measuring tool-touch (between cuts), and lasers for measuring workpiece dimensions have been studied.

A list of some of the needs for improvements in machining, including implementation of high-speed machining, is presented in Table 4. The goal is to increase metal removal rates as well as to increase use of automated machining. The list includes not only sensors and diagnostics but also better cutting tools, faster response machine tool controls and greater tool rigidity. These needs are, by and large, technical in nature. On the other hand, significant reductions in non-cutting time can be achieved by procedures which to a large extent involve improved management. These procedures include setup, part load/unload, queing, chip cleanup, tool change, maintenance, scheduling, and general operator efficiency. Until improvements in these procedures are achieved, cutting times will remain only a small fraction of the overall manufacturing sequence and the full advantages of high-speed machining will not be realized.

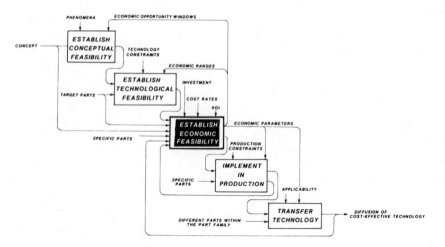

Figure 19.   Stages in the Introduction of New Manufacturing Processes.

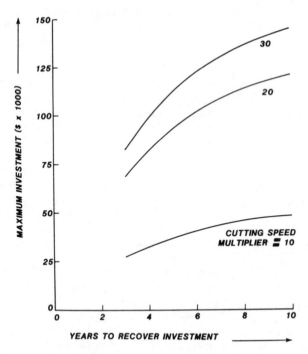

Figure 20.   Analysis of Economic Feasibility for Aluminum High-
Speed Machining.

Table 4. Future Needs in Advanced Machining

o Integrated sensing techniques for monitoring machining
   processes generally.

o Sensors and diagnostics for detecting tool breakage specifically (including incipient tool breakage).

o Linkage of signal analysis to tool wear mechanisms and
   machining variables.

o Self-teaching adaptive control systems.

o Faster response in machine tool controls when machining
   aluminum.

o Better cutting tools (composition and geometry) for
   machining titanium.

o Water-base cutting fluids for machining titanium.

o More reliability and predictability in cutting tools.

o Greater machine tool rigidity for increased productivity.

o DCL-notch resistant cutting tools for superalloys.

ACKNOWLEDGEMENTS

    Most of the work discussed in this paper was performed under
contract F33615-79-C-5119 sponsored by the Defense Advanced Research
Projects Agency and managed by the Manufacturing Technology Division
of the Air Force Wright Aeronautical Laboratories. The support of
both organizations is gratefully acknowledged.

    The author also wishes to thank the many subcontractors to
General Electric who contributed so significantly to the program.
Specifically, the principal investigators from those organizations
included B. F. von Turkovich - University of Vermont; R. F. Recht -
Denver Research Institute; R. T. Sedgwick - S-Cubed; T. S. Stelson -
TRW; and V. A. Tipnis - Tipnis Associates, Inc.

    Thanks are due also to the many within General Electric who
have contributed their talents to this effort. These include F. W.
Gorsler, S. Haque, S. R. Hayashi, J. Hazra, R. Komanduri, D. Lee,
M. Lee, W. R. Reed, Jr., T. A. Schroeder, R. A. Thompson, and A.
W. Urquhart.

REFERENCES

1. B. F. von Turkovich, "Influence of Very High Cutting Speed on
    Chip Formation Mechanics", Proc. NAMRC-VII, pp. 241-247,
    1979.
2. D. G. Flom, ed., "Advanced Machining Research Program (AMRP)",
    Semiannual Technical Report, Air Force Contract No.
    F33615-79-C-5119, GE Report No. SRD-80-018, February 15,
    1980.
3. D. G. Flom, ed., "Advanced Machining Research Program (AMRP)",
    Annual Technical Report, Air Force Contract No. F33615-
    79-C-5119, GE Report No. SRD-80-118, August 16, 1980.
4. D. G. Flom, ed., "Advanced Machining Research Program (AMRP)",
    Semiannual Technical Report, Air Force Contract No.
    F33615-79-C-5119, GE Report No. SRD-81-018, February 15,
    1981.
5. D. G. Flom, ed., "Advanced Machining Research Program (AMRP)",
    Annual Technical Report, Air Force Contract No. F33615-
    79-C-5119, GE Report No. SRD-81-062, August 17, 1981.
6. D. G. Flom, ed., "Advanced Machining Research Program (AMRP)",
    Semiannual Technical Report, Air Force Contract No.
    F33615-79-C-5119, GE Report No. SRD-82-027, February 15,
    1982.
7. D. G. Flom, ed., "Advanced Machining Research Program (AMRP)",
    Annual Technical Report, Air Force Contract No. F33615-
    79-C-5119, GE Report No. SRD-82-070, August 16, 1982.
8. D. G. Flom, ed., "Advanced Machining Research Program (AMRP)",
    Semiannual Technical Report, Air Force Contract No.
    F33615-79-C-5119, GE Report No. 83-SRD-012, February 15,
    1983.
9. "The 12th American Machinist Inventory of Metalworking Equipment
    1976-78", American Machinist, pp. 133-148, December 1978.
10. D. G. Flom, R. Komanduri, and M. Lee, "Review of Past Work in
    High-Speed Machining", TMS-AIME Meeting, Louisville, KY,
    October 15, 1981.
11. J. P. Kottenstette, and R. F. Recht, "An Ultra-High Speed
    Machining Research Facility", Proceedings of NAMRC-IX,
    The Pennsylvania State University, University Park, PA,
    May 19-22, 1981, published by SME.
12. R. Komanduri, and J. Hazra, "A Metallurgical Investigation of
    Chip Morphology in Machining an AISI 1045 Steel at Vari-
    ous Speeds up to 10,100 SFPM", Ibid.
13. M. Lee, and D. G. Flom, "Metallurgical Aspects of the Chip
    Formation Process at Very High Cutting Speed", Ibid.
14. R. Komanduri, and B. F. von Turkovich, "New Observations on the
    Mechanism of Chip Formation When Machining Titanium
    Alloys", WEAR, 69, 179-188, 1981 .
15. R. Komanduri, "Titanium - A Model Material for Studying the
    Mechanism of Chip Formation in High-Speed Machining",
    TMS-AIME Meeting, Louisville, KY, October 15, 1981.

16. B. F. von Turkovich, and D. R. Durham, "Machining of Titanium and Its Alloys", Ibid.

17. M. Lee, "The Failure Characteristics of Cutting Tools Machining Titanium Alloys at High Speed", Ibid.

18. R. Komanduri, "Some Clarifications on the Mechanics of Chip Formation when Machining Titanium Alloys", WEAR, 76, 15-34, 1982.

19. R. F. Recht, "Catastrophic Thermoplastic Shear", Trans. ASME, Vol. 86, p. 186, June 1964.

20. R. Komanduri and R. H. Brown, "The Mechanics of Chip Segmentation in Machining", ASME Journal of Engineering for Industry, Vol. 103, pp. 33-51, February, 1981.

21. R. Komanduri, T. Schroeder, J. Hazra, B. F. von Turkovich and D. G. Flom, "On the Catastrophic Shear Instability in High-Speed Machining of an AISI 4340 Steel", ASME Journal of Engineering for Industry, Vol. 104, pp. 121-131, May 1982.

22. T. A. Schroeder and J. Hazra, "High Speed Machining Analysis of Difficult-to-Machine Materials", Proceedings of NAMRC-IX, The Pennsylvania State University, University Park, PA, May 19-22, 1981, published by SME.

23. M. C. Shaw, P. A. Smith, and N. H. Cook, "The Rotary Cutting Tool", Trans. ASME, 1065 1952 .

24. V. A. Tipnis, "Economic Models for Process Development", ASM Fall Meeting, Cleveland, Ohio, October 28-30, 1980.

25. V. A. Tipnis, S. J. Mantel, and G. L. Ravignani, "Sensitivity Analysis for Macroeconomic and Microeconomic Models of New Manufacturing Processes", Annals of CIRP, 30/1, 401, 1981.

26. S. J. Mantel, Jr., V. A. Tipnis and G. L. Ravignani, "Economic Evaluation of Potential Process Innovations", Joint National Meeting of The Institute of Management Sciences/ The Operations Research Society of America, Houston, TX, Fall-1981.

27. V. A. Tipnis, G. L. Ravignani and S. J. Mantel, Jr., "Microeconomic Model for Process Development", TMS-AIME Meeting, Louisville, KY, October 15, 1981.

28. S. J. Mantel, Jr., V. A. Tipnis, and G. L. Ravignani, "Adaptive Technological Planning Under High Uncertainty", Joint National Meeting of the Institute of Management Science/ The Operations Research Society of America, Detroit, MI, April 18-21, 1982.

ADVANCES IN PRECISION GRINDING

Robert L. Mahar

Norton Company
1 New Bond Street
Worcester, MA 01606

The army is not alone in its quest for producibility, reliability and affordability. Industry in general has the same concerns and is searching for producivity and total cost per part improvements. The economic situation since the late 1970's has significantly changed manufacturing management's attitude toward change - especially the larger world class manufacturers like Saginaw Steering Gear, Garrett Turbine, Caterpillar Tractor - to not only be receptive to new processing methods but to be pro-active in searching for them, providing engineering resources to work on them, and implementing them in their manufacturing procedures.

Lagging productivity in a competitive world economy, double digit inflation, declining world competitiveness, and recession have all helped focus national attention on the search for productivity improvements. High energy and material costs have accelerated the development of new forming processes to make parts closer to near-net-shape so less material has to be removed. Utilization of improved materials that are stronger, lighter, and smaller produced to closer tolerances, better finishes, and metallurgically sound improves part performance capability, reliability and durability.

These trends are making the grinding process more important than ever as a machining process and advancements in the grinding process are causing it to become a process of choice instead of one of necessity. Advances in abrasives, bonding systems, machine design, machine control systems, and process understanding are responsible for these changes. Manufacturing and process engineers must become more knowledgable about capabilities of the grinding

process. Grinding machine builders and abrasive manufacturers
must do more to help engineers assess the grinding process as one
of their part processing options.

I will address two major advances in the abrasive process.
The first is creep feed grinding and the second is cubic boron
nitride, a super abrasive and referred to as CBN hereafter. I will
also touch on coated abrasive advancements; computer numerical
controls on grinders; adaptive controls, and process understanding
which is lifting the grinding process out of the "black art"
category into the understandable, predictable, and stable process
that it is.

But first I would like to discuss grinding from a historical
viewpoint. It is fair to say that the grinding process is generally
perceived as first a finishing process - i.e., light stock removal,
second - an unstable/unpredictable mysterious process, and third
a costly process. This leads process and manufacturing engineers
to the conclusion that grinding should be chosen only as a last
resort process when a part cannot be processed any other way. Now,
I would like to show why these perceptions "miss the mark".

It is true that grinding is a complex process with many variables which must and can be controlled. It is capable of abrasive
machining - i.e., heavy material removal - not just finishing
because it can do both, and finally it may be a costly process,
but if the part processing plan is designed around grinding the <u>total</u>
cost per part can be significantly reduced. The traditional part
processing method is to turn or mill in the soft state - heat treat
- then finish grind. However, better abrasives and abrasives
systems are currently capable of machining the part in the hardened
state thus eliminating the need to turn or mill as a perliminary
step in the soft state. Later, I will present a case history to
illustrate this.

"Abrasive machining" is not a new concept and to convince some
of you who may not be aware of the abrasive process' capability as
a metal removal tool I would like to mention two applications. The
first is steel conditioning. While definitely not a precision
application, it clearly demonstrates the abrasive systems' metal
removal capability when the entire system is designed and optimized
around the grinding process. Here 24 x 3 inch coarse grit resin
bond wheels on three hundred horsepower machines rip off carbon
and stainless steel at the rate of 2400 pounds per hour. That
converts to 140 cubic inches per minute or 47 cubic inches per
minute per inch of wheel width. Specific grinding energy is a very
efficient 2 HP/in 3/minute. Clearly this is a roughing operation
on soft steel, but is as severe as they come.

More toward precision grinding, however, is fluting of high

speed steel drills, end mills, and taps on specialized equipment
employing high wheel speeds and high pressure oil as a lubricant/
coolant. Here in a single pass at full depth of a high speed steel
drill flute metal is removed at a rate of 6 cubic inches per minute
per inch of wheel width at a traverse rate of up to 53 inches per
minute with no metallurgical burn. Depth of cut on a 1/2" drill
is .180 inches per flute. This application is not new, but clearly
demonstrates the grinding wheel's machining capability on hardened
tool steels.

So much for the historical viewpoint and I hope I have con-
vinced you of the grinding process metal removal capability. Now
let's discuss the two most notable advances in precision grinding
which are creep feed grinding - better termed full depth/slow
traverse grinding (and not always so "slow" at that): - and cubic
boron nitride - (CBN for short) a superabrasive for ferrous based
materials. Creep feed and CBN are separate developments, but can
be teamed up to magnify their separate advantages. Neither of
these advances are really new either because creep feed grinding has
been used in Europe for over twenty years, but because of some very
unfortunate situations when first introduced to the United States
some fifteen years ago the process was totally discredited and it
has taken that long to begin regaining credibility in the U.S.A.
CBN has been around since 1969.

While most creep feed grinding systems utilize conventional
grinding wheels made from aluminum oxide or silicon carbide abrasive,
an increasing number of situations involve the combining of the
creep feed mode with a CBN or diamond grinding wheel. Most situations where these advanced techniques or tools are used is upgrading
an existing abrasive operation, but an increasing number of
instances involve replacing a turning or milling operation with
grinding or eliminating the turn or mill step in an operation and
grinding to finished part dimensions from the solid and hardened
state.

Creep Feed Grinding

Creep feed grinding is a high efficiency process for mass
production where high metal removal rates and high surface integrity
are required. It is most often used in grinding deep slots and
grinding profiles in a surface grinding mode especially in difficult to grind materials. Several of these creep feed pictures are
the courtesy of Dr. P. G. Werner, University of Bremen, Bremen,
W. Germany. /1/

Creep feed grinding is generally characterized by single pass
grinding depths of from forty thousandths of an inch (.040") to over
1 inch and traverse rates from 1 to 40 inches per minute. Metal
removal rates of 2 to 3 times conventional grinding are achievable.

Table I

	MAXIMUM REMOVAL RATE Z' (MM³/MM/S)							
	CYLINDRICAL GRINDING				SURFACE GRINDING			
	CONVENTIONAL		CREEP FEED		CONVENTIONAL		CREEP FEED	
COOLANT	V_S (M/S)		V_S (M/S)		V_S (M/S)		V_S (M/S)	
	25-30	60-80	25-30	60-80	25-30	60-80	25-30	60-80
DRY	>0,5[+]	—	—	—	>0,5[+]	—	—	—
WATER	[5]	—	[10][+]	—	4	—	[8][+]	—
EMULSION (2%)	20	50	[50]	150	[15]	[40]	40*	[120]
OIL	[60]	[180]	[150]	600	[40]	[120]	100*	[400]

DR. P.G. WERNER, BREMEN, WEST GERMANY

[+] WITHOUT PRACTICAL RELEVANCY

[] ASSESSED VALUES

*) HIGHER VALUES ARE ATTAINABLE BY CONTINUOUS DRESS

Special cases like drill fluting obviously exceed these general-
izations. The capability of creep feed grinding vs. conventional,
as well as the effects of wheel speed and coolant are shown in this
chart. /1/

To achieve success with creep feed grinding requires that the
entire grinding system - that is machine, truing/dressing, fluid
application, etc. be designed specifically for this mode of
grinding. Retrofit attempts on conventional grinders are almost
never successful.

Creep feed systems are characterized by:

. Machines with high rigidity
. Increased horsepower - three to five times conventional
. Variable wheel speed capability
. Numerical control
. Accurate, consistent, stick-slip free table drives (usually
  ball screw driven)
. High pressure coolant and wheel cleaning systems
. Integrated wheel truing and dressing systems

Most systems today are of European manufacture like Elb, Magerle, Blohm, and Aba, but U.S. builders are getting interested. It's ironic in a way that the first creep feed system is credited to the Thompson Grinder Company some thirty years ago. Apparently Thompson could not develop market interest in the U.S.A. European machine builders were obviously more successful. Thompson is renewing its interest in creep feed grinding as an emerging U.S.A. market develops.

It is very important that all elements of the system be well maintained and balanced. Negligence of any part of the system will usually cause a poor work result. It is necessary that the operating environment and personnel be attuned to the system's needs. The most successful applications to date have been in the aerospace and tool manufacturing industries or in grinding job shops specializing in creep feed where a high quality and precision mentality exists.

I have already mentioned one specialty application of creep feed abrasive machining that has been in existence for over 25 years in the U.S.A. which is fluting high speed steel drill, taps, and end mills from solid and hardened M1, M2, and M7 high speed steel blanks. The flute forms are fairly simple and geometric accuracy is not stringent, but the principle of creep feed at its best is demonstrated. Metal removal rates of 6 cubic inches per minute per inch of wheel width are common. 12 to 14 cubic inches per minute have been achieved. Traverse rates on straight fluting of taps go as high as 300 inches per minute taking three passes at .030 depth to reach final flute depth on a 3/8" tap. Drills and end mills are single pass operations full depth at traverse rates of 53 inches per minute. Straight grinding oil is used at pressures of 120 spi and 50 gallons per minute. Attention to the fluid delivery system is very important for a satisfactory work result.

As the geometric form and accuracy complexity increase lower metal removal rates at slower traverse speeds are generally found. All elements of the system must be balanced for optimum results.

Most creep feed jobs today use conventional aluminum oxide or silicon carbide abrasives with vitrified or ceramic bonding systems. Wheels are reformed with diamond truing tools - usually by formed rotary diamond cutters.

Two recent advances on the creep feed state of the art are continuous dressing of conventional abrasive vitrified wheels and the use of CBN or diamond super-abrasives in bonded (multi-layer) wheels or in the single layer plated wheel design. I will discuss superabrasives in creep feed grinding shortly.

In the continuous dressing mode a rotary diamond form roll is in continuous contact with the grinding wheel and continuously

feeding into the grinding wheel a few millionths of an inch per
revolution of the grinding wheel. This assures that the wheel
profile and cutting sharpness are always optimal and metal removal
rates up to 25 times higher than conventional grinding are
achievable.

The following chart developed as part of Dr. Stuart Salmon's
doctoral thesis gives a second comparison of continuous dressing
capability. /2/

Table II

Material	Wheel Grade	Creep-Feed Grinding		With Continuous Dressing	
		Feed Rate (in./min.)	Limitation	Feed Rate (in./min.)	Limitation
C 1023	WA 60 KV	0.400	Burn	41.75	Burn
MAR M002		0.400	Burn	46.06	Burn
C 1023	WA 6080F P2V	2.362	Burn	41.73	Wheel Breakdown
MAR M002		2.165	Burn	53.15	Wheel Breakdown

Depth of Cut — 0.120"
Wheel Dia. — 24"

Ref. Dr. Stuart C. Salmon. Ph.D Thesis 1979

While not technically creep feed grinding it should be mentioned
that abrasive machining of softer materials using coated abrasive
belts is a proven technique. High horsepower rigid machines using
zirconia alundum abrasives applied to high strength polyester
backings have significantly upgraded coated abrasives' metal removal
and productivity capability. Shown is a two headed abrasive
machiner with 150 horsepower on the grinding head and 75 horsepower
on finishing head. Metal removal rates of 30 to 60 cubic inches
per minute are typical. Flatness to .001 (one thousandth) inches
is possible. This process is geared for high production of flat
parts made of cast iron, die cast aluminum, etc.

Finally, creep feed grinding can be applied to most hardened
ferrous formed parts as well as non-ferrous materials like
tungsten carbide. It is quite a versatile yet technically sophisticated metal removal system that can significantly improve

productivity. A final reminder - the word "creep" is a bit of a misnomer because as the drill fluting and continuous dressing examples show the linear feed rate is hardly "creeping".

CBN

Now, let's explore the capabilities of the super-abrasive cubic boron nitride - CBN for short. This material is very nearly an ideal precision abrasive. It performs superbly on ferrous based materials and its existence will permit part designers to upgrade their materials choices to stronger, lighter, tougher materials whose inherent characteristics make them difficult to machine and grind with conventional cutting tools and abrasives.

CBN is a man-made abrasive from the same high temperature high pressure process technology that produces man-made diamond. For the grinding of ferrous based materials CBN is a near ideal abrasive. Its hardness, edge durability, chemical and thermal stability, and low co-efficient of friction make it an excellent cutting tool. It is the second hardest material known, diamond being the hardest.

CBN is expensive - about $4,500/lb. Industry's challenge is how to effectively apply this excellent cutting tool. Because of its cost, bonded CBN wheels are made with only a 1/16 to 1/4 inch radial depth of CBN on the wheel periphery. It is also possible to plate a single layer of CBN on a precision steel preform. And because of its slow wearing characteristic diameter reduction is minimal through the life of the wheel which maintains the wheels peripheral speed and keeps its cutting efficiency at a high optimum level.

CBN can be bonded with resin, metal, or vitrified (ceramic) systems as well as plated in a single layer to a precision preform. Plated wheels are very aggressive metal removal tools that do not need truing (which is making the wheel round and concentric with the spindle) or dressing (which is a cleaning operation to expose the abrasive particles from the surrounding bond matrix). When the CBN particles are dull, they can be stripped off and a new layer plated on. Diamond abrasive can be used on carbide, glass, ceramics and other non-ferrous materials.

CBN's primary utility today in production grinding is on those materials that are difficult to grind with conventional abrasives - the high speed tool steels, particularly M and T series steels, the high nickel superalloys - like Inconel, Rene, etc. and hardened alloy structural steels, and to repeat - any material where conventional abrasives have difficulty. Although the abrasive cost per part may double on average the productivity improvement comes from shorter cycle times which significantly lowers labor and overhead

costs. Improved part geometry, reduced part rejections, no metallurgical damage, and improved machine utilization are additional
productivity gains from CBN. Total cost savings per part are
typically in the 30 to 50% range.

The benefits offered by CBN are many:

. Shorter cycle times
. Improved productivity
. Lower labor/overhead costs
. Lower total grinding costs per part
. Improved part geometry
. Lower rejection rates
. Improved surface integrity
. Improved machine utilization

Let's look at two examples where CBN resin bonded wheels
significantly improved productivity versus conventional wheels.

CASE #1 - Cylindrical Grinding CPM 10V Fuel Pump Gear Shafts

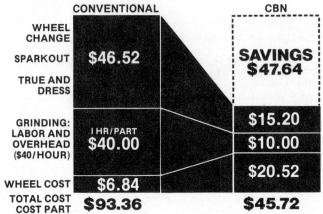

CYLINDRICAL GRINDING OF A CPM IOV FUEL PUMP GEAR SHAFT
CBN RESIN BOND WHEEL VS CONVENTIONAL AI$_2$O$_3$ WHEEL

CASE #2 - Double Disc Grinding M50
Power Steering Pump Vanes
   Wheel: 30" diameter CBN - resin bond disc wheel
   Results:
      1) Rejections cut from 25/30% to 5%
      2) Productivity doubled - from 9000 to 18,000 pieces
          per hour
      3) Total cost savings - several thousand dollars per
          month

This and the next case history show the electro-formed single layer CBN wheel capability vs. conventional wheels.

CASE #3 - Internal Grinding Cast Iron
Valve Seal Rings

APPLICATION: I. D. GRINDING EIGHT VALVE SEAL RINGS
SIMULTANEOUSLY FROM CAST IRON
ALLOYS (RC 38-40) IN AN AS-CAST STATE.

MACHINE: MAJOR AUTOMOTIVE COMPANY, MICHIGAN.

MATERIAL: HEALD CONTROLLED FORCE GRINDERS.

TEST RESULTS:

	PROCESS E CBN WHEEL	CONVENTIONAL WHEEL
CYCLE TIME TO REMOVE .040''	1½ MINUTES	3 MINUTES
RMS FINISH	17	20
PARTS GROUND PER WHEEL	30,000 (SET OF TWO WHEELS, COURSE & FINE)	120
APPROX. WHEEL COST PER PART	$.01 (ONE CENT $326 FOR BOTH WHEELS)	$.007 (7/10 OF ONE CENT @ $.85 PER WHEEL)
APPROX. PER PART TOTAL GRINDING COST ($40./HR. USED AS LABOR & OVERHEAD COST)	$1.01	$2.01

CONCLUSIONS:

- • CYCLE TIME CUT IN HALF
- • GRINDING MACHINE CAPACITY DOUBLED
- • FINISH NOTICEABLY IMPROVED
- • WHEEL LIFE INCREASED 250 TIMES
- • REDUCED TOTAL GRINDING COST BY 50%
- • 300 FEWER WHEEL CHANGES

CASE #4 — Sizing Slot Width in Premachined
Hardened Steel Forging

APPLICATION: SIZING SLOT WIDTH FROM PREMACHINED,
THEN HARDENED PART. WHEEL IS FED
DOWN INTO SLOT THEN MOVED UP AGAINST
ONE SIDE OF THE SLOT, THEN THE OTHER SIDE.

LOCATION: 10 HP EXCELLO O.D. GRINDER.

MACHINE: RC 45-48 STEEL FORGING.

TEST RESULTS:

	PROCESS E CBN WHEELS	CONVENTIONAL WHEELS
APPROXIMATE FLOOR TO FLOOR TIME/PART	15 MINUTES	1 HOUR
PARTS GROUND PER WHEEL	1.100	50
COST PER WHEEL	$500.00 REPLATE	$30.00
TRUING/DRESSING	NONE REQUIRED	SINGLE POINT—EVERY PART
WHEEL COST PER GOOD PART	$.45	$.60
APPROXIMATE LABOR AND OVERHEAD PER PART ($40./HR. USED AS LABOR AND OVERHEAD COST)	$10.00	$40.00
APPROXIMATE TOTAL COST PER PART	$10.45	$40.60
SAVINGS PER PART	$30.15 (74% LOWER COST PER PART)	

CONCLUSIONS:

PROCESS E WHEEL

1. LOWERED PER PART TOTAL GRINDING COST BY $30.15 OR 74%.

2. REDUCED CYCLE TIME BY 75%.

3. INCREASED OPERATOR PRODUCTIVITY 4 TIMES.

4. INCREASED CAPACITY OF EXISTING EQUIPMENT 4 TIMES.

5. INCREASED MACHINE UP TIME BY REDUCING THE NUMBER OF WHEEL CHANGES NEEDED.

6. ELIMINATED DOWNTIME FOR DRESSING AND TRUING.

7. ELIMINATED TRUING TOOL COST (NOT INCLUDED IN ABOVE SAVINGS).

This case history and the one following show creep feed grinding and CBN teaming up to provide substantial productivity benefits and in one case replace a pre-milling operation.

CASE #5 – Creep Feed Groove Grinding of Inconel 718 Material
on a Surface Grinder

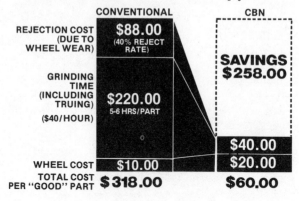

**CREEP FEED GROOVE GRINDING OF INCONEL 718
ON A SURFACE GRINDER –
PROCESS E CBN WHEEL VS CONVENTIONAL Al₂O₃ WHEEL**

	CONVENTIONAL	CBN
REJECTION COST (DUE TO WHEEL WEAR)	$88.00 (40% REJECT RATE)	
		SAVINGS $258.00
GRINDING TIME (INCLUDING TRUING) ($40/HOUR)	$220.00 5-6 HRS/PART	
		$40.00
WHEEL COST	$10.00	$20.00
TOTAL COST PER "GOOD" PART	$318.00	$60.00

CASE #6 – Creep Feed Surface Grinding of
RC60 L6 Tooling Parts
  Wheel: 12" resin bond CBN
  Results: 1) Eliminated milling operation
        2) CBN grind cycle from 25% to 75% shorter than
           mill plus conventional grind
        3) CBN tooling costs 22% lower
        4) Labor and overhead costs 46% lower
        5) Total costs/part 41% lower
        6) Savings/yr = over $20,000

The conditions that need to exist today for successful CBN
production grinding are:

  1) A problem to be solved
  2) Customer commitment – has to want CBN to work
  3) Manufacturing management involvement
  4) Total cost accounting
  5) Suitable machinery
  6) Proper grinding fluid
  7) Proper process/system parameters

Machines designed to use CBN are becoming more numerous. Existing honing machines and jig grinders can use CBN without modification. Tool and cutter machines for resharpening cutting tools is also an easy substitute for CBN from aluminum oxide and a well developed CBN application where nearly half of the CBN wheels sold in the U.S. are consumed.

The Japanese have been aggressively working on the implementation of CBN in production grinding and have a CBN growth rate estimated at nearly twice that of either the U.S.A. or Europe. The challenge for American industry is to technically and economically qualify CBN in production applications. CBN has proven its utility in grinding modes like cylindrical, surface, internal, and disc grinding, but is also working very well in the creep feed mode we talked about earlier – both in bonded and plated wheels.

Before I close I would like to comment on computer numerical control (CNC), adaptive control, and process understanding. In the last four years CNC has rapidly been adopted by the grinding machine builder industry. Stepping motors, ball screws, programmable controllers, and microprocessors have improved the control, accuracy, and capability of grinding machines. They are easier to set up, change setups, and easier for an operator to learn and control. Such precision control when combined with slow wearing CBN wheels makes the system capable of automatic operation with excellent part size and geometry consistency over long intervals. Consequently the operation does not require an operator in attendance on a continual basis. With the trends moving toward flexible machine systems and unmanned machines computer numerical control and CBN can make that dream a reality.

Even fast wearing conventional abrasive wheels can provide highly productive work results when teamed with appropriate adaptive controls in the machine. Just now emerging is such a system designed by Adaptive Energy, Inc., Rockford, IL. By always keeping the grinding wheel sharp high metal removal rates without metallurgical damage are achievable. It makes sense that a constantly sharp cutting tool will out produce one that is permitted to dull. Work is currently going on in diesel crankshaft and camshaft parts processing where metallurgical damage must be avoided to assure long part life.

The objective is shorter grinding cycles without grinding burn. This translates to more pieces per hour, lower total cost per part, and higher asset (machine) utilization.

Finally, considerable knowledge has been gained about grinding process behavior. The basic parameters of metal removal rate, wheel wear rate, horsepower, forces, wheel to work conformity, G-ratio, and wheel cutting sharpness can all be measured and documented for any grinding system.

This one slide is worth an hour's talk all by itself and I don't have time to elaborate on it, but I did want to show it to indicate that the grinding system can be measured, predicted, and optimized if one is willing to study the specific job in question and make performance measurements.

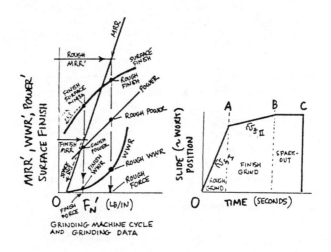

Figure 1.

Dr. Richard Lindsay of Norton Company is one of the foremost advocates of this approach to assessing the grinding system efficiency. /3/

There is no magic or black art about the grinding system. Like all processes knowledge about it comes from adequate study of the system to determine and measure the fundamental behavioral relationships and the output results. This is basic to the system understanding and instrumental in any attempts by the machine tool industry and the abrasives industry to advance the grinding process metal removal capability. Cooperative work is occurring at an

increasing rate which will keep grinding process technology advancing and continually improve its value to the manufacturing process.

REFERENCES

1. P. G. Werner, Increased Removal Rates and Improved Surface
        Integrity by Creep Feed Grinding., A.E.S. 21st Annual
        Conference, April 1983.
2. S. C. Salmon, Taking Advantage of the Latest Abrasive Manufac-
        turing Technology. A.E.S. 21st Annual Conference, April 1983.
3. R. P. Lindsay and others, Grinding....Putting Control in the
        User's Hands - round table discussion - five experts, Manu-
        facturing Engineering (SME), June 1983.

ION IMPLANTATION OF METAL AND NON-METAL SURFACES

Charles Levy
Army Materials and Mechanics Research Center

and

James K. Hirvonen
Zymet, Incorporated

ABSTRACT

For material surface modification, the ion implantation process
has been found to have beneficial effects on surface-sensitive
properties such as hardness, wear, coefficient of friction,
fatigue, aqueous corrosion, oxidation, adhesion, optical
characteristics and catalytic activity. These are in part due
to the fact that ion implantation is inherently a non-
equilibrium phenomenon which can substantially alter surfaces of
metals and non-metals to impart new characteristics. From the
chemistry standpoint, conventional alloys can be formed on the
surface without affecting the bulk material. More importantly,
since elements can be introduced in concentrations far beyond
equilibrium solid and compound solubility limits, the process may
be used to form unusual alloys, metastable phases, and morphologies
which are impossible to prepare by other methods.

The technique involves directing a beam of high energy ions,
in the one hundred kiloelectron volt and higher range in energy,
at a substrate surface, under vacuum. An analyzing magnet is used
to separate the ion or ions of interest, and thus, a pure isotope
species is energetically injected into the surface to a depth up
to several thousand angstroms.

Recent variations in direct beam implantation include ion
beam mixing, simultaneous deposition and implantation, and ion cluster
beam deposition. These processes result in thicker alloy deposits
and higher additive concentrations, along with implantation.

Discussion of applications of implantation will be limited to friction and wear and general corrosion. Results in the literature document wear improvement factors of between two and twelve times for treated cutting tools and dies. In corrosion, one program has shown substantial improvement of M50 steel bearings in resistance to chloride in lubricating oil.

Applications of implantation to non-metals is much more limited. Investigators have examined effects of implantation on silicon carbide, aluminum oxide and titanium diboride. Miscellaneous investigations on non-metals include studies on polymers and polymer films, and on glasses.

INTRODUCTION

Ion implantation is a process for modifying the near surface region of solid materials by bombardment with energetic ions. To date the process has been applied principally in the electronics industry for the doping of silicon with elements such as boron, arsenic, antimony, and phosphorus for semi-conductors. It is believed that approximately 1,500 implanters are in production use worldwide for this purpose. Outside the electronics field, ion implantation shows promise of beneficial effects on hardness, wear, coefficient of friction, fatigue, corrosion, and other surface-dominated effects.[1]

The implanted species can be any chemical element which can be ionized into a plasma and the substrate can be any solid. Classes of substrates which have been implanted and reported in the literature include metals, ceramics, glasses, and plastics. The typical ion energy of implantation is 20 to 200 kiloelectronvolts to achieve a dosage or fluence in the 10^{15} to 10^{17} ions per square centimeter range. These parameters result in implant element concentrations of 0.1 to 30 atomic percent, at depth levels of 1 to 10 microinches (0.025 to 0.25 micrometers) beneath the solid surface.

IMPLANTATION PROCESSES

Figure 1 shows the components of an ion implanter. Beginning with the ion source, the gasified element or compound is drawn into an extraction electrode at high voltage to form the ion beam. The beam is then purified by the analyzer magnets which remove unwanted elements and isotopes. An acceleration tube raises the beam to higher energy, followed by the scanner plates to focus the beam onto the target. Figure 2 shows the Gaussian sub-surface distribution of the implanted ions, which have now become elemental in nature, and assumed substitutional, interstitial, or vacancy sites in a metal target substrate. The applied voltage controls

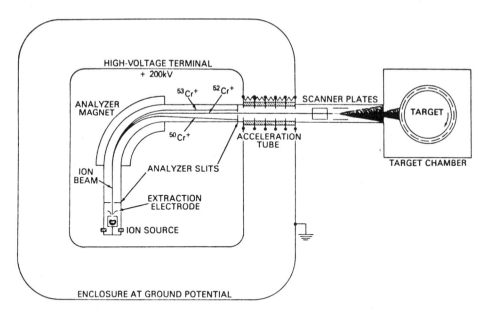

FIGURE 1 - Schematic of an Ion Implanter

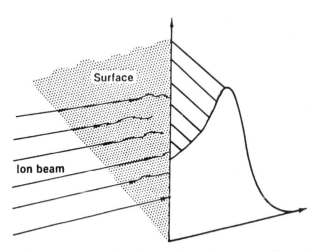

FIGURE 2 - Gaussian Sub-Surface Distribution of Implanted Ions

the penetration depth, while the current establishes the time of
treatment to achieve a desired concentration. Since these
parameters can be controlled very closely, the implantation
process becomes highly reproducible from sample to sample of the
same source and target.

Figure 3 depicts the relative kinetic energy levels in volts
for various types of vacuum coating processes. Ion implantation
is on the highest energy end of the scale, while modifications of
the direct beam implant methods require somewhat less energy. For
example, ion beam mixing and simultaneous deposition and the
implantation require several magnitudes less energy for similar
source materials while cluster beam deposition requires an even
lower voltage range, comparable to that for ion plating and ion
beam deposition.

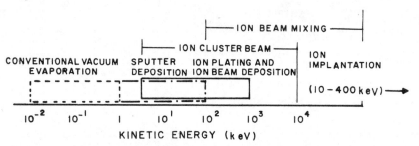

FIGURE 3 - Kinetic Energy Ranges for
some Vacuum Deposition Processes

The principle advantages of the direct ion implantation
process can be seen in Table I. While most of the advantages
cited are self-explanatory, it should be particularly noted that
there is no interface created by this form of coating, and therefore no potential for major voids which can cause spalling of
coatings. On the other hand, major limitations of direct beam
implantation are shown in Table II. It has a line-of-sight beam
and requires a scanning and/or mechanical manipulation device to
uniformly bombard most surfaces. Since a high vacuum is required
and there are sophisticated electrical, magnetic and electronic
components in the implanter, purchase and maintenance costs are
high. Last but not least, there is a limitation to the energy
level which may be applied to a substrate during implantation,
since significant sputtering of the substrate may occur.

Taking a look at recent advances beyond the direct implantation
process, these all involve combinations of previously developed
coating techniques in conjunction with the ion beam. In ion beam
mixing,[2] or recoil implantation, a thin coating is first applied to
the substrate, following which an ion beam of an inert gas drives
the coating atoms into the substrate, as shown in Figure 4. In

TABLE I: ADVANTAGES OF ION IMPLANTATION

1. No sacrifice of bulk properties
2. Solid solubility limit can be exceeded
3. Alloy preparation independent of diffusion constants
4. No coating adhesion problems since there is no interface
5. No macroscopic change in dimensions
6. Depth concentration distribution controllable
7. Room temperature process
8. Precision control
9. Automatic handling possible
10. Clean vacuum process

TABLE II: MAJOR LIMITATIONS OF DIRECT ION IMPLANTATION

- LINE-OF-SIGHT PROCESS
- SHALLOW PENETRATION
- HIGH TECHNOLOGY (MAINTENANCE)
- RELATIVELY HIGH CAPITAL COSTS
- SPUTTERING

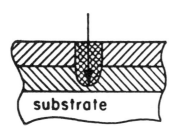

• collision cascade volume
  around ion track

• ion induced mixing and
  enhanced diffusion

• fast quench rate

Formation of: 1) single phase (silicides)

metastable   2) amorphous (metal/Si)

                     3) solid solution (metal/metal)

FIGURE 4 – Micro-Alloying by Ion Beam Mixing

another variation, a plasma of atoms from an ion or electron beam
source is implanted into a target substrate by simultaneous
deposition (by evaporation, sputtering, ion beam, electron beam,
etc.) and ion beam implantation of the deposited layer by bombardment with another metallic ion species or inert gas ions, as shown
in Figure 5. These combinations result in improvements shown in
Table III. An even more recent development, ion cluster beam (ICB)
deposition,[3] takes advantage of a radio frequency bias at low voltage

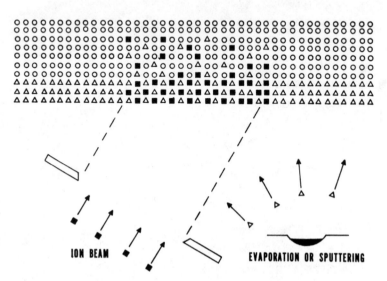

FIGURE 5 – Simultaneous Deposition and Ion Implantation

TABLE III: IMPROVEMENTS BY COMBINATION PROCESSES

- Ion Beam Enhanced Deposition (IBED)

or

- Ion Beam Assisted Deposition (IBAD)

or

- Dynamic Recoil Implantation (DRI)

    - No limit to thickness

    - Controllable stoichiometry

    - Superior adhesion

    - High density coatings

    - Low temperature process

    - Improved reactivity

to form neutral clusters of atoms, some of which are ionized and
accelerated to form a broad ion beam with neutral and ionized
clusters. A simplified schematic diagram is shown in Figure 6.
The various phenomena which take place during ion cluster beam
deposition are illustrated in Figure 7. Note that sputtering of
the substrate and re-evaporation of the deposit take place while
the neutral clusters are being deposited, and migration condensation
or splattering deposition is occurring rather uniformly along the
substrate surface. By proper control, a coating thickness variation within + 3% can be attained over a substrate surface. All
the while, implantation of the highly energized ions are penetrating
into the substrate as described above for direct implantation. The
principal advantages of ICB deposition are the low energy requirement and the potential relative simplicity of the equipment;
this development is in its early stages. Major features of ICB are
listed in Table IV, several of the which were discussed above.

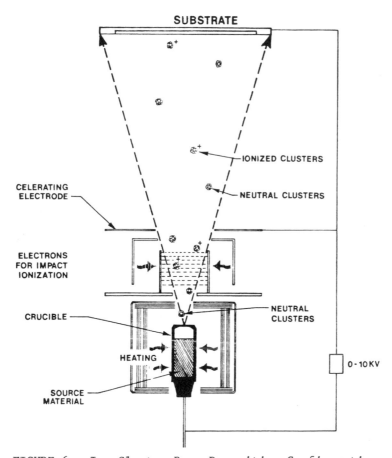

FIGURE 6 – Ion Cluster Beam Deposition Configuration

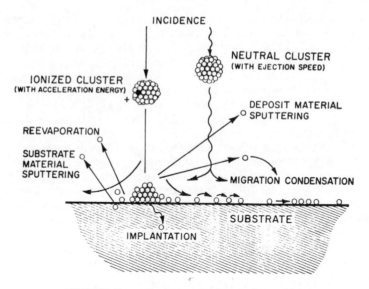

FIGURE 7 – Cluster Impact Phenomena

TABLE IV: MAJOR FEATURES OF ICB

- Ability to adjust average energy per deposition
  atom over 0.01-100 eV range

- Effective conversion of cluster kinetic energy to adatom
  surface energy due to snowball effect

- Inherent cleansing action by sputtering and micro-scale
  heating (improve adhesion)

- Enhanced reactive processes due to ionic charge presence

- Reduce stress in film

APPLICATIONS TO METAL SUBSTRATES

     Some of the surface characteristics of metals which can be
substantially improved by the aforementioned ion implantation
techniques include hardness, coefficient of friction, adhesion,
optical and catalytic activity and resistance to wear, fatigue and
aqueous corrosion. Data given in this paper will be limited to
improvements in wear and aqueous corrosion resistance.

     Friction and wear has been extensively investigated at the
United Kingdom Atomic Energy Research Establishment (Harwell),
U.S. Naval Research Laboratory, and Westinghouse Electric Corpora-
tion. At Harwell, results shown in Figure 8 demonstrate the

reduction in friction and displacement afforded by ion implantation with nitrogen using the spin-on-disc test for evaluation.[4]   Tables V and VI give more specific results for components evaluated at Harwell[4] and Westinghouse.[5]   It can be seen that a minimum of three times the normal service life can be expected under the test conditions applied.

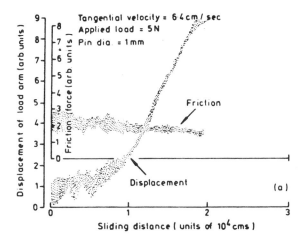

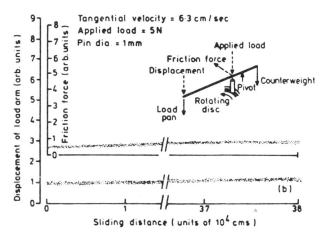

FIGURE 8 - Reduction in Friction and Displacement after Nitrogen Implantation

TABLE V: HARWELL INDUSTRIAL TRIAL RESULTS

APPLICATION	TOOL	RESULT
Steel and copper wire drawing	Co-cemented WC die	3X-5X normal life, improved surface finish
Plastic injection molds, nozzles, and runner blocks	Cr-plated steel	4X-6X normal life

TABLE VI: WESTINGHOUSE COMPONENT RESULTS

APPLICATION	TOOL	RESULT
Index slotting of silicon steel	Co-cemented WC punch and die	6X normal lifetime
Steel rolling mill rolls	H-13 steel	Negligible wear at 3X normal life
Cutting of sheet scrap	Co-cemented WC scrap chopper blade	Over 3X normal life

Corrosion protection experimentation using ion implantation
has been underway at the Naval Research Laboratory for several years.
The components of interest are bearings, fabricated of M50 steel, for
which the corrosion protective methods must result in no loss of
hardness, no degradation of performance characteristics, no reduced
fatigue life, and no significant dimensional changes. Figure 9 shows
the results of simulated field service tests of corrosion, where the
specimen ion implanted with chromium, molybdenum and nitrogen was not
attacked.[6]   Figure 10 shows an arrangement of bearings in a carousel
mode to achieve uniformity of implantation by rotation and rastering.

POTENTIAL FOR IMPLANTATION IN NON-METALS

Many non-metal surfaces have been ion implanted. Those
examples which are cited here are from papers presented at the
Third International Conference on Ion Beam Modification of Materials,
Grenoble, France, September 6-10, 1982. These papers represent the
state-of-the-art areas of investigation on a world-wide aspect. In
less recent work, implantation of nitrogen in tungsten carbide has
been examined by Dearnaley et al at AERE Harwell, United Kingdom,
as mentioned previously, with dramatic results.

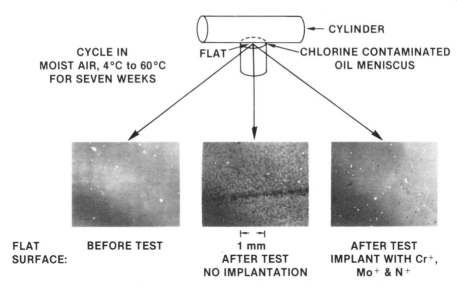

CYCLE IN
MOIST AIR, 4°C to 60°C
FOR SEVEN WEEKS

CYLINDER

FLAT — CHLORINE CONTAMINATED
OIL MENISCUS

FLAT
SURFACE:

BEFORE TEST

⊢ ⊣
1 mm
AFTER TEST
NO IMPLANTATION

AFTER TEST
IMPLANT WITH Cr$^+$,
Mo$^+$ & N$^+$

FIGURE 9 – Simulated Field Service Test of
Corrosion of M50 Bearings

FIGURE 10 – Carousel of M50 Bearings for Ion Implantation

A great deal of interest has been shown in implantation of insulating materials.[7] These include silicon dioxide, rhenium nitride, diamond, garnet, lithium fluoride and calcium fluoride, principally for electronic applications up to the present time.

In ceramic substrates, a great deal of implantation into aluminum oxide has been done at Oak Ridge National Laboratory (ORNL).[7a] The implanted ions include chromium, titanium and zirconium, in concentrations between 25 and 45 atomic percent at the surface. Other ceramic substrates implanted include silicon carbide (at ORNL[8] and University of Cambridge, UK) for amorphization by nitrogen; zirconium oxide (at Orsay, France) with calcium for conductivity;[9] lead lanthanum zirconate titanate (PLZT) (at Sandia Laboratories) with argon, neon or helium for photo-imaging;[10] and quartz (at University of New South Wales, UK) with helium for stress determination.[11]

Implantation into glass was studied at Sandia Labs, where lead was put into silicate glass to stabilize radioactive waste;[12] at University of Padova, Italy, where argon was implanted into silicate glass for optical effects;[13] and at State University of New York, Albany where various types of glasses where implanted with xenon for reactions with water.[14]

Finally a few investigations of polymers have been carried out. At Massachusetts Institute of Technology, arsenic, krypton or bromine[15] were implanted into p-phenylene sulfide for electrical conductivity. In an adhesion program at the University of Missouri, Rolla,[16] polymethane was implanted with carbon and argon.

CONCLUSIONS

1. In the past ten years, ion implantation experimentation has resulted in laboratory improvements in mechanical and chemical surface-sensitive properties of materials.

2. Industrial applications have been growing very slowly, excluding semi-conductors.

3. Future developments will involve simplification of the processing technique as well as combination with other deposition processes.

REFERENCES

1. Hirvonen, J.K., and Preece, C.M., Ion Implantation Metallurgy, The Metallurgical Society of AIME, New York, 1979.

2. Mayer, J.W., "Materials Modification by Ion Beam Mixing", USAMMRC. Ion Implantation for Army Needs Workshop, July 1981.

3. Takagi, T., "Role of Ions in Ionized Cluster Beam Deposition", Thin Solid Films, 92, 1-17 (1982).

4. Dearnaley, G., "Practical Applications of Ion Implantation", Ion Implantation Metallurgy, The Metallurgical Society of AIME, New York, 1979, 1-20.

5. Fromson, R.E. and Kossowsky, R., Proc. 9th North American Manufacturing Research Conference, Pennsylvania State University, 1981.

6. Valori, R. and Hubler, G., "Ion Implantation of Bearings for Improvement of Corrosion Resistance", Naval Research Laboratory Memo Report 4527, June 1981.

7. Davenas, J., Dupuy, C., Long, X.X., Maitrot, M., Andre, J.J., Fracois, B., "A Soliton Model of the Nonmetal-Metal Transition in Implanted Organic (polyacetylene) and Inorganic (lithium fluoride) Materials", Radiation Effects 4(1-4), 1983, 209-217.

7a. Narmoto, H., McHarque, C.J., White, C.J., Williams, J.M., Holland, O.W., Abraham, M.M., and Appleton, B.R., "Near Surface Modification of α-Al$_2$O$_3$ by Ion Implantation followed by Thermal Annealing", Nuclear Instruments and Metals in Physics Research 209/210, May 1983, 1159-1166, North Holland Publishing Co.

8. Williams, J.M., McHargue, C.J., and Appleton, B.R., "Structural Alterations in SiC as a Result of Cr+ and N+ Implantation", Ibid, 317-323.

9. Schnell, J.P., Croset, M., Dieumegard, D., Velasco, G., and Siejka, J., "Characterization of Calcium Implanted Zirconia Thin Films", Ibid, 1187-1191.

10. Percy, P.S., and Land, C.E., "Photographic Image Storage in Ion Implanted PLZT Ceramics", Ibid, 1167-1178.

11. King, B.V., Kelly, J.C., and Dalglish, R.L., "Stress and Strain in Quartz and Silica Irradiated with Light Ions", Ibid, 1135-1143.

12. Arnold, G.W., and Petit, J.C., "Ion Beam Analysis of Implanted Simulated Nuclear Waste Glasses", Ibid 1071-1077.

13. Mazzoldi, P., "Properties of Ion Implanted Glasses", Ibid, 1089-1098.

14. Langford, W.A., and Burman, C., "Effects of Ion Implantation on the Reaction Between Water and Glass", Ibid 1099-1103.

15. Mazurek, H., Day, D.R., Maby, E.W., Abel, J.S., Senturia, S.D., Dresselhaus, M.S., Dresselhaus, G., "Electrical Properties of Ion-Implanted Poly (p-Phenylene Sulfide)", Journal of Polymer Science, Polymer Physics Edition 21, 1983.

16. Hale, E.B., et. al. "Use of Ion Implantation to Improve Adhesion Between a Polymer and a Metal", Ion Implantation into Metals, Ashworth, V., Grant, W.A., and Procter, R.P.M., Pergamon Press, 1982, 167-171.

SURFACE MODIFICATION BY PLASMA POLYMERIZATION

Nicholas Morosoff

Research Triangle Institute
P. O. Box 12194
Research Triangle Park, NC   27709

Plasma polymers are surface coatings formed in a plasma by the fragmentation of "monomer" molecules, the formation of reactive sites on surfaces (including newly formed plasma polymer) in contact with the plasma, and the reaction of monomer fragments with the surface and each other. Such films are formed in a glow-discharge, a plasma (electrically neutral ionized gas) formed (and sustained) by an electric field in a partial vacuum (less than 10 Torr pressure). As a consequence of the low pressure, electrons are characterized by electron temperatures of $10,000^0C$, average electron energies of 1-10 eV and non-equilibrium with gas molecules.[1] The temperature of the latter is therefore close to ambient.

Inelastic collisions of the electrons with the monomer molecules result in a relatively high concentration of chemically reactive species (e.g., radicals, ions, excited molecules) in the plasma phase and at surfaces in contact with the plasma. As a result, a variety of organic and organometallic monomers will polymerize. In addition to unsaturated organic monomers subject to conventional free radical polymerization, saturated and normally insert compounds (e.g., methane, silane, nitrogen, benzene) may also be induced to form polymers and/or copolymers which deposit on or are grafted to a surface (glass, metal, polymer) in the form of thin films (which are usually densely crosslinked) or, under some conditions, as powders. Typically, e.g., in the plasma polymerization of hydrocarbons, the structure of the product bears no direct relationship to that of the monomer; no repeat unit is recognizable. This is illustrated by the structure of an ethylene plasma polymer,[2] shown in Figure 1, as deduced from infrared and NMR data. The reason for this is, of course, because of the fragmentation of the monomer that occurs in the plasma. As shown below, the degree of fragmentation and cross-linking will depend on reaction conditions.

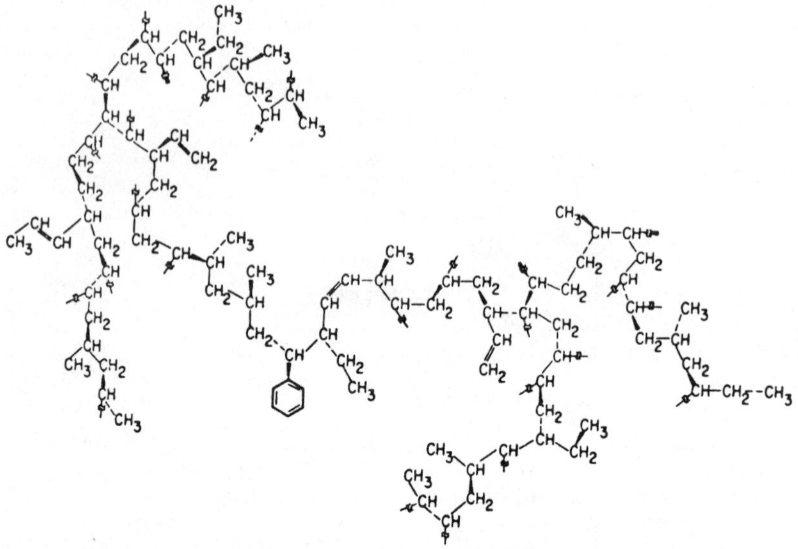

Figure 1. Molecular structure for a section of plasma polymerized ethylene film as derived from infrared and NMR data of plasma polymerized films and oils (from reference 2). Broken bonds indicate crosslink points (1 crosslink per 6-7 backbone carbon atoms).

In the following we will examine the extent to which the chemical and physical properties of plasma polymers may be controlled by the process parameters and the variety of materials and properties made accessible by this technology.

Plasma polymerization for the formation of useful coatings is normally carried out in a flow system with the continuous introduction of monomer into the reactor accompanied by evacuation of gaseous by-products and unreacted monomer. The electric field can be supplied via electrodes in contact with the plasma or electrodes or a coil external to the reaction vessel. Some typical reactors are shown in Figure 2. The field may be DC or may alternate at frequencies up to the microwave region. External electrodes require frequencies exceeding 1 MHz.[3,4,5,6] The process parameters include the frequency of excitation, the excitation power, the flow rate, the plasma pressure, the substrate temperature, and the choice of monomer. As mentioned above, a glow-discharge requires pressures of 10 Torr or less to avoid elevated molecule and molecular fragment temperatures, thus avoiding pyrolysis of the monomer and plasma

polymer. The electron temperature, T_e, decreases with the parameter E/p, where E is the electric field and p the pressure.[1,7]
An increase in pressure therefore leads to a decrease in electron
temperature with a concomitant increase in gas temperature as the
molecules in the gas phase approach equilibrium with the electrons.
Such equilibrium is achieved at pressures exceeding 100 Torr and
the resulting "hot" plasma is called an arc, with applications in
metallurgy.

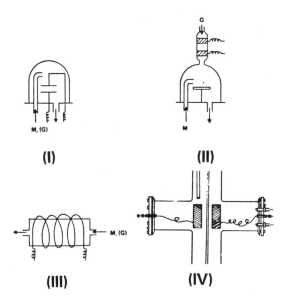

Figure 2.   Schematic representation of some typical arrangements for glow discharges showing flow of monomer, M, and carrier
gas, G. Arrangement I shows a capacitively coupled reactor in a
bell jar, arrangement IV is similar but allows deposition on a
moving substrate located between the electrodes. Arrangement II
involves capacitive coupling of the plasma by electrodes outside
the reactor. The carrier gas is excited by the electric field
between the electrodes. The monomer gas is introduced into the
tail flame of the glow-discharge directly above the site of the
plasma deposition. This arrangement minimizes monomer contact with
the glow-discharge and has been used to deposit reverse osmosis
films[32] where some degree of plasma polymer films flexibility is
desirable. Arrangement III illustrates inductive coupling where
the glow discharge is induced by the field across the ends of the
coil wound around the outside of the cylindrical reactor.

The basic plasma parameters have been given as the electron
density, n_e, the distribution of electron energies, $f(\varepsilon)$, the gas
density, N, and the residence time, τ, for gas molecules in the
plasma volume.[8] Only the gas density and residence time are

directly obtainable from easily measured process parameters.
However the dependence of the electron density and the distribution
of electron energies on the process parameters is such that the
quantity W/F (at constant pressure) is expected to determine the
properties of the plasma polymer deposited using a given monomer(s).
Here W signifies power input, F the monomer flowrate, and W/F has
the units of energy/mass.

The electron density, n_e, will increase with current through
the plasma. The electron energy distribution will resemble that
shown in Figure 3, with the mean electron energy increasing with the
electric field in the plasma, if the pressure is held constant
(E/p increases). The residence time, τ, for molecules in the plasma
is given by the expression, pV/F, where p is the pressure, V is the
volume of the plasma and F is the flow rate. The gas density, N,
is proportional to the pressure. It may be noted that the distribution of electron energies is such (Figure 3) that only electrons
in the tail end of the distribution are energetic enough to break
bonds. The probability of any given process occurring is therefore
given by the intersection of the crosssection for that process
(shown in Figure 3), with the electron energy distribution.[9]
Obviously, the process will not occur if no electrons are available
above the threshold energy for that process.* It would appear that
the chemical nature of

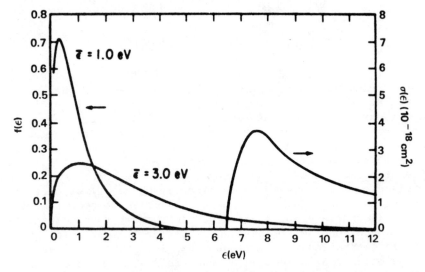

Figure 3. The distribution of electrons, $f(\varepsilon)$ and reaction
cross-section, $\tau(\varepsilon)$, as a function of electron energy (after
reference 9). Electron energy distribution is shown for average
energies of 1 and 3 eV.

*This is not strictly true, but is an assumption used for the
sake of simplicity.

the plasma polymer resulting from a given plasma polymerization
would be affected by the ratio

$$\frac{n_{e(\varepsilon > \varepsilon \ min)} \times \tau}{N} \qquad\qquad (1)$$

This expression defines the probability of a given molecule (in the
plasma phase) undergoing an inelastic collision with an electron
of energy exceeding ε min. Here $n_{e(\varepsilon > \varepsilon \ min)}$ signifies the density
of electrons of energy exceeding ε_{min}. As W/F is equal to $W\tau/pV$,
an increase in W/F results in an increase in expression (1) above.
This is because $n_{e(\varepsilon > \varepsilon \ min)}$ will increase with W, the power, which
is effected by an increase in either current or electric field
(proportional to voltage). An increase in current increases n_e
(i.e., the density of all electrons); an increase in electric field
(E/p increases) increases the average electron energy, thereby
increasing the proportion of electrons in the high energy tail of the
distribution. Either way $n_{e(\varepsilon > \varepsilon \ min)}$ is increased.

   The importance of the interaction of the plasma with the
surface on which the plasma polymer is deposited can be illustrated
by the observation that surfaces can be modified by the interaction
of non-polymerizing plasmas with the surfaces. The modification of
polyethylene surfaces with an oxygen plasma to render them hydroph-
ilic (and therefore printable) and to improve adhesion[10] and the
crosslinking of the inside surface of poly(vinyl chloride) tubing
to inhibit the leaching of plasticizer may be cited.[11] In addition,
fluorocarbons (and other fluorine containing compounds) are used in
the etching of silicon. Obviously such processes can contribute to
(and compete with) the deposition of plasma polymer coatings.
Studies of the kinetics of plasma polymer deposition suggest that
the actual propogation and termination steps occur at the surface.[12]
During plasma polymerization, the extent of modification of a newly
laid down element of plasma polymer surface would be expected to
increase with $n_{e(\varepsilon > \varepsilon \ min)}$ and decrease with the deposition rate.
As the deposition rate is proportional to flow rate over a wide
range of conditions, this again leads to dependence on W/F. Thus
it has been shown that the extent of interaction of energetic
electrons with both monomer molecules in the gas phase and plasma
polymer segments in contact with the plasma is expected to increase
with an increase in W/F.

   One example of the effect of W/F on the nature of the resultant
plasma polymer is given by the plasma polymerization of tetrafluoroethylene. Here plasma polymerization competes with etching with the
former predominant at low W/F and the latter predominant at high
W/F.[13] This is shown by Figure 4, which shows the deposition rate
as a function of power at various flow rates. Deposition rates are
seen to increase with power to a plateau level at high flow rates
(low W/F) but decrease with power at low flow rates (high W/F).

ESCA spectra obtained at high W/F show that the plasma polymer
deposited is fluorine poor at high W/F, fluorine rich (containing
CF, CF_2, and CF_3 groups) at low W/F.[6]

In the preceeding the effect of pressure, flow rate and power
on the chemistry of the resultant plasma polymer has been described.
A proposed free radical mechanism of plasma polymerization that
takes into account both gas phase reactions and reactions at the
surface in contact with the plasma is shown below.[14]

1. $e + M \rightarrow 2R_g + e$,

2. $P_s \xrightarrow{e, I, hv} 2R_s$,

3. $R_g + P_s \rightarrow R_s$,

4. $R_g + R_s \rightarrow P_s$.

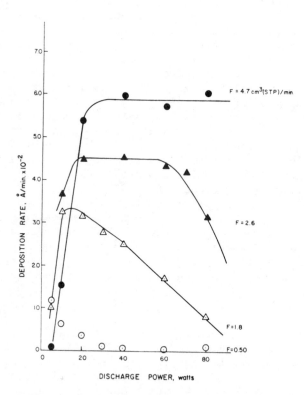

Figure 4.   The deposition rate of tetrafluoroethylene plasma
polymer as a function of rf discharge power at various flow rates
in an inductively coupled plasma reactor (after ref. 13).

Here R signifies a radical, M monomer, P polymer, and the subscripts g and s signify the gas and solid phases, respectively.

In Figure 1, we have shown the structure of the plasma polymer formed from ethylene, a highly crosslinked structure. However, a wide range of chemical and physical properties can be obtained by plasma polymerization by manipulation of the process parameters. For the case of ethylene, it has been shown that oils, films or powders can be laid down by plasma polymerization, by control of the process parameters.[15] Oils are formed at high pressures (which would mean a low E/p and consequently a low average electron energy) and at high flow rates (low W/F and short residence time). Powders reflect formation of plasma polymer entirely in the gas phase and require long residence times (low flow rates). The effect of pressure on powder formation depends on the reaction rate for a given plasma polymerization. Thus acetylene plasma polymer powder is formed under a wide range of conditions because of its extremely rapid rate of polymerization. Ethylene can also polymerize rapidly and is formed at low pressures (0.5 - 1 Torr). Ethane, which cannot polymerize by conventional free radical mechanisms, polymerizes more slowly and requires somewhat higher pressures for the onset of powder formation.[16] High pressures increase the frequency of gas-gas collisions as compared to gas-surface collisions and increase residence time (favoring powder formation) but also decrease average electron energy (which would tend to decrease the rate of reaction and of powder formation). Obviously the first two effects are more important than the latter in the case of ethane.

The formation of oils in the plasma polymerization of hydrocarbons implies that the degree of crosslinking can be controlled, as the oils are obviously not crosslinked. A similar effect is seen for fluorocarbon plasma polymers, for which solubility in fluorocarbon solvents has been reported.[17]

The range of products that can be formed by plasma polymerization can be appreciated by considering that the process is related to the organic plasma synthesis at one extreme and inorganic plasma deposition or plasma assisted chemical vapor deposition at the other. These are listed in Figure 5 in order of increasing probability of interaction of energetic electrons with molecules in the gas phase and with the surface. Organic plasma synthesis is a process involving minimal interaction of the reactants with the plasma. The reactants flow through a reactor similar to the kind used for plasma polymerization, but at flow rates 1 to 2 orders of magnitude higher.[18,19] As a result of the extremely short residence times, little or no solid products are formed and the product is collected in a cold trap(s) between the reactor and the pump. The recently reported cyanation of toluene by cyanogen in a plasma is given as an example.[19] It may be noted that the degree of fragmentation is less than that shown in Figure 1; the structure of the reactants (the benzene or toluene entity and the cyano group) is clearly recognizable in the product.

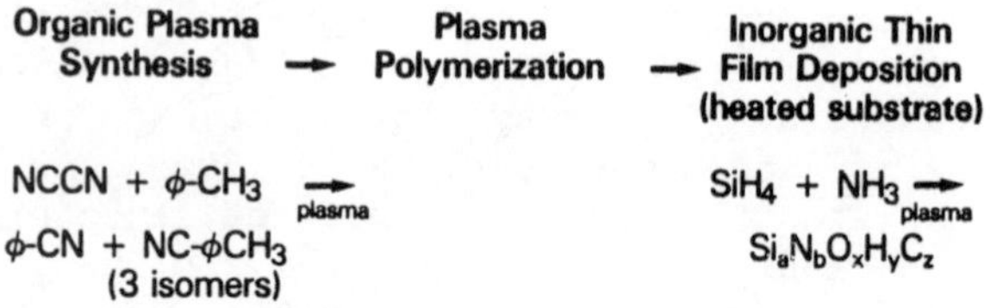

Organic Plasma Synthesis	→	Plasma Polymerization	→	Inorganic Thin Film Deposition (heated substrate)

NCCN + φ-CH₃ $\xrightarrow[\text{plasma}]{}$

φ-CN + NC-φCH₃
(3 isomers)

SiH₄ + NH₃ $\xrightarrow[\text{plasma}]{}$

Si$_a$N$_b$O$_x$H$_y$C$_z$

Figure 5. The relationship of plasma polymerization to the related processes of organic plasma synthesis and inorganic thin film deposition with examples of the latter. Processes are listed (left to right) in order to increasing exposure of reactants and products to energetic electrons (and other energetic components) of the plasma.

Inorganic thin film deposition involves deposition on a heated substrate, which reduces the deposition rate and allows for extended contact between the plasma and any given molecular building block in a growing film. The effect of heat (driving off normally involatile fragments and pyrolysis) and extended contact with plasma is to remove organic components and hydrogen, leaving only inorganic elements behind. The reaction of silane and ammonia to yield silicon nitride[20] (with traces of O, H, C) or the deposition of amorphous hydrogenated silicon[21] by plasma assisted chemical vapor deposition may be cited as examples.

Plasma polymerizations which approximate both of these extremes have been reported. To approximate organic plasma synthesis, the polymerization must be predominantly a conventional polymerization (thus preserving the structure of the reactant) as has been reported for the plasma polymerization of vinyl ferrocene.[22,23] This plasma polymer has been used for the fabrication of surface modified electrodes. The ferrocene moiety is left largely intact allowing reduction and oxidation between the ferrocene and ferricenium states on the surface of the electrode. Some degradation of the ferrocene structure does occur and the resultant coating swells, but does not dissolve in acetonitrile.[22]

At the other extreme, the plasma polymerization of hexamethyl-disiloxane may be cited. By using oxygen as a comonomer, and manipulating W/F, a predominantly inorganic product may be deposited.[24] The infrared spectrum reported for the hexamethyldisiloxane plasma polymer is shown in Figure 6. Evidence for an organic siloxane structure is found in the peaks at 1250 and 800 cm^{-1} for

Si–CH$_3$ bonds and the peak at 1030 cm^{-1} indicating the Si–O–Si group. Introduction of oxygen into the plasma and an increase in the W/F ratio results in diminution of infrared bands associated with the Si–CH$_3$ group and the C–H bond. The process can be considered a combination of plasma polymerization and plasma etching with the oxygen plasma acting as a carbon scavenger. The product formed is increased in inorganic character and has been found to be effective as an anti–abrasion coating.[24] It begins to approximate inorganic thin film deposition in the tendency to preserve only the silicon and oxygen atoms in the structure of the final product.

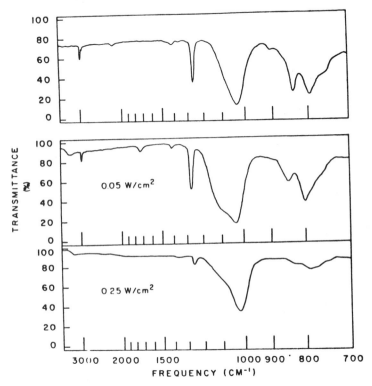

Figure 6. Infrared spectra of hexamethyldisiloxane plasma polymer films. Top spectrum: monomer feed is hexamethyldisiloxane, rf power is 5W (in a capacitively coupled plasma reactor). Middle and bottom spectrum: monomer feed is hexamethyldisiloxane + O$_2$ (in a 1/4.5 molar ratio), power is 5W (middle) and 25W (bottom spectrum). (After reference 24).

Potential applications of plasma polymerization make use of the ability to lay down extremely thin, highly adherent and coherent, crosslinked and amorphous organic films by this method and the ability to vary their chemical character and degree of crosslinking. They include abrasion resistant coatings[24] and barrier coatings, either for control of leaching rate from the polymer[25,26] or as a protection from corrosive agents. In the latter category the ability of plasma polymers to protect metal from corrosion by alkaline solutions[27], of rubber seals[28] from corrosion by sour gas in an oil drilling application and of infrared windows (alkali halides) from attack by moisture have been reported.[29] Other applications making use of control of diffusion through plasma polymers include permselective membranes (for separation of components of a gaseous mixture[30]) and reverse osmosis membranes (for purification of saline[31,32] or waste water). Plasma polymers have been deposited on fillers or rein-forcing components of composites as adhesion promoters between the filler and the matrix.[33] Here use is made of the excellent adhesion between the plasma polymer and the filler, while the surface energy of the plasma polymer is tailored to achieve a close match to that of the matrix. The use of vinyl ferrocene plasma polymer in making chemically modified electrodes has been cited; other plasma polymeric systems have also been shown to be useful.[34] The compatibility of plasma polymerization with an industrial process has been recognized in the development of a plasma polymerization process (using styrene and nitrogen as comonomers) in the development of the RCA videodisc.[35] Although plasma polymerization is not used in the fabrication of the commercial product, the plasma polymer coating that was developed met all technical requirements for both deposition of and performance of a dielectric layer. It was one of several steps discarded as simpler and more economic ways of making the product were found.

REFERENCES

1. Bell, A. T., Fundamentals of Plasma Chemistry, in "Techniques and Applications of Plasma Chemistry", Hollahan, J. R., Bell, A. T., eds., J. Wiley, New York (1974).
2. Tibbit, J. M., Shen, M., Bell, A. T., J. Macromol. Sci., Chem., A10, 1623-48 (1976).
3. Maissel, L., Applications of Sputtering to the Deposition of Films, in "Handbook of Thin Film Technology", Maissel, L. I., Glang, R., eds., McGraw-Hill, N.Y. (1970).
4. Morita, S., Bell, A. T., Shen, M., J. Polym. Sci., Polym. Chem. Ed., 17, 2775-83 (1979).
5. Morosoff, N., Newton, W., Yasuda, H., J. Vac. Sci. Technol., 15, 1815-22 (1978).
6. Morosoff, N., Yasuda, H., ACS Symp. Ser., 108, 163-79 (1979).
7. Lam, D. K., Baddour, R. F., Stancell, A. F., J. Macromol. Sci., Chem., 10, 421-50 (1976).
8. Kay, E., Coburn, J., Dilks, A., Top. Curr. Chem., 94, 1-42 (1980).
9. Bell, A. T., J. Macromol. Sci., Chem., A10, 369-81 (1976).
10. Hall, J. R., Westerdahl, C. A. L., Bodnar, M. J., Levi, D. W., J. Appl. Polym. Sci., 16, 1465-77 (1972).

11. Morosoff, N., Yasuda, H., Final report, NIH contract N01-HV-3-2913 (1980).
12. Lam, D. K., Baddour, R. F., Stancell, A. F., J. Macromol. Sci., Chem., A10, 421 (1976).
13. Yasuda, H., Hsu, T. S., Brandt, E. S., Reilley, C. N., J. Polym. Sci., Polym. Chem. Ed., 16, 415-25 (1978).
14. Bell, A. T., Top. Curr. Chem., 94, 43-68 (1980).
15. Kobayashi, H., Shen, M., Bell, A. T., J. Macromol. Sci., Chem., 8, 373-91(1974).
16. Shen, M., Bell, A. T., ACS Symp. Ser., 108, 1-33 (1979).
17. Pavlath, A. E., Pittman, A. G., ACS Symp. Ser., 108, 181-92 (1979).
18. Suhr, H., Angew. Chem., Internat. Ed., 11, 781-92 (1972).
19. So, Y. H., Miller, L. L., J. Am. Chem. Soc., 103, 4204-9 (1981).
20. Hollahan, J. R., Rosler, R. S., Plasma Deposition of Inorganic Thin Films, in, "Thin Film Processes", Vossen, J. L., Kern, W., eds., Academic Press, N. Y. (1978).
21. Brodsky, M. H., Thin Solid Films, 50, 57-67 (1978).
22. Nowak, R. H., Schultz, F. A., Umana M., Lam, R., Murray, R. W., Anal. Chem., 52, 315-21 (1980).
23. Dautartas, M. F. and Evans, J. F., J. Electroanal. Chem. Interfacial Electrochem., 109, 301-312 (1980).
24. Hays, A. K., Proc.-Electrochem. Soc., 82(6) (Proc. Symp. Plasma Process., 3rd, 1981), 75-87 (1982).
25. Chang, F. Y., Shen, M., Bell, A. T., J. Appl. Polym. Sci., 17, 2915-8 (1973).
26. Colter, K. D., Shen, M., Bell, A. T., Biomater., Med. Devices, Artif. Organs, 5, 13-24 (1977).
27. Schreiber, H. P., Tewari, Y. B., Wertheimer, M. R., Ind. Eng. Chem. Prod. Res. Dev., 17, 27-30 (1978).
28. Arnold, C., Bieg, R. W., Cuthrell, R. E., Nelson, G. C., J. Appl. Polym. Sci., 27, 821-37 (1982).
29. Wydeven, T., Johnson, C. C., Polym. Eng. Sci., 21, 650-7 (1981).
30. Stancell, A. F., Spencer, A. T., J. Appl. Polym. Sci., 16, 1505-14 (1972).
31. Yasuda, H., Marsh, H. C., Tsai, J., J. Appl. Polym. Sci., 19, 2157-66 (1975).
32. Hollahan, J. R., Wydeven, T., J. Appl. Polym. Sci., 21, 923-35 (1977).
33. Schreiber, H. P., Wertheimer, M. R., Sridharan, A. U., ACS Symp. Ser., 108, 287-98 (1979).
34. Morosoff, N., Patel, D. L., Lugg, P. S., Crumbliss, A. L., J. Appl. Polym. Sci., Appl. Polym. Symp., in press (1983).
35. Ross, D. L., RCA Review, 39, 136-161 (1978).

ADVANCES IN CARBURIZING - VACUUM CARBURIZING

Samuel J. Hruska

School of Materials Engineering
Purdue University
West Lafayette, Indiana   47907

INTRODUCTION

Steel components having appropriate combinations of wear resistance, mechanical strength, impact strength, etc., especially
at higher temperatures (above 500^0F) can be produced by employing
the well-known conventional carburizing process or a suitably
modified process. In carburizing a carbon-rich layer ranging
from $\sim$ .005 inch to more than $\sim$ .100 inch is produced on some or
all of the surfaces of a relatively low carbon component. By
suitable post-carburizing treatment the carbon-rich layer can be
caused to be quite hard (> 60 R_c) but also relatively brittle
while the low carbon interior (< 35 R_c) retains adequate strength
and ductility.

Recent advances in carburizing processing allow purposeful
development of carbide distributions in the carbon-rich layer
This causes the component to have a layer of relatively high
elastic modulus and superior wear resistance. Such processing can
be carried out utilizing conventional carburizing facilities
where appropriate temperature cycling and atmosphere cycling cause
development of carbides; the process is termed cyclic carburizing.
Vacuum carburizing permits much better control of atmosphere variations and consequently the carbide development process.

Carburization of low carbon (modified) tool steels offers
an avenue for producing carburized components having resistance to
loss of properties at higher temperatures. Conventional carburizing atmospheres maintain an oxidizing potential that produces
oxides of alloying elements, inhibiting carbon penetration.
Vacuum carburizing atmospheres usually have negligible oxidizing

potential and allow maximum rate of carbon penetration.

CONVENTIONAL CARBURIZING

Most commonly a $CO-CO_2$ atmosphere is employed to establish a carbon potential. However, these gases are produced by controlled air combustion of CH_4 (or other hydrocarbon) producing a gas mixture approximately 20% $CO-CO_2$, 40% N_2, and 40% $H_2 - H_2O$, the CO_2 and H_2O levels being less than 1%. At typical carburizing temperatures ($\sim 1700^0 F$) chemical equilibrium is rapidly achieved and the carbon potential obeys the relation [1]

$$K_1 = \frac{a_c \ P_{CO_2}}{P_{CO}^2} \tag{1}$$

where K_1 = equilibrium constant at temperature T for the
      equilibrium 2 CO = C + CO_2.
  a_c = carbon activity.
  P_{CO}= CO pressure.
  P_{CO_2}= CO_2 pressure.

For a gas mixture at a given temperature having a fixed carbon percentage, i.e., a fixed % $P_{CO} + P_{CO_2}$ control of P_{CO_2} permits control of a_c (carbon potential).
Such a gas mixture can also be controlled by control of oxygen pressure (oxygen activity) that obeys

$$K_2 = \frac{P_{O_2}^{\frac{1}{2}} \ P_{CO}}{P_{CO_2}} \tag{2}$$

where K_2 = equilibrium constant at temperature T for the
      equilibrium CO_2 = ½ O_2 + CO.
  P_{O_2} = O_2 pressure.

For $P_{CO} + P_{CO_2}$ fixed a fixed P_{O_2} fixes P_{CO} and P_{CO_2} thereby fixing a_c by relation (1). If the gas mixture arises from combustion of CH_4 and C_3H_8 mixtures of unknown relative amounts (as frequently created by suppliers of natural gas for heating purposes), the carbon potential can be controlled only by control of two gas pressures, e.g., P_{CO} and P_{CO_2} or P_{CO_2} and P_{O_2}. This circumstance is emphasized because close control of carbon potential is required not only for satisfactory execution of the cyclic carburizing process and similar processes where temperature is varied, but also for high quality commercial carburizing.

In a typical conventional carburizing process the component, during the carburizing stage, is maintained at a relatively fixed temperature and first exposed to an atmosphere that maintains the carbon potential at near the maximum, avoiding soot formation. This step, called the "carburize" step, permits maximum absorption of carbon. Then the atmosphere is changed to one having a lower carbon potential; a lower carbon concentration at the surface and a region of uniform carbon near the surface is desirable for optimum final properties. During this step, called the "diffuse" step, there can be an initial loss of carbon but the overall effect is to produce some additional absorption of carbon and a carbon distribution (carbon profile) as shown in Figure 1. During

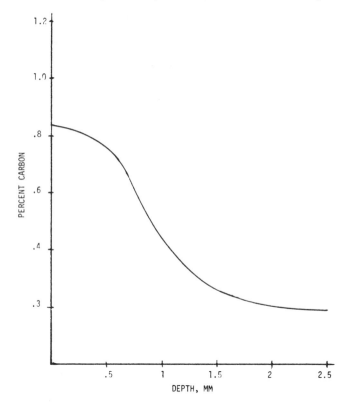

Figure 1.

the "carburize" step the carbon potential was 1.2% carbon; during the "diffuse" step the potential was .82% carbon.

An engineering analysis of this isothermal process requires consideration of three distinct reactions [2]: gas phase reactions, the gas-surface reaction and the diffusion process. At temperatures normally used (for other reasons) gas equilibrium is readily achieved; relation (1) dictates the maximum carbon potential attainable by the component surface. The time-dependent carbon potential at the surface is controlled by the remaining competing reactions: the gas-surface reaction and the diffusion process. The appropriate differential equation is Fick's Second Law in one dimension,

$$\frac{\partial c}{\partial t} = \frac{\partial}{\partial x}\left[D \frac{\partial c}{\partial x} \right]$$ (3)

where D = carbon diffusivity, dependent on c and
     c = carbon concentration.

    The engineering control of the carburizing process arises by transforming relation (3) into a finite difference equation [2,4], specifying initial (usually uniform) carbon level of the component, and formulating the boundary conditions in terms of the gas-surface reaction rate constant and the time dependent carbon potential of the atmosphere. The carbon distribution can be computer-micro-processor controlled. The most notable features of this work are the importance of the gas-surface reaction (a one to two hour time constant) and the variation of diffusivity with composition (a variation of about a factor of three). Figure 2 shows the

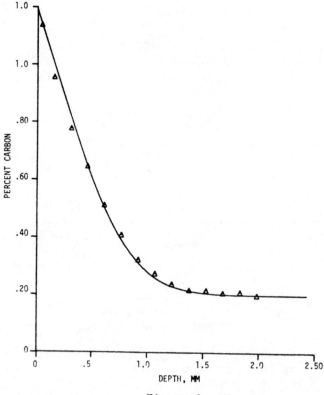

Figure 2.

agreement between a calculated carbon profile and an experimental
profile [3] for AISI 8620 (.55% Ni, .50% Cr, 20% Mo) carburized 4
hours at 928^0C.

CYCLIC CARBURIZING

The objective of cyclic carburizing [5] is to develop a useful
high concentration of alloy carbides (10% to 50%) in the near-
surface region of a component. Normally the temperature range
employed has its maximum at that employed for conventional carburizing. Carbides develop during temperature cycling because a
carbon supersaturated region will precipitate carbides during
cooling, depleting the surroundings, and upon heating, the under-
saturated surroundings absorb additional carbon from adjacent richer
regions, while the previously precipitated carbides are dissolving.
Such carbide development is accentuated by slow cooling followed
by rapid heating and by carbide-stabilizing alloying elements.

In a typical cyclic carburizing process the component is exposed to the "carburize" step as in conventional carburizing, then
it is cooled from $\sim 1700^0$F to $\sim 1400^0$F and held at $\sim 1400^0$F to
precipitate carbides. This is followed by heating to $\sim 1700^0$F for
further carburization after which the cooling-hold-reheating is
repeated as desired.

The engineering analysis of this process [6] requires consideration of temperature variation effects and the precipitation-
growth-dissolution of carbides in addition to the gas-surface
reaction and the diffusion process; oxide kinetics are neglected
for simplicity here. The appropriate differential equation in
this case is a modified form of Fick's Second Law that incorporates
carbide kinetics,

$$f \frac{\partial c}{\partial t} = \frac{\partial}{\partial x} \left(fD \frac{\partial c}{\partial x} \right) + K(1-f) \ (c-c_e) \ (c_c-c) \qquad (4)$$

where c_e = composition of austenite in equilibrium with carbide,
      c_c = composition of. carbide,
      f = volume fraction austenite, and
      K = conglomerate constant related to the composition
             linearized form of the growth equation.

$$\frac{\partial f}{\partial t} = (f-1) \ K(c-c_e) \qquad (5)$$

Suitable simple assumptions for carbide kinetic behavior appear
satisfactory; carbides readily precipitate on cooling and do not
dissolve at all on heating. However, the gas-surface reaction is
sufficiently slow that the carbon activity in the surface of the
component cannot be varied as rapidly as may be appropriate. The
effectiveness of vacuum carburizing to obviate this constraint is

now being studied. Work by Mittal [3], Kim [7] and Park [8] aided
by Fairfield Manufacturing Co. and Cummins Engine Co. on cyclic
carburization taking into account the three reactions shows that
the "carbiding" process is adequately described. Figure 3 shows

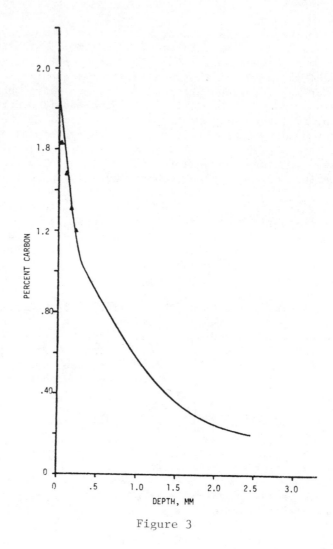

Figure 3

the agreement between the calculated carbon profile and data in
the highly carbided region for AISI 8620 given a guadruple carburizing treatment cycling between 927^0C and 837^0C. Note that the
carbon content is well in excess of the solubility limit, $\sim$ 1.30%
carbon. Figure 4 is a 200X micrograph of carbided AISI 8620.

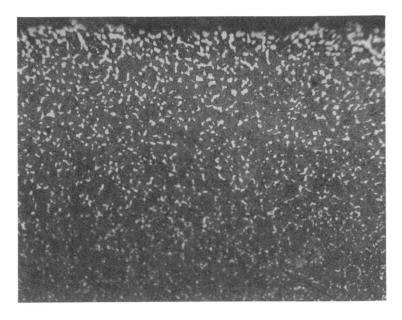

Figure 4

VACUUM CARBURIZING

Vacuum carburizing as described herein is of a rather specialized but essentially simple kind: exposure of a component, at
temperature, in an evacuated space to a hydrocarbon, CH_4, in our
studies. In comparison to conventional carburizing there is no
readily controllable gas phase equilibrium. Decomposition of the
gas occurs within the heated space and especially on contact with
the component to be carburized. The net result found experimentally is that the surface of the component very rapidly (< 5 min)
saturates with carbon. An attendant developmental problem is the
formation of soot, fine graphite particles. In the heated space
the gas mixture accordingly is CH_4 and H_2, primarily; the amount
of H_2 being determined by carbon absorption of the component and
by soot formation. Since virtually no O_2 is introduced, no
appreciable oxidation occurs; carburization of tool steels is not
prohibited.

The vacuum carburizing process usually involves two steps,
similar to conventional carburizing. In the first step, again
termed "carburize" step, the previously evacuated space is filled
to ∿ .30 atm flowing CH_4. Following rapid evacuation to ∿ 0.100
torr the second step commences, termed the "diffuse" step.
Analysis of this process is relatively straightforward and simple.
The boundary conditions are approximately instantaneous saturation
for the "carburize" step, found experimentally, and zero carbon
flux for the "diffuse" step, due to the relatively low gas

pressure; there is no need to consider gas reactions or the gas-surface reaction as in the conventional carburizing case. For isothermal treatments of steels where carbides do not form only diffusion effects need to be accounted for. Exposure of AISI 9310 (3.25% Ni, 1.2% Cr, .12% Mo) to a 1 hour "carburize" at 1900°F shows excellent agreement [9], Figure 5, with the finite difference

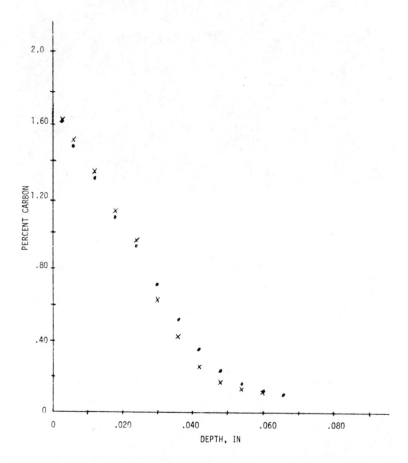

Figure 5

solution employing the simple boundary condition above; the dots are calculated values. When strong carbide forming elements are present, as in Vasco X-2 (5% Cr, 1.4% Mo, 1.3% W, .4% V, .15% C), internal carbiding occurs even for isothermal treatments; carbide precipitation-growth-dissolution kinetics must also be considered. Specialized processes are currently under study at AMMRC as are the conditions relating to the maintenance of the zero carbon flux boundary condition.

SUMMARY

The advances in vacuum technology, microprocessor controls, computer design and heat treating equipment are allowing incorporation of fundamental concepts into the processing of superior performance steel components. Vacuum carburizing, in particular, is being developed to facilitate new product development via cyclic carburizing and to substantially widen the range of alloys that can be used for high temperature wear environments.

REFERENCES

1. ASM Metals Handbook, Vol. 4, 9th Ed., 123.
2. Collin, R., Brachaczek, M. and Thulin, D., J. Iron Steel Inst., 210, 1122 (1969).
3. Mittal, S. K., M.S. Dissertation, School of Materials Engineering, Purdue University, W. Lafayette, IN, July 1979.
4. Goldstein, J. I. and Moren, A. E., Metall. Trans., 9A, 1515 (1978).
5. Cullen, O. E., Patent No. 610,554, Canadian Patent Office, 1 (1960).
6. Winchell, P. G. (now deceased) private communication.
7. Kim, S., M.S. Dissertation, School of Materials Engineering, Purdue University, W. Lafayette, IN, Dec. 1980.
8. Park, Y. G., M.S. Dissertation, School of Materials Engineering, Purdue University, W. Lafayette, IN, Dec. 1982.
9. Hruska, S. J., J. Heat Treating, in press.

ELECTROPLATING OF REFRACTORY METALS

Georges J. Kipouros and Donald R. Sadoway

Materials Processing Center and Department of Materials
Science and Engineering
Massachusetts Institute of Technology
Cambridge, MA 02139

ABSTRACT

Fused salt electroplating offers the prospects for producing
refractory metals in a variety of technologically important forms:
coatings of metals, alloys, and compounds; thin films and graded
interfaces; and ceramic to metal interlayers. Previous results
of fused salt electroplating of refractory metals are reviewed.
New techniques of studying the electroplating process, process
control, and evaluation of the deposit are discussed.

INTRODUCTION

Virtually every engineering endeavor becomes materials-limited.
Surface quality is a very important consideration in the development of materials processing operations to produce high performance
materials. Carefully prepared surfaces can offer improved mechanical properties such as better abrasion or wear resistance and
enhanced corrosion resistance. The refractory metals (elements of
groups IVB, VB, and VIB) with their high melting points, good
thermal and electrical conductivities, and excellent mechanical
properties, are very attractive as surface coatings. Electroplating
of refractory metals from fused salts is a technique that offers
the prospect of making uniform coatings of controlled thickness in
a way which may be superior to chemical vapor deposition or
sputtering.

A related issue is that of bonding dissimilar materials, in
particular, metals to ceramics. The ability to produce thin
reactive layers of refractory metal or alloy would advance the
entire joining effort in this area.

493

TABLE I

Previous Work in Fused Salt Electroplating of Refractory Metals

Electrolyte Metal	Chlorides	Fluorides	Oxides
Ti	−	−	−
V	−	M[8]	−
Cr*	−	−	−
Zr	A(Zr-Al)[8,14]	C(ZrB$_2$)[6,15-17]	−
Nb	−	M[8,21] C(Nb$_x$Ge$_y$)[26]	−
Mo	M[2] A(Ni-Mo)[i] C(Mo$_2$C)[24,25]	M[8]	M[2]
Hf	−	M[8] C(HfB$_2$)[6,15-17]	−
Ta	−	M[8,22] C(Ta$_2$C)[23]	−
W	−	M[8]	M[4,13]

*Electroplated from aqueous solutions
−No successful electroplating/not reported
Successful electroplating: M − metal product
                                  A − alloy product
                                  C − compound product

The preparation of refractory metal coatings on base metals has been the subject of considerable research.[1,27,28] The goal of this work was to take advantage of many of the properties of the refractory metal without incurring the high cost of manufacturing the entire component from it. With the exception of chromium, for which electrodeposition from aqueous solutions has been a successful industrial operation for almost 50 years, none of the refractory metals can be electroplated from aqueous media, due to their reactivity with water.[18,19] This reactivity also impedes the complementary operation: plating onto refractory metals.[20]

Electroplating of refractory metals in fused salts had limited success. Table I is a sampling of such efforts as reported in the literature. The electroplating of refractory metals or alloys, as well as electrosynthesis of refractory metal compounds on the surface of base metals, are included. There appears to have been no major effort with the goal of codepositing refractory metal alloys. The few alloy products noted in Table I were produced coincidentally during the course of attempts to improve the refractory metal plating process.

Although fused salt electroplating of refractory metals had been achieved at the laboratory scale as long ago as the sixties, there has been no major commercial application. Among the reasons for this are the following:

1. Chemical attack of the base metal by the fused salt plating bath.
2. Reactivity of the plating bath with the electroplated metal itself.
3. Poor reproducibility of the electrodeposition process due to improper surface preparation of the base metal.
4. Poor adhesion of the deposit to the substrate.
5. Poor deposit morphology: porous, powdery, denritic, rarely smooth.

Subsequent discussion will address the control and optimization of fused salt electroplating processes with reference to items 1 through 5. Attention will be given to the use of what for this field are advanced instrumentation and novel techniques of investigation. As well, some of the parameters of the various processes will be described. While much can be borrowed from the related field of aqueous electroplating, fused salt electroplating has its own special requirements.

THE ELECTROPLATING PROCESS

In electroplating a soluable electrode, the <u>anode</u>, is dissolved in the electrolyte by the passage of an electric current and is deposited on the surface of the other electrode, the <u>cathode</u> or

Table II. Characteristics of electroplating in fused salts

ELECTRODES		CELL		Feed	Type	DEPOSIT
Anode	Cathode	Container	Plating melt			Minimum common requirements
Pure metal to be electrodeposited Inert to plating melt	Metal or substrate to be covered Electronic* conductor Electropositive with respect to electrodeposited metal Nonreactive with plating melt Clean surface - physically - chemically - mechanically**	Inert to plating melt High purity	High purity salts Free of oxygen & moisture Non-corrosive to container, electrode & diaphragm materials Good solubility for additives	Anode only in the presence of a compound of the metal to be plated	Pure metal coatings (0.1–10 μm) Alloy coatings Compound coatings Thin films Graded interfaces Composites	Surface: 1. Smoothness 2. Continuity Plated film: 1. Texture 2. Low porosity 3. Ductility 4. Thick 5. Uniformity of thickness Interface: 1. Adherence - bonding 2. Internal stress Specific[5]: 1. Appearance 2. Color 3. Hardness 4. Contact resistance

*   Coating on non-electronic substrates is possible via electrophoretic deposition or electroless deposition, which are not Faradaic processes.

** Free of stressed, smeared, or torn surface or other surface defects.

[5] Required only for coatings designed for specific service.

substrate, where it forms the electroplated film. It is the end-
product usage that sets the specifications and determines the type
of testing to be performed. The plating parameters that can be
controlled to produce the required electroplated film may be chosen
from among the following: overpotential, current density, temperature, composition of electrolyte, agitation and the addition or not
of substances to control the texture, the so-called additives. It
is the correct combination of these parameters and the proper
sequence of operations that will produce the desirable form of the
plate. Electroplating in aqueous solutions as well as advances in
controlling the process have been recently described.[29]

    Table II displays the characteristics of the electroplating
process pertinent to fused salt systems. It should be noted that
each salt constitutes an individual solvent system serving the
same purpose as water does for the aqueous systems and requiring
the same degree of understanding. This explains the slow evolution
of fused salts as common industrial solvents where precise operations
such as electroplating can be performed.

    The central part of an electroplating operation is the plating
bath. For fused salt electroplating, melts of high purity are
essential; in particular, moisture and oxygen must be kept at the
parts-per-million level. The same applies to the inert atmosphere
of the cell. Care must also be given to the selection of materials
for construction of the cell container, diaphragms and electrodes
to ensure that they will not be attacked by the electrolyte to
produce harmful electroactive species.

    As indicated in Table II, the cathode must be an electronic
conductor. Plating on non-electronic substrates is possible via
electrophoretic deposition or electroless deposition, processes
which are not truly Faradaic and consequently are not studied here.
Electroless plating of a base metal with a compound of a refractory
metal in fused salts has been recently reported.[30] The cathode
should also be more electropositive than the electrodeposited metal.
This is very restrictive in regard to the base metals that can be
plated with refractory metals. For example, direct electroplating
of molybdenum on steel in molten chlorides, as shown in Table I,
is not possible because molybdenum is displaced from the melt by
all the commonly used metals. Electroplating in this case is only
possible after the application of an underplating -- often called
"strike" -- of a metal of the platinum group (Pt, Ir, Pd) or gold.

    Another requirement for the cathode is that it not react with
the plating bath during electroplating. This refers both to the
cathode substrate and to the deposit itself. Of the electrolytes
shown in Table I, the molten chlorides are the least corrosive
toward most of the base metals (in the absence of moisture) and
allow the lowest operating temperatures. For these reasons,

their function is not completely understood, their properties
cannot be correlated to the properties of other candidate materials.
However, studying inorganic materials which may function as additives
in fused salt electroplating baths is an open research area.
Potential reagents in this regard are the sulfides, oxides and
borides of the corresponding refractory metals. It has long been
known that activated alumina, for example, acts as a levelling
agent in fused salt electrolysis. The need for additives for
electroplating of refractory metals in fused salts cannot be over-
emphasized given the fact that the deposit produced presently are
powdery, dentritic, or at best, highly faceted.

The most crucial issue facing the fused salt electroplating of
refractory metals is the uncertainty as to what species is
responsible for producing a smooth deposit. Refractory metals are
known to exist in more than one valence. During electrolysis
higher valent species are converted to lower valent species, and
electroplating of crystalline refractory metals begins only after
"conditioning" of the plating bath. Recent electrodeposition
studies of refractory metals in fused salts[21,22,31] have indicated
that the "average valence" of the refractory metal in the fused
salt is an important requirement for the deposition of good quality
metal at high current efficiencies. Those studies have been conducted using indirect electrochemical methods of investigation.

However, powerful new direct techniques of investigation of
structural entities in fused salts are now available. In situ
fast Raman spectroelectrochemical techniques are being applied to
laboratory scale representations of the conventional electrolysis
cells for the production of light metals such as magnesium and
aluminum. The results indicate that these spectroelectrochemical
techniques are very powerful and capable of elucidating many aspects
of the local chemistry as the process occurs.

Application of fast Raman spectroelectrochemical techniques
for studying the more complicated fused salt refractory metal
electroplating process is seen as the natural extension of the light
metals investigations. The goals of such studies would be to
identify the refractory metal subvalent species which form during
electroplating, to define the proper "average valence" of the
refractory metal in the plating melt, and to formulate the proper
sequence of operations for the electroplating of good quality
refractory metal.

Fused salt electrolysis of refractory metal has been successful. Titanium electrodeposition by step-wise electroreduction of
$TiCl_4$ in an alkali chloride melt compares favorably[32] with the
existing technology and in some aspects of metal quality is superior.
Direct spectroelectrochemical techniques may be very helpful in
unravelling the mechanism of electrodeposition of titanium and

chlorides are the preferred electrolytes for many fused salt plating operations.

Another very important requirement of the electroplating process is that the substrate surface be clean. Investigators who have paid little attention to this aspect of the process have had difficulty reproducing their results. The surface must be free of dirt, gas bubbles and inclusions, which often initiate the formation of pores. Chemical purity of the surface requires the absence of grease, oxidized spots or any other chemically absorbed substance which will prevent the electrolyte from wetting the electrode. Methods developed for cleaning the substrate for aqueous electroplating processes such as electropolishing may not be applicable to refractory metals due to their reactivity with water. Other surface treatment techniques such as ion bombardment, ion implantation and vacuum metallization are known[20] to produce satisfactory surfaces for the electroplating of refractory metals.

It should be noted that electroplating and electrodeposition, in general, are non-equilibrium processes, the success of which largely depends on the proper mechanical surface preparation. Surface defects or a stressed, smeared or torn surface introduce electroactive spots that give rise to dendrite or needle plating texture and to the lack of adhesion of the electrodeposit. For electroplating of refractory metals in fused salts, the presence of such defects is detrimental given the tendency of refractory metals in molten salts to form dendritic and powdery deposits. Good mechanical surface preparation along with the application of advanced cleaning techniques, such as ultrasonic cleaning, is expected to remove the non-reproducibility associated with the electroplating of refractory metals from fused salts. In contrast to what can be done in aqueous electroplating where the mechanical imperfections may be corrected by the application of a gold and/or nickel underplate or strike layer, in the case of the fused salt electroplating of refractory metals this possibility has not been adequately studied.

PROCESS OPTIMIZATION

The electroplating parameters are interrelated in a complicated manner which is at present not well understood, although in principle the operation appears simple. In aqueous electroplating the introduction of additives, organic compounds with beneficial effects on the properties of the deposit, allows greater variation of the plating parameters.

In fused salt electroplating of refractory metals the use of additives has not been studied at all. This is due to the fact that usually the additives are organic macromolecular compounds unstable in the high temperature environments of fused salts. As

in addition may produce a method for electroplating the metal.

In general, the data base for fused salt electroplating is inadequate. For the design of new processes and for the optimization of existing processes as well, it is necessary to know accurate values of the relevant thermodynamic and electrochemical properties. Without this knowledge it is difficult to understand the fundamentals that serve as the basis for changing the process.

PROCESS CONTROL

Closely related to the topic of process optimization is the topic of process control. With reference to fused salt electroplating, the goal of process control is to correlate the operation of the cell at any moment to the morphology of the electrodeposit being produced. On-line monitoring of cell operating parameters is necessary. The great gains in computer technology have made possible real-time management of multiple parameters in a complicated process algorithm. However, the inputs must be electrical. In electroplating operations this becomes a problem of developing a sensor capable of converting a specific chemical potential or ion concentration into a voltage. The solid electrolyte oxygen sensor with its rapid and continuous emf response has been a boon to those concerned with monitoring oxygen levels in liquid metals. Given a suitable halide probe, new avenues of dynamic control would open in fused salt electroplating.

One promising area of control that does not require new sensors exploits digital signal processing. The fast Fourier transform (FFT) of fluctuating cell parameters is computed on-line in real time. Results aid diagnostics and may prove to be useful in control of the process as our understanding of the relationships between various malfunctions and the cell's electrical signature improves. This technique has the additional advantage of being relatively easy to retrofit.

The adaptation of FFT to fused salt electrolysis operations is currently under study. The application to fused salt electroplating of refractory metals seems a logical extension for this kind of research.

TYPES AND EVALUATION OF DEPOSITS

The purpose of this section is to examine briefly the possible types of deposits of refractory metals electroplated in fused salts. In addition, the research needed for the commercialization of fused salt electroplating as well as the issue of evaluation of the deposit are discussed.

Table II shows the types of deposits of refractory metals which may be produced in fused salts.

   Pure metal coatings. Attempts to produce elemental deposits
of refractory metals have been described in Table I. Fused salt
electroplating of pure refractory metals is capable of producing
thick deposits due to higher current densities than those encountered
in aqueous electroplating. The cost and quality are comparable[1]
to chemical vapor deposition.

   Alloy coatings. Codeposition of refractory metals in fused
salts is possible given the narrow variation of their decomposition
potentials. Since electrodeposition is a non-equilibrium process,
refractory metal codeposition is capable of producing alloys not
shown in equilibrium phase diagrams and may produce structures with
properties superior to equilibrium phases. Electroplating of refractory alloys in fused salts has not been studied systematically,
but only coincidentally during the course of attempts to improve
the refractory metal plating process. Electroplating of refractory
alloys is an area that needs attention.

   Compound coatings. As the use of materials is extended to
higher temperatures and more aggressive environments compounds of
refractory metals, rather than the metals themselves, are becoming
favored by designers and engineers. The electroplating of some of
these ceramic compounds, such as borides and carbides, has been
demonstrated in the laboratory, as shown in Table I. Future needs
may require electroplating of nitrides or carbonitrides; this is
also an expanding area for fused salt electroplating.

   Thin film technology of other metals by aqueous electroplating
is well-established, but for refractory metal thin films formation
in fused salts appears to be virtually unexplored.

   Graded interfaces and composites. Joining of dissimilar
materials and preparation of composite materials are two rapidly
expanding areas. The problem of reliable bonding of ceramics to
metals may be solved by producing multilayered (graded) interfaces
between the two materials. Refractory metals would be part of
these graded interfaces which can be produced by epitaxial electroplating, a natural tendency of refractory metal electrodeposition
in fused salts.

   Evaluation of deposits. Table II gives the minimum common
requirements of electrodeposits classified according to the specific
components they refer to: surface, plated or main deposit and
interface. Despite recent[33] attempts to classify minimum
required testing, the question remains unresolved.

   Testing and evaluation are determined by the final application
of a particular coating. Thus given the diversity of the electro-

plated products, it seems that standardization of testing is unlikely.

Another complicating factor is the fact that some of the requirements refer to subjectively evaluated properties which should then be combined with quantitatively assessed properties to determine the acceptability of the final product.

The applications of the refractory metals are also rapidly expanding; this prevents developing standard and commonly accepted testing methods for this category of metal coatings.

SUMMARY

We have shown that as an innovation in materials processing, fused salt electroplating offers exciting possibilities for producing refractory metals in a variety of technologically important forms: coatings of metals, alloys and compounds; thin films and graded interfaces; and ceramic-to-metal interlayers. Although a number of technical issues must still be addressed before the process can be fully commercialized, it is felt that fused salt electroplating will be a major choice among surface treatment techniques.

REFERENCES

1. Senderoff, S., in "The Metal Molybdenum", ed. J. J. Harwood, American Society for Metals, Cleveland (1958).
2. Senderoff, S. and Brenner, A., J. Electrochem. Soc. 101, 16 (1954).
3. Senderoff, S. and Brenner, A., U. S. Pat. 2,715,093 (1955).
4. Couch, D. E. and Senderoff, S., Trans. Metallurg. Soc. AIME 212, 320 (1958).
5. Mellors, G. W. and Senderoff, S., J. Electrochem. Soc. 112, 266 (1965).
6. Mellors, G. W. and Senderoff, S., J. Electrochem. Soc. 113, 60 (1966).
7. Mellors, G. W. and Senderoff, S., Plating 51, 976 (1964).
8. Mellors, G. W. and Senderoff, S., Can. Pat. 688,546,(1964).
9. Senderoff, S. and Mellors, G. W., J. Electrochem. Soc. 112, 841 (1965).
10. Senderoff, S. and Mellors, G. W., J. Electrochem. Soc. 113, 66 (1966).
11. Senderoff, S. and Mellors, G. W., Science 153, 1475 (1966).
12. McCawley, F. X., Kenahan, C. B. and Schlain, D., U. S. Bur. Mines, RI 6455,(1964).
13. Schlain, D., McCawley, F. X. and Wyche, C., J. Metals 17, 92 (1965).

30. Oki, T. and Tanikawa, J., "Film Formation of Tantalum Carbide by Disproportionation in Molten Salts," paper presented at the First International Symposium on Molten Salt Chemistry and Technology, Kyoto, Japan, April 20-21, 1983

31. Kipouros, G. J., "Electrorefining of Zir onium Metal in Alkali Chloride and Alkali Fluoride Electrolytes and Thermodynamic Properties of Some Alkali Metal Hexachlorozirconate and Hexachlorohafnate Compounds," Ph.D. Thesis, University of Toronto, (1982).

32. Poulsen, E. R. and Hall, J. A., J. of Metals 35$^{(6)}$ 60,(1983).

33. Sard, R., Leidheiser, H. and Ogburn, F., eds., "Properties of Electrodeposits - Their Measurements and Significance," The Electrochemical Society, Princeton, NJ, (1975).

14. Pint, P. and Flengas, S. N., Trans. Inst. Min. and Met. 87, C29 (1978).

15. Mellors, G. W. and Senderoff, S., J. Electrochem. Soc. 118, 220 (1971).

16. Anthony, K. E. and Welch, B. J., Aust. J. Chem. 22, 1593 (1969).

17. Frazer, E. J., Anthony, K. E. and Welch, B. J., Electrodep. and Surf. Treatment 3, 169 (1975).

18. Senderoff, S., Metall. Rev. 11, 97 (1966).

19. Inman, D. and White, S. H., in "Molten Salt Electrolysis in Metal Production", International Symposium organized by the Inst. of Mining and Metallurgy, Grenoble, pp. 51-61 Sept. 19-21, (1977).

20. Wiesner, H. J., AES Symp. Plat. Difficult-Plate Met., Proc. Am. Electroplat. Soc., Winter Park, Fla., pp. 1-21, (1980).

21. Schulze, K. K., J. Metals 33, 33 (1981).

22. Ahmad, I., Spiak, W. A. and Janz, G. J., "Proceedings of the Third International Symposium on Molten Salts", ed. by G. Mamantov, M. Blander and G. P. Smith, The Electrochemical Society Inc., Pennington, NJ p. 558 (1981).

23. Stern, K. H. and Gadowski, S. T., J. Electrochem. Soc. 130, 300 (1983).

24. Heinen, H. J., Barber, C. L. and Baker, D. H., Jr., U. S. Bur. Mines, RI 6590,(1965).

25. Suri, A. K., Mukherjee, T. K. and Gupta, C. K., J. Electrochem. Soc. 120, 622 (1973).

26. Cohen, U., J. Electrochem. Soc. 130, 1480 (1983).

27. Holt, M. L., J. Electrochem. Soc. 98, 33c,(1951).

28. Campbell, T. T. and Jones, A., U. S. Bur. of Mines, Information Circular 7723,(1955).

29. Landau, U., in "Electrochemistry in Industry - New Directions," eds. U. Landau, E. Yeager and D. Kortan, Plenum Press, NY pp. 215 - 245, (1982).